Exploring Data Science with R and the Tidyverse

This book introduces the reader to data science using R and the tidyverse. No prerequisite knowledge is needed in college-level programming or mathematics (e.g., calculus or statistics). The book is self-contained so readers can immediately begin building data science workflows without needing to reference extensive amounts of external resources for onboarding. The contents are targeted for undergraduate students but are equally applicable to students at the graduate level and beyond. The book develops concepts using many real-world examples to motivate the reader.

Upon completion of the text, the reader will be able to:

- Gain proficiency in R programming
- Load and manipulate data frames, and "tidy" them using tidyverse tools
- Conduct statistical analyses and draw meaningful inferences from them
- Perform modeling from numerical and textual data
- Generate data visualizations (numerical and spatial) using `ggplot2` and understand what is being represented

An accompanying R package "`edsdata`" contains synthetic and real datasets used by the textbook and is meant to be used for further practice. An exercise set is made available and designed for compatibility with automated grading tools for instructor use.

Exploring Data Science with R and the Tidyverse

A Concise Introduction

Jerry Bonnell and Mitsunori Ogihara

CRC Press
Taylor & Francis Group
Boca Raton London New York

CRC Press is an imprint of the
Taylor & Francis Group, an **informa** business

A CHAPMAN & HALL BOOK

Designed cover image: © Jerry Bonnell and Mitsunori Ogihara

First edition published 2024
by CRC Press
6000 Broken Sound Parkway NW, Suite 300, Boca Raton, FL 33487-2742

and by CRC Press
4 Park Square, Milton Park, Abingdon, Oxon, OX14 4RN

CRC Press is an imprint of Taylor & Francis Group, LLC

Library of Congress Cataloging-in-Publication Data
Names: Bonnell, Jerry, author. | Ogihara, Mitsunori, 1963- author.
Title: Exploring data science with R and the tidyverse : a concise introduction / Jerry Bonnell and Mitsunori Ogihara.
Description: First edition. | Boca Raton : CRC Press, 2024. | Includes bibliographical references. | Identifiers: LCCN 2023003151 (print) | LCCN 2023003152 (ebook) | ISBN 9781032329505 (hardback) | ISBN 9781032341705 (paperback) | ISBN 9781003320845 (ebook)
Subjects: LCSH: Data mining. | R (Computer program language)
Classification: LCC QA76.9.D343 B66 2024 (print) | LCC QA76.9.D343 (ebook) | DDC 006.3/12--dc23/eng/20230424
LC record available at https://lccn.loc.gov/2023003151
LC ebook record available at https://lccn.loc.gov/2023003152

ISBN: 978-1-032-32950-5 (hbk)
ISBN: 978-1-032-34170-5 (pbk)
ISBN: 978-1-003-32084-5 (ebk)

DOI: 10.1201/9781003320845

Typeset in Latin Modern
by KnowledgeWorks Global Ltd.

Publisher's note: This book has been prepared from camera-ready copy provided by the authors.

Access the Instructor and Student Resources: Support Material URL
https://routledgetextbooks.com/textbooks/instructor_downloads/?_gl=1*amqjdp*_ga*MjYwMDY4MTU0LjE2Nzk0N
Tk3MzY.*_ga_0HYE8YG0M6*MTY4NDkzMTI2OS43LjAuMTY4NDkzMTI2OS4wLjAuMA..

To our students and families

Contents

Welcome

This is the first edition of "Exploring Data Science with R and the Tidyverse: A Concise Introduction", published by CRC Press.

This textbook was used to teach the courses *Data Science for the World* (CSC100) and *Introduction to Data Science in R* (CSC200) at the University of Miami in Fall 2021 and Fall 2022.

Colophon

This book was written in bookdown[1] inside RStudio[2]. The accompanying website is hosted with Netlify[3], and built from source[4] by GitHub Actions[5]. It is made freely available to interested readers at https://ds4world.cs.miami.edu/.

---

[1] http://bookdown.org/
[2] http://www.rstudio.com/ide/
[3] https://www.netlify.com
[4] https://github.com/jerrybonnell/ds4everyone
[5] https://github.com/features/actions

Programming Fundamentals

1

Data Types

Data scientists work with different kinds of data. We have already seen two different types in the previous chapters: *numerical* data (e.g., 14 or 1.5) and *text* data (e.g., "you found yourself in a wide, low, straggling entry with old-fashioned wainscots").

These different representations of data are often referred to as *data types*. In this chapter we will learn about four fundamental data types in R and how the tidyverse can be used for working with them. They are:

- a whole number, which we call the *integer* type
- a real number, which we call the *double* type
- a truth value representing TRUE or FALSE, which R calls the *logical* type. These are also often called the *boolean* type, especially in other programming languages
- a character sequence, which R calls the *character* type. These are commonly referred to as the *string* type.

We will also study two important structures for *storing* data known as the *vector* and the *list*.

1.1 Integers and Doubles

In this section, we turn to our first two principal data types: the *integer* and the *double*.

1.1.1 A primer in computer architecture

The vacuum tubes we mentioned earlier as one of the principal ideas in the early modern computers had the role of regulating flow of electricity. The architects of the machines used two conditions, the existence of current and the non-existence of current, for information coding.

We used the word *bit* to present the dichotomy, 1 for having the current and 0 for not having the current. All the computer models that followed used the principle.

An *integer* type consists of some fixed number of (e.g., 32) bits. In its simplest representation, one of the bits is used to represent the sign (1 for negative numbers and 0 for non-negative numbers) and the rest are to represent the absolute value of the number with the power-of-2 denominations.

The power-of-2 denominations consist of the powers of 2 in increasing order starting from the 0th power (which is equal to 0): $2^0 = 1, 2^1 = 2, 2^2 = 4, 2^3 = 8, 2^4 = 16, ...$

We call this the *binary representation* as opposed to the *decimal representation* as we write numbers in our daily life.

DOI: 10.1201/9781003320845-1

For example,

$$0 \cdots 011101$$

in the representation is to represent by reading the bits from right to left,

$$1 \cdot 2^0 + 0 \cdot 2^1 + 1 \cdot 2^2 + 1 \cdot 2^3 + 1 \cdot 2^4$$

which is equal to $1 + 4 + 8 + 16 = 29$.

This implies that there is a limit to how large a positive whole number or how small a negative whole number R can accurately represent. If a number should go out of the range of accurately presentable whole numbers, R switches to a double number using some approximation.

To represent a *double* number, computers split the number of bits it can use into three parts: the sign (1 bit), the *significand*, and the *exponent*. How R uses the significand and exponents in representing a real number is beyond the scope of this text.

$$\text{(sign) (number significand represents)} \cdot \text{(base)}^{\text{number exponent represents}}$$

The representation in the significand part uses the inverse powers of 2: $1/2$, $1/4$, $1/8$, ..., down to a certain value $1/2^m$, where m is the length of the significand part.

By combining these fractional quantities, we can obtain an approximation for a number between 1 and 2. Like integer, the double type thus has a range of numbers it can record an approximation.

If you are interested in learning more about digital representations of numbers, sign up for a course in computer organization and architecture!

1.1.2   Specifying integers and doubles in R

Our brief dive into how integers and doubles are represented boils down to two basic ideas:

- Integers are called `integer` values in R. They can only represent whole numbers (e.g., positive, negative, or zero).
- Real numbers are called `double` values in R. They can represent whole numbers and fractions, but there are limitations by how well it approximates some values.

Here are some examples of `double` values.

```
1.2
```

```
## [1] 1.2
```

```
3.0
```

```
## [1] 3
```

```
1.5 + 2.2
```

```
## [1] 3.7
```

We can confirm these are doubles using the typeof function.

```
typeof(3.0)
```

```
## [1] "double"
```

However, for numbers without a decimal point, i.e., integers, R will still treat them as double.

```
3   # here is a value that looks like an integer
```

```
## [1] 3
```

```
typeof(3) # ..but is actually a double!
```

```
## [1] "double"
```

Hence, to create an integer, we must specify this explicitly by placing an L directly after the number. Here are some examples.

```
# Some integer values
2L
```

```
## [1] 2
```

```
1L + 3L
```

```
## [1] 4
```

```
-123456789L
```

```
## [1] -123456789
```

We can confirm these are integers using the typeof function.

```
typeof(-123456789L)
```

```
## [1] "integer"
```

Check out what happens when we try placing an L after a decimal value!

```
3.2L
```

```
## [1] 3.2
```

When a mathematical expression contains a double and an integer, the result is always a double. The third expression 4L + 1 adds an integer 4 and a double together and so is double. In the fourth one, 1 is also an integer, and so the result is an integer.

```
3.5 + 1.2
```

```
## [1] 4.7
```

```
3L + 1.3
```

```
## [1] 4.3
```

```
4L + 1
```

```
## [1] 5
```

```
4L + 1L
```

```
## [1] 5
```

We can confirm the types using typeof().

```
typeof(3.5 + 1.2)
```

```
## [1] "double"
```

```
typeof(3L + 1.3)
```

```
## [1] "double"
```

```
typeof(4L + 1)
```

```
## [1] "double"
```

```
typeof(4L + 1L)
```

```
## [1] "integer"
```

1.1.3 Finding the size limit

How big numbers (or negative with big absolute value) can a double represent? To specify a double number with a large absolute value, we can use the e expression. If you type a literal that goes beyond the range, R gives you Inf or -Inf to mean that the number is out of range.

Let's start with e1000.

```
-1.0e1000
```

```
## [1] -Inf
```

```
1.0e1000
```

```
## [1] Inf
```

So, using whether the result is `Inf` or not, we can explore around where between the boundary between presentable numbers and non-presentable numbers:

```
1.0e500
```

```
## [1] Inf
```

```
1.0e400
```

```
## [1] Inf
```

```
1.0e300
```

```
## [1] 1e+300
```

```
1.0e310
```

```
## [1] Inf
```

```
1.0e305
```

```
## [1] 1e+305
```

```
1.0e306
```

```
## [1] 1e+306
```

```
1.0e307
```

```
## [1] 1e+307
```

```
1.5e308
```

```
## [1] 1.5e+308
```

```
1.6e308
```

```
## [1] 1.6e+308
```

```
1.7e308
```

```
## [1] 1.7e+308
```

```
1.8e308
```

```
## [1] Inf
```

```
1.790e308
```

```
## [1] 1.79e+308
```

So, around `1.79e308` is the limit. The quantity is large enough so as to accommodate the computation we will do in this text.

How about integers? We can do the same exploration, this time appending the character `L` at the end of each number literal. We leave this as an exercise for the reader. Keep in mind that if you supply an integer that is too big, R will present a warning that the value must be converted to a double.

1.2 Strings

A *string* is a sequence of characters, which takes its place as our third principal data type. Formally, R calls such sequences *character vectors*, where the word "vector" should seem like gibberish jargon at this point. In this section, we examine the string type.

1.2.1 Prerequisite

There is a variety of operations that are available in base R for manipulating strings. A part of the `tidyverse` super-library is the `stringr` library. We will use a few functions from `stringr` here, so let us load this package.

1.2.2 Strings in R

We use a matching pair of double quotations or a matching pair of single quotation marks (a matching pair of apostrophes) to mark the start and the end of the character sequence we specify as its contents. An advantage of using the double quotation marks is that single quotation marks can appear in the character sequence.

Here are some examples of strings.

```
"This is my first string."
```

```
## [1] "This is my first string."
```

```
"We love R!"
```

```
## [1] "We love R!"
```

```
'"Data Science" is a lot of fun :-)'
```

```
## [1] "\"Data Science\" is a lot of fun :-)"
```

Notice that the R substituted the single quotation marks we used for the last literal with double quotation marks. We can inspect these types as well.

```
typeof("We love R!")
```

```
## [1] "character"
```

1.2.3 Conversions to and from numbers

First of all, characters are not compatible with numbers. You cannot apply mathematical operations to strings even if their character contents are interpretable as numbers.

In other words, "4786" is *not* the number 4786, but a character sequence with the four characters, "4", "7", "8", and "6". Knowing that the string can mean the number, we can generate an integer representing the number using the function as.integer.

```
as.integer("4786")
```

```
## [1] 4786
```

The as functions are useful also when you want to interpret a string as a double and when you want to interpret a number as a string.

The functions as.double and as.character convert a string to a double number and a number to a string, respectively.

```
as.double("3.14")
```

```
## [1] 3.14
```

```
as.double("456")
```

```
## [1] 456
```

```
as.character(3.14159265)
```

```
## [1] "3.14159265"
```

```
as.character(-465)
```

```
## [1] "-465"
```

1.2.4 Length of strings

If you know the contents of the string, you can count the characters in the sequence to obtain its length. However, if you do not know or if the string is very long, you can rely on R to do the counting for you using str_length from the stringr package.

```
str_length("310A Ungar, 1365 Memorial Drive, Coral Gables, Florida")
```

```
## [1] 54
```

1.2.5   Substrings

A string is a sequence of characters. Each character composing a string receives a unique position. Let us consider the following string.

```
my_string <- "song"
```

The positioning starts from the left end of the sequence.

- The first position has value 1.
- The four characters of the string, "s", "o", "n", and "g", have positions 1, 2, 3, and 4, respectively.
- You can specify a string and two positions and obtain a new string consisting of the characters between the two positions.

For example, the substring from position 2 to position 3 of the string "song" is "on". We can use the function `str_sub` from `stringr` to retrieve substrings.

```
str_sub("song", 2, 3)
```

```
## [1] "on"
```

The syntax is `str_sub(some_string, start, end)` where `some_string` from which we will build a substring, `start` is the staring position, and `end` is the ending position.

You can omit the ending position if it is the last position of the string.

```
str_sub("song", 2)
```

```
## [1] "ong"
```

You can also retrieve substrings by searching from the *right*. If the starting number is less than -1, R looks at the index starting from the right. For instance, the following extracts the substring from the third-to-last to last position.

```
str_sub("song", -3, -1)
```

```
## [1] "ong"
```

If the ending position is smaller than the starting position, the substring is empty.

```
str_sub("song", 2, 0)
```

```
## [1] ""
```

1.2.6   String concatenation

We may want to combine multiple strings into a single string. We call such action *concatenation*.

Let us consider three strings.

```
str1 <- "data"
str2 <- "science"
str3 <- "rocks"
```

There are two types of concatenation. One type connects strings with no gap, and the other connects strings with one white space inserted in between. We can perform these actions using the `str_c` function from `stringr`.

Below, we think of concatenating three strings "data", "science", and "rocks" with the two functions individually.

```
str_c(str1, str2, str3)
```

```
## [1] "datasciencerocks"
```

```
str_c(str1, str2, str3, sep = " ")
```

```
## [1] "data science rocks"
```

In the second, we provide a whitespace character " " to the argument `sep`. The effect obtained is that each individual word is separated by a space.

1.3 Logical Data

As we stated before, boolean is the data type for logical values, true and false. `TRUE` is the value representing true, and `FALSE` is the one representing false. When you use a number where R is expecting to see a boolean, it interprets 0 as `FALSE` and any non-zero as `TRUE`.

Here are the two values.

```
TRUE
```

```
## [1] TRUE
```

```
FALSE
```

```
## [1] FALSE
```

1.3.1 Comparisons

While we can specify a boolean value with a boolean literal, we can use comparisons to generate boolean values. In the case of numbers, we can compare the values of two mathematical expressions. Here is an example.

```
1 + 4 > 7 - 4
```

```
## [1] TRUE
```

This asks if the value of 1 + 4 is strictly greater than the value of 7 - 4. The former is 5 and the latter is 3, and so the answer to the comparison is in the affirmative. Therefore, the value of the comparison expression is TRUE.

They represent "is equal to", "is not equal to", "is greater than", "is greater than or equal to", "is smaller", and "is smaller than or equal to".

There are six possible types of comparisons. We will define these in the following table using an example with the numbers 6 and 3. Before looking, can you try to guess them all? Compare your guesses against the table.

Operator	Meaning	True example	False example
<	Smaller than	3 < 6	6 < 3
>	Greater than	6 > 3	3 > 3
<=	Smaller than or equal to	3 <= 3	6 <= 3
>=	Greater than or equal to	6 >= 6	3 >= 6
==	Equal to	3 == 3	6 == 3
!=	Not equal to	6 != 3	3 != 3

An expression can chain together multiple comparisons with the AND operator &, and they all must hold in order for the whole expression to be True.

For example, we can express that 1+1 is between 1 and 3 using the following expression.

```
1 < (1 + 1) & (1 + 1) < 3
```

```
## [1] TRUE
```

You may recall the functions max and min for obtaining the maximum and the minimum from a group of numbers, respectively. One thing to remember is that the minimum is greater than or equal to the average, and the maximum is greater than or equal to the average. The equality holds when the two numbers in the group are all identical. Let us set some numbers to variables x and y and see what the maximum and the minimum functions produce.

```
x <- 12
y <- 5
min(x, y) <= (x + y)/2
```

```
## [1] TRUE
```

```
max(x, y) >= (x + y)/2
```

```
## [1] TRUE
```

How about the equality? Assuming that we have executed the previous section of the code, we can reuse x and y in the following computation.

```
x <- 17
y <- 17
min(x, y) == (x + y)/2
```

```
## [1] TRUE
```

```
max(x, y) == (x + y)/2
```

```
## [1] TRUE
```

1.3.2 Boolean operations

There are three fundamental boolean operations. They are *negation*, *disjunction*, and *conjunction*. Negation flips the value of a boolean. Disjunction tests if at least one of boolean values appearing on a list is true, and conjunction tests if all values appearing on a list are. R uses symbols !, |, and & for them. Here are some examples of using the boolean operations.

```
a <- TRUE
b <- FALSE
c <- TRUE
!a
```

```
## [1] FALSE
```

```
a | b | c
```

```
## [1] TRUE
```

```
a & b & c
```

```
## [1] FALSE
```

The roles these operations play are analogous to the roles −, +, and ∗ play in the numbers.

1.3.3 Comparing strings

R can compare strings for equality and non-equality using == and !=. R can also compare two to see if one is greater than the other and if one is smaller than the other. To compare two strings, R compares their characters position-wise, starting from the beginning. The position-wise comparison continues until it reaches a position where no comparison is possible because either at least one string has no characters remaining or the two strings showing non-identical pair of characters.

- In the first case, if R has run out of characters on both sides, it asserts that the two strings are equal to each other; otherwise, the one that has just run out of characters is smaller than the other.
- In the second case, R examines the *character code* of the two characters.

R (and any programming language) uses a table of characters where each character has a unique number. The result of comparing two characters is that if the two characters are not equal to each other, then the character with a lower position than the other is smaller than the other as character.

```
print("Bach" > "Back")
```

```
## [1] FALSE
```

```
print("Darth Vader" > "Dark Chocolate")
```

```
## [1] TRUE
```

```
print("09:00 AM" > "Nine in the morning")
```

```
## [1] FALSE
```

```
print("data science" > "Data Science")
```

```
## [1] FALSE
```

```
print("abc" > "ABC" )
```

```
## [1] FALSE
```

The examples below show how R interprets numbers to boolean values.

```
3.0 == TRUE
```

```
## [1] FALSE
```

```
-5 == FALSE
```

```
## [1] FALSE
```

```
0 == TRUE
```

```
## [1] FALSE
```

```
0 == FALSE
```

```
## [1] TRUE
```

1.4 Vectors

You might have been wondering about the meaning of the [1] that appears when you inquire about the value of an expression. The square brackets [] in R means the position in a series of distinct elements known as a *vector*.

1.4.1 Sequences

The [1] indicates that the value that follows is the first element of a sequence that contains the value. Meaning, the value is not a standalone value "per se" but appears as an element in a sequence. Encompassing a value in a sequence is a distinctive feature of R. In more technical terms, R uses *vectorization* to put objects in vectors. Let us see an example of this in action.

Suppose we have assigned a value of 10 to an object my_vec.

```
my_vec <- 10
```

Then a is a *vector* containing one element, whose value is 10.

```
my_vec
```

```
## [1] 10
```

The [1] 10 appearing as the output states exactly that. Visually, this looks like:

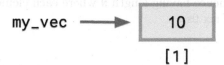

The way we access an element of a vector is to state the position of the element in the sequence comprising the vector inside square brackets and attach it after the name of the vector. So, let us see what my_vec[1] returns. The expression means to refer to the first element of my_vec, which we know to be 10.

```
my_vec[1]
```

```
## [1] 10
```

Wait a second, it still says [1] 10! Because a vector is the most primitive information structure that R uses, there is no smaller structure. This means that you can apply [1] as many times you want to a.

```
my_vec[1][1]
```

```
## [1] 10
```

```
my_vec[1][1][1]
```

```
## [1] 10
```

```
my_vec[1][1][1][1]
```

```
## [1] 10
```

```
my_vec[1][1][1][1][1]
```

```
## [1] 10
```

Since the number inside the brackets specifies a position, you can try a number other than 1. It is only that in the case of a, positions other than 1 do not exist.

```
my_vec[2]
```

```
## [1] NA
```

```
my_vec[0]
```

```
## numeric(0)
```

What are these? NA is short-hard for "not available" and means that there is no such thing. numeric(0) means that it is a vector with no elements. Once you get to length 0, [1] becomes NA but [0] produces numeric(0).

numeric(N) with N produces a sequence having length N where each element is 0. Here is an example which creates a vector of six 0s.

```
my_vec2 <- numeric(5)
```

We can examine its contents and even change the values. Below we change the values of my_vec2 at positions 2 and 4 to 2 and 4, respectively.

```
my_vec2[2] <- 2
my_vec2[4] <- 4
my_vec2
```

```
## [1] 0 2 0 4 0
```

Here is a visualization of the situation.

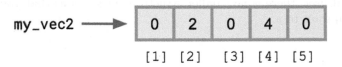

Let's play with these vectors a bit more. In addition to specify a single position, you can give a range of positions, "from here to there". The syntax for a range specification is FROM:TO where FROM is the starting position and TO is the ending position.

The code below changes the values for the five positions of series_n and then provides examples of some range indexing.

```
my_vec2[1] <- 1
my_vec2[3] <- 3
my_vec2[5] <- 5
my_vec2[1:5]
```

```
## [1] 1 2 3 4 5
```

```
my_vec2[2:4]
```

```
## [1] 2 3 4
```

```
my_vec2[2:2]
```

```
## [1] 2
```

If the END value is smaller than the START value, the elements appear in the reverse order.

```
my_vec2[4:1]
```

```
## [1] 4 3 2 1
```

You can use the negative sign in the range and position specification. The - sign means "all positions other than". If the negative sign appears with the FROM or the TO index, then the other index must have the negative sign. In the case of a negative range, the order between FROM and TO does not matter.

```
my_vec2
```

```
## [1] 1 2 3 4 5
```

```
my_vec2[-3]
```

```
## [1] 1 2 4 5
```

```
my_vec2[-2:-4]
```

```
## [1] 1 5
```

```
my_vec2[-4:-2]
```

```
## [1] 1 5
```

1.4.2   The combine function

We can create sequences with element specification. The creation of this uses the function c, which means to "combine".

The syntax is quite simple. Within the parentheses following the initial c, state the elements of the series with a comma in between.

```
c(6, 2, 3)
```

```
## [1] 6 2 3
```

```
c("data", "science", "rocks", "my", "socks")
```

```
## [1] "data"    "science" "rocks"    "my"        "socks"
```

```
c(3.0, 4.0, 2.0, 2.2, -4.5, -25.7)
```

```
## [1]    3.0    4.0    2.0    2.2   -4.5 -25.7
```

Using the c construction, you can obtain a subseries of a mother sequence with specific positions.

```
my_vec2[c(1,3,4)]
```

```
## [1] 1 3 4
```

```
my_vec2[c(4,4,3,3,5,3,5,3)]
```

```
## [1] 4 4 3 3 5 3 5 3
```

R retrieves the elements individually, and whether the numbers repeat or whether the numbers are in order do not matter.

1.4.3   Element-wise operations

Let us look at the combine function more closely. We can define two sequences having the same lengths using the combine function.

```
a <- c(2, 3, 4, 5, 1, 6)
b <- c(9, 8, 7, 1, 2, 1)
```

A very common and useful set of operations when working with vectors is known as *element-wise* operations, which work on each element of a vector. Here is an example with element-wise addition.

```
a + 1/2
```

```
## [1] 2.5 3.5 4.5 5.5 1.5 6.5
```

To explain what happened, here is a visualization.

2				2 + 1/2			2.5
3				3 + 1/2			3.5
4				4 + 1/2			4.5
5	+	1/2	=	5 + 1/2	=		5.5
1				1 + 1/2			1.5
6				6 + 1/2			6.5
a				a			a

We can also apply element-wise subtraction, multiplication, division, and remainder.

```
a - 7
```

```
## [1] -5 -4 -3 -2 -6 -1
```

```
a * 3
```

```
## [1]  6  9 12 15  3 18
```

```
b / 2
```

```
## [1] 4.5 4.0 3.5 0.5 1.0 0.5
```

```
b %% 3
```

```
## [1] 0 2 1 1 2 1
```

These element-wise operations can be applied to two sequences that have the same length.

```
a + b
```

```
## [1] 11 11 11  6  3  7
```

```
a - b
```

```
## [1] -7 -5 -3  4 -1  5
```

```
a * b
```

```
## [1] 18 24 28  5  2  6
```

```
a / b
```

```
## [1] 0.2222222 0.3750000 0.5714286 5.0000000 0.5000000
## [6] 6.0000000
```

```
a %% b
```

```
## [1] 2 3 4 0 1 0
```

You can access some properties of the vector. `length`, `max`, and `min` provide the length of a vector, the maximum among the elements in a vector, and the minimum among the elements in it, respectively.

```
length(a)
```

```
## [1] 6
```

```
max(a)
```

```
## [1] 6
```

```
min(a)
```

```
## [1] 1
```

In the case of numbers, the summation of all elements is possible.

```
sum(a)
```

```
## [1] 21
```

Another important feature of the function `c` is you can connect two vectors with `c`. Here we recall the vectors `a` and `b` from earlier and then present the difference between the component-wise addition `a + b` and the sequence connection `c(a, b)`.

```
c(a, b)
```

```
##  [1] 2 3 4 5 1 6 9 8 7 1 2 1
```

Since a single value is a vector, connecting a vector with a conspicuously single value are actually vector concatenation.

```
c(a, 10)
```

```
## [1]  2  3  4  5  1  6 10
```

```
c(78, a)
```

```
## [1] 78  2  3  4  5  1  6
```

You can connect more than two elements.

```
c(79, a, 17)
```

```
## [1] 79  2  3  4  5  1  6 17
```

1.4.4 Booleans and element-wise operations

Not only can we apply mathematical operations, but also, we can apply comparisons.

```
a > b
```

```
## [1] FALSE FALSE FALSE  TRUE FALSE  TRUE
```

```
a >= 3
```

```
## [1] FALSE  TRUE  TRUE  TRUE FALSE  TRUE
```

Once you have a sequence of boolean whose length is equal to the vector at hand, you can use that boolean sequence (or vector) to select elements to generate subvectors.

```
a[a > b]
```

```
## [1] 5 6
```

```
a[a >= 3]
```

```
## [1] 3 4 5 6
```

We can also use element-wise comparisons to count the number of occurrences of a certain element in a vector. For instance, consider this vector of greetings.

```
greetings <- c("hello", "goodbye", "hello", "hello", "goodbye")
```

We can count the number of `hello`'s that occur as follows.

```
greetings == "hello"
```

```
## [1]  TRUE FALSE  TRUE  TRUE FALSE
```

```
sum(greetings == "hello")
```

```
## [1] 3
```

1.4.5 Functions on vectors

R provides programmers with convenient and powerful functions for creating and manipulating vectors.

The `mean` of a collection of numbers is its average value: the sum divided by the length. Each of the examples below performs a computation on the vector called `temps`.

```
temps <- c(87.5, 87.5, 66.5, 90.0, 65.5, 71.0)
length(temps)
```

```
## [1] 6
```

```
sum(temps)
```

```
## [1] 468
```

```
mean(temps)
```

```
## [1] 78
```

The `diff` function computes the difference between each adjacent pair of elements in an array. The first element of the `diff` is the second element minus the first.

```
diff(temps)
```

```
## [1]    0.0 -21.0   23.5 -24.5    5.5
```

Following are some more commonly used functions that work over vectors. The list includes many functions that we have not gone over yet, especially those that work on *character* vectors. Learning this vocabulary is an important part of learning R, so refer to this list often as you work through examples and problems.

Please note: you **do NOT need to memorize these!**

Each of these functions takes some vector x as an argument and returns a single value.

Function	Description
`sum(x)`	Add elements together
`prod(x)`	Multiply elements together
`all(x)`	Test whether all elements are true values
`any(x)`	Test whether any elements are true values
`sum(x != 0)`	Number of non-zero elements

Each of these functions takes some vector x as an argument and returns a vector of values.

Function	Description
`diff(x)`	Difference between adjacent elements
`round(x)`	Round each number to nearest integer
`cumprod(x)`	For each element, multiply all elements so far (cumulative product)
`cumsum(x)`	For each element, add all elements so far (cumulative sum)
`exp(x)`	Computes the exponential function
`log(x)`	Computes the natural logarithm of each element
`sqrt(x)`	Computes the square root of each element
`sort(x)`	Sort the elements

The `stringr` package from the tidyverse provides us a collection of useful functions for working with character vectors. A full cheat sheet can be found here[1], but listed here are

---

[1]https://evoldyn.gitlab.io/evomics-2018/ref-sheets/R_strings.pdf

some of the commonly used ones. Following are functions that take a character vector x as an argument and return a vector.

Function	Description
str_to_lower(x)	Lowercase each element
str_to_upper(x)	Uppercase each element
str_trim(x)	Remove spaces at the beginning and/or end of each element

The following function takes a character vector x with additional arguments but also returns a vector.

Function	Description
str_sub(x, start, end)	Extracts a substring from x given by start position and end position
str_detect(x, "[:alpha:]")	Whether each element is only letters (no numbers or symbols)
str_detect(x, "[:digit:]")	Whether each element is only numeric (no letters)

The following functions take both a character vector x and a *pattern string* to search for. Pattern strings can be more general like "[:alpha:]" or "[:digit:]" from the above list (these are also called regular expressions[2] or regexp for short, which we won't cover in detail :-). Each of these functions returns a vector.

Function	Description
str_count(x, pattern)	Count the number of times a pattern appears among the elements of an array
str_which(x, pattern)	The indexes of vector x where the pattern is found
str_replace_all(x, pattern)	Replace all matched patterns in each string
str_detect(x, ^pattern)	Whether each element starts with the pattern

The following function also takes a vector x of strings and a pattern string. However, unlike the above table, this function returns a *list*. We will cover lists in the next section.

Function	Description
str_locate_all(x, pattern)	The positions within each element that a pattern is found

The following function is a helpful diagnostic tool to view matched patterns.

---

[2]https://github.com/rstudio/cheatsheets/blob/main/regex.pdf

Function	Description
str_view_all(x, pattern)	Visualizes all pattern matches in vector x

1.5   Lists

Like vectors, lists also group values together. However, unlike vectors, lists can hold values that are of *different* types. For instance:

```
mixed <- list("apple", 1.5, 2L, TRUE)
mixed
```

```
## [[1]]
## [1] "apple"
##
## [[2]]
## [1] 1.5
##
## [[3]]
## [1] 2
##
## [[4]]
## [1] TRUE
```

It can be helpful to examine the structure inside the list. We use str for this.

```
str(mixed)
```

```
## List of 4
##  $ : chr "apple"
##  $ : num 1.5
##  $ : int 2
##  $ : logi TRUE
```

Lists hold just about anything; they can even contain vectors...

```
mixed2 <- list(c("asparagus", "arrowroot", "tomato"),
               c("mango", "kumquat"),
               3.14159)
str(mixed2)
```

```
## List of 3
##  $ : chr [1:3] "asparagus" "arrowroot" "tomato"
##  $ : chr [1:2] "mango" "kumquat"
##  $ : num 3.14
```

...or more lists!

```
omg <- list(list(1,1), list(2,2,2), "hello world")
str(omg)
```

```
## List of 3
##  $ :List of 2
##   ..$ : num 1
##   ..$ : num 1
##  $ :List of 3
##   ..$ : num 2
##   ..$ : num 2
##   ..$ : num 2
##  $ : chr "hello world"
```

1.5.1 Working with lists

Let's examine the mixed2 list more closely.

```
mixed2
```

```
## [[1]]
## [1] "asparagus" "arrowroot" "tomato"
##
## [[2]]
## [1] "mango"   "kumquat"
##
## [[3]]
## [1] 3.14159
```

[extracts a sub-list. The result is always a list.

```
str(mixed2[2])
```

```
## List of 1
##  $ : chr [1:2] "mango" "kumquat"
```

We can also *subset* a list the same way we do with vectors. Remember that the result is still a list.

```
str(mixed2[1:2])
```

```
## List of 2
##  $ : chr [1:3] "asparagus" "arrowroot" "tomato"
##  $ : chr [1:2] "mango" "kumquat"
```

If we wish to extract the vector *inside* mixed2[2], we must use [[. This extracts a single component from a list. We can use it to retrieve, for example, the vector of fruits:

```
mixed2[[2]]
```

```
## [1] "mango"   "kumquat"
```

What if we only wanted the mango?

```
mixed2[[2]][1]
```

```
## [1] "mango"
```

Yum!

1.5.2 Visualizing lists

The difference between `[` and `[[` is important, but too easy to confuse. The following visual will clarify the point.

Suppose that we have the following list `a`.

```
a <- list("A big cat", c(1,2,3), 3.14159)
```

Then we can imagine operations on `a` as the following:

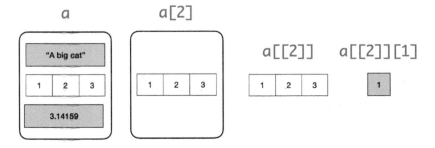

- a is a list
- a[2] is also a list, this time containing a single element
- a[[2]] is the second component of the list, a vector
- a[[2]][1] is the first element of that vector

Groovy! The "pepper shaker photos" in R for Data Science[3] provides another nice visualization of the different ways we can extract pieces from a list.

1.6 stringr Operations

As we mentioned earlier, `tidyverse` is a collection of packages. One of the packages `tidyverse` contains is `stringr`, which offers a variety of methods for manipulating strings. Like numbers, a string is a vector of strings with just one string. So, applying a function to a string is the same as applying a function to a string vector.

Depending on what we are interested in doing, we can classify the functions in `stringr` into the following categories:

---

[3] https://r4ds.had.co.nz/vectors.html#lists-of-condiments

- Detect Matches: Finding and locating matches of a pattern in each string in a string vector
- Subset Strings: Extracting substrings and subvectors from a string vector matching a pattern
- Manage Lengths: Inquiring about the lengths and padding/trimming the strings
- Mutate Strings: Mutating the strings appearing in a string vector
- Join and Split: Changing the structure of a vector
- Order Strings: Sorting

Be sure to bookmark the `stringr` cheatsheet[4].

1.6.1 Prerequisites

To set up, let us load the `tidyverse`.

```
library(tidyverse)
```

1.6.2 Regular expressions

The pattern matching functions of `stringr` accept *regular expressions*.

A regular expression is a string that specifies a collection of strings in a possibly compact manner. Here are some examples of regular expressions and how one of the `stringr` functions `str_detect` uses it find patterns in string vector c("May 17, 2019", "Certified mail", "FL 33333", "Oppa Locka", "to Mr. Haan", "arrived"). The function `str_detect` receives two arguments. The first is a string (or a string vector) and the second is the pattern. The function returns for each element in the vector, whether the pattern appears. Let us first define a string vector with the five elements.

```
s <- c("May 17, 2019", "Certified mail", "FL 33333",
       "Oppa Locka", "to Mr. Haan", "arrived")
s
```

```
## [1] "May 17, 2019"   "Certified mail" "FL 33333"
## [4] "Oppa Locka"     "to Mr. Haan"    "arrived"
```

Now you see a bracket with a number other than 1. The [5] states that "to Mr. Haan" is the fifth element of the vector.

The pattern strings appearing in the following have the following meaning

pattern	meaning
"a"	any appearance of "a"
"a[iy]"	any "a" then one of "i" or "y"
"a$"	"a" at the end of string
"^a"	"a" at the start of string
"^a.*d$"	"a" at the start, any string, and then a "d" at the end

---

[4]https://github.com/rstudio/cheatsheets/blob/master/strings.pdf

pattern	meaning
"ppa"	"ppa"
"3{4}"	"3" repeated four times in sequence
"[aeiou].[aeiou]"	one from "aeiou", some character, and one from "aeiou"

```
s
```

```
## [1] "May 17, 2019"    "Certified mail" "FL 33333"
## [4] "Oppa Locka"      "to Mr. Haan"    "arrived"
```

```
str_detect(s, "a")
```

```
## [1]  TRUE  TRUE FALSE  TRUE  TRUE  TRUE
```

```
str_detect(s, "a[iy]")
```

```
## [1]  TRUE  TRUE FALSE FALSE FALSE FALSE
```

```
str_detect(s, "a$")
```

```
## [1] FALSE FALSE FALSE  TRUE FALSE FALSE
```

```
str_detect(s, "^a")
```

```
## [1] FALSE FALSE FALSE FALSE FALSE  TRUE
```

```
str_detect(s, "^a.*d$")
```

```
## [1] FALSE FALSE FALSE FALSE FALSE  TRUE
```

```
str_detect(s, "ppa")
```

```
## [1] FALSE FALSE FALSE  TRUE FALSE FALSE
```

```
str_detect(s, "3{4}")
```

```
## [1] FALSE FALSE  TRUE FALSE FALSE FALSE
```

```
str_detect(s, "[aeiou].[aeiou]")
```

```
## [1] FALSE  TRUE FALSE FALSE FALSE  TRUE
```

So, the syntax is:

- use the square brackets to specify a list of characters
- use the curly brackets to specify the number of repetitions
- use the period to specify any character

- use the caret "\^" and the dollar sign to specify the start and the end of string, respectively
- use * to specify any number of repetitions, including 0 repetitions.

A useful diagnostic tool to visualize matches against a pattern is the function str_view_all, which highlights any matches hit by the regular expression. For instance:

```
str_view_all(s, "[aeiou].[aeiou]", html = TRUE)
```

May 17, 2019

Certified mail

FL 33333

Oppa Locka

to Mr. Haan

arrived

In the square brackets you can specify a range using a dash and the caret to mean "not". By putting a pair of numbers inside a pair of curly brackets, you can specify a permissible range of the number of repetitions.

The patterns below mean the following:

pattern	meaning
"[a-z]"	"a" through "z" at least one
"[A-Z]{2,5}"	"A" through "Z" between 2 and 5 times
"[^a-zA-Z]"	a non-alphabet
"[0-9]"	a numeral

```
s
```

```
## [1] "May 17, 2019"   "Certified mail" "FL 33333"
## [4] "Oppa Locka"      "to Mr. Haan"    "arrived"
```

```
str_detect(s, "[a-z]")
```

```
## [1]  TRUE  TRUE FALSE  TRUE  TRUE  TRUE
```

```
str_detect(s, "[A-Z]{2,5}")
```

```
## [1] FALSE FALSE  TRUE FALSE FALSE FALSE
```

```
str_detect(s, "[^a-zA-Z]")
```

```
## [1]  TRUE  TRUE  TRUE  TRUE  TRUE FALSE
```

```
str_detect(s, "[0-9]")
```

```
## [1]  TRUE FALSE  TRUE FALSE FALSE FALSE
```

1.6.3   Detect matches

There are functions in this category.

- str_detect(STRING, PATTERN): As we have seen previously, the function returns a vector of boolean representing whether the elements of STRING matching PATTERN.
- str_which(STRING, PATTERN): The function works like str_detect but instead of boolean vector, returns the indexes at which the pattern appears; in other words, at which indexes, the value of str_detect(STRING, PATTERN) is true.
- str_count(STRING, PATTERN): The function returns for each element of STRING, at how many different positions the pattern aligns.
- str_locate(STRING, PATTERN): The function finds for each element of STRING, the first (or the closest to the start of string) match of the pattern and provides the start and end character positions of the first match as a pair of integers (that is, a length-2 vector of integers). If there are no matches for an element of STRING, the function returns a pair of NA.

Let us recall s.

```
s
```

```
## [1] "May 17, 2019"   "Certified mail" "FL 33333"
## [4] "Oppa Locka"     "to Mr. Haan"    "arrived"
```

Here is the result of finding "[0-9]" in the strings.

```
str_which(s, "[0-9]")
```

```
## [1] 1 3
```

Here is finding one of "aeiou" and any alphabet and counting the occurrences. Note that in "to Mr. Haan" there are two places that match the pattern "[aeiou][A-Za-z]", that is "aa" and "an". The pattern counting goes by repeating the action of finding the first match and then asserting the characters leading to the end of the match unusable for finding matches. With the feature, after finding the first, "aa", the prefix "to Mr. Haa" is no longer available, and so "an" does not qualify as a match.

```
str_count(s, "[aeiou][A-Za-z]")
```

```
## [1] 1 4 0 1 1 3
```

So, if we find three numerals in sequence with no gap, the elements 1 and 3 have exactly one match each, though there are multiple possibilities for aligning the pattern.

```
str_count(s, "[0-9]{3}")
```

```
## [1] 1 0 1 0 0 0
```

1.6.4 Subset strings

Here are the functions in this category.

- str_sub(STRING, START, END): Creates a new vector consisting of the substrings of the elements of STRING with START and END as the starting and ending positions, respectively.
- str_subset(STRING, pattern): Creates a new vector keeping only those elements in s matching the pattern.
- str_extract(STRING, pattern): The same as str_subset concerning what to find, but the function returns a table not a one-dimensional vector. See the examples below to examine the differences among str_subset, str_extract, and str_match.

```
str_sub(s, 4, 7)
```

```
## [1] " 17," "tifi" "3333" "a Lo" "Mr. " "ived"
```

```
str_subset(s, "[a-zA-Z][^a-zA-Z]+[a-zA-Z]")
```

```
## [1] "Certified mail" "Oppa Locka"     "to Mr. Haan"
```

```
str_extract(s, "[a-zA-Z][^a-zA-Z]+[a-zA-Z]")
```

```
## [1] NA    "d m" NA    "a L" "o M" NA
```

```
str_match(s, "[a-zA-Z][^a-zA-Z]+[a-zA-Z]")
```

```
##        [,1]
## [1,] NA
## [2,] "d m"
## [3,] NA
## [4,] "a L"
## [5,] "o M"
## [6,] NA
```

Both str_extract and str_match have their version for "apply to all matches", like str_locate_all. Their names are str_extract_all and str_match_all.

1.6.5 Manage lengths

The following functions belong to this category.

- str_length(STRING): Returns the length of each element.
- str_pad(STRING, WIDTH, side=OPTION, pad=X):Appends the string X to both or either side of each element of STRING as many times as necessary so as to inflate the length of element to at least WIDTH. The OPTION is one of "left", "right", and "both" to indicate

where the padding should occur. They respectively represent "the left side only", "the right side only", and "both sides so as to center the original". If the original has length greater than or equal to WIDTH no padding occurs. In the case of "both", if the number of necessary padding is an odd number to make the length equal to WIDTH, the right side receives one more than the left side. The padding sTring X must be a single character. You may omit the specification part pad= in pad=X.

- str_trunc(STRING, WIDTH, side=OPTION, ellipsis=E): This is in some sense at the opposite of str_pad. The function shrinks each string to length WIDTH with the single-character string E for replacement. The side= specification states where the replacement occurs and should be one of "left", "right", and "center". You may omit the ellipsis= prefix.
- str_trim(STRING, side=OPTION): Trims the white space at the end. The option value is one of "left", "right", and "both", which correspond to "the left side only", "the right side only", and "both sides".

```
str_length(s)
```

```
## [1] 12 14  8 10 11  7
```

```
str_pad(s, 11, side="both", pad=".")
```

```
## [1] "May 17, 2019"   "Certified mail" ".FL 33333.."
## [4] "Oppa Locka."    "to Mr. Haan"    "..arrived.."
```

```
str_pad(s, 10, side="left", ".")
```

```
## [1] "May 17, 2019"   "Certified mail" "..FL 33333"
## [4] "Oppa Locka"     "to Mr. Haan"    "...arrived"
```

```
str_trunc(s, 3, side="right", ellipsis="_")
```

```
## [1] "Ma_" "Ce_" "FL_" "Op_" "to_" "ar_"
```

```
str_trunc(s, 5, side="center", "#")
```

```
## [1] "Ma#19" "Ce#il" "FL#33" "Op#ka" "to#an" "ar#ed"
```

```
str_trim("   abc   ", side="right")
```

```
## [1] "   abc"
```

```
str_trim("   abc   ", side="both")
```

```
## [1] "abc"
```

1.6.6 Mutate strings

The first three functions execute substitutions to parts of each string.

- str_sub() <- VALUE: Here the str_sub() part follows the syntax of the method we earlier discussed - the substrings with specific range of indexes inside strings. After this action, each substring will become VALUE.

```
s <- c("5/17/2019", "Certified mail", "FL 33333",
       "Oppa Locka", "to Mr. Haan", "arrived")
```

Since the str_sub() <- VALUE modifies the original, let us copy s to scopy and then execute the action on the copy, not the original. In that manner, we can preserve the original. Note the use of 0 and 0 in the second instance, the position range of 0 through 0 corresponds to the part before the start of the string.

```
scopy <- s
str_sub(scopy, 1, 3) <- "I am "
scopy
```

```
## [1] "I am 7/2019"     "I am tified mail"
## [3] "I am 33333"      "I am a Locka"
## [5] "I am Mr. Haan"   "I am ived"
```

```
scopy <- s
str_sub(scopy, 0, 0) <- "I am "
scopy
```

```
## [1] "I am 5/17/2019"     "I am Certified mail"
## [3] "I am FL 33333"      "I am Oppa Locka"
## [5] "I am to Mr. Haan"   "I am arrived"
```

```
s
```

```
## [1] "5/17/2019"      "Certified mail" "FL 33333"
## [4] "Oppa Locka"     "to Mr. Haan"    "arrived"
```

- str_replace(STRING, PATTERN, REPLACEMENT): Replaces the first occurrence of PATTERN with REPLACEMENT.

- str_replace_all(STRING, PATTERN, REPLACEMENT): Replaces all occurrences of PATTERN with REPLACEMENT. Like str_locate_all the function finds a set of non-overlapping occurrences and then replaces all the occurrences it has identified.

  Both functions return the vector resulting from their action. The original remains intact.

```
t <- s
str_replace(t, "a", "oo")
```

```
## [1] "5/17/2019"      "Certified mooil" "FL 33333"
## [4] "Oppoo Locka"    "to Mr. Hooan"    "oorrived"
```

```
t <- s
str_replace_all(s, "a", "oo")
```

```
## [1] "5/17/2019"       "Certified mooil" "FL 33333"
## [4] "Oppoo Lockoo"    "to Mr. Hoooon"   "oorrived"
```

```
t <- s
str_replace_all(s,"[0-9][0-9]", "##")
```

```
## [1] "5/##/####"       "Certified mail" "FL ####3"
## [4] "Oppa Locka"      "to Mr. Haan"    "arrived"
```

There are three more string mutation functions.

- str_to_lower(STRING): This converts each uppercase letter to its corresponding lower-case letter.

- str_to_upper(STRING): This converts each lowercase letter to its corresponding upper-case letter.

- str_to_title(STRING): This converts each string to a title-like format.

  Each of the three methods returns a new vector.

```
t <- s
str_to_lower(s)
```

```
## [1] "5/17/2019"       "certified mail" "fl 33333"
## [4] "oppa locka"      "to mr. haan"    "arrived"
```

```
t <- s
str_to_upper(s)
```

```
## [1] "5/17/2019"       "CERTIFIED MAIL" "FL 33333"
## [4] "OPPA LOCKA"      "TO MR. HAAN"    "ARRIVED"
```

```
t <- s
str_to_title(s)
```

```
## [1] "5/17/2019"       "Certified Mail" "Fl 33333"
## [4] "Oppa Locka"      "To Mr. Haan"    "Arrived"
```

1.6.7 Join and split

We only show two functions in this category.

- str_dup(STRING, TIMES): Create from a vector STRING, a new vector where each element appears consecutively, without a gap, TIMES times.
- str_c(STRING): Collapse the elements column-wise into a single dimensional vector.
- str_c(STRING, collapse=X): Concatenate the elements column-wise with X in between and then concatenate the elements row-wise with X in between, thereby generating a single-element vector.

Below are examples.

```
t <- s
str_dup(t, 3)
```

```
## [1] "5/17/20195/17/20195/17/2019"
## [2] "Certified mailCertified mailCertified mail"
## [3] "FL 33333FL 33333FL 33333"
## [4] "Oppa LockaOppa LockaOppa Locka"
## [5] "to Mr. Haanto Mr. Haanto Mr. Haan"
## [6] "arrivedarrivedarrived"
```

```
str_c(t)
```

```
## [1] "5/17/2019"      "Certified mail" "FL 33333"
## [4] "Oppa Locka"     "to Mr. Haan"    "arrived"
```

```
str_c(t, collapse=":")
```

```
## [1] "5/17/2019:Certified mail:FL 33333:Oppa Locka:to Mr. Haan:arrived"
```

1.6.8 Sorting

There are two functions.

- str_sort(STRING,decreasing=X,na_last=Y,numeric=Z): Sorts the elements as the string. The arguments after the first one are optional. X, Y, and Z are boolean. With decreasing=TRUE the ordering is reverse. The na_last=FALSE, all NA move to the start. With numeric=TRUE, the function treats numerical sequences as numbers; otherwise, it treats them as character sequences.
- str_order(...): The same arguments as str_sort, but instead of actually sorting, returns the sequence of indexes such that ordering the elements according to the index list produces the same result as str_sort.

```
u <- str_c(t)
str_sort(u)
```

```
## [1] "5/17/2019"      "arrived"        "Certified mail"
## [4] "FL 33333"       "Oppa Locka"     "to Mr. Haan"
```

```
u <- str_c(t)
str_sort(u, decreasing=TRUE)
```

```
## [1] "to Mr. Haan"    "Oppa Locka"     "FL 33333"
## [4] "Certified mail" "arrived"        "5/17/2019"
```

```
u <- str_c(t)
str_order(u, decreasing=TRUE)
```

```
## [1] 5 4 3 2 6 1
```

1.6.9   An example: `stringr` and lists

We end this section with an educational example of when you might need to work with the
`stringr` package.

Recall that `str_locate_all` is a method for working with strings; it returns the positions
within each element that a pattern string is found as a *list*. Let's try to use what we know
to extract common bird names.

```
birds <- c("Black-crowned Night-Heron-Nycticorax",
           "Little-egret-Egretta garzetta")
```

The common names are `Black-crowned Night-Heron` and `Little-egret`. We can extract this
by looking for the *last* occurrence of the dash (-). Here they are:

```
str_view_all(birds, "-", html=TRUE)
```

Black-crowned Night-Heron-Nycticorax

Little-egret-Egretta garzetta

The `str_locate_all` function can tell us the starting and ending positions of all occurrences
of the dash.

```
dash_positions <- str_locate_all(birds, "-")
dash_positions
```

```
## [[1]]
##      start end
## [1,]     6   6
## [2,]    20  20
## [3,]    26  26
##
## [[2]]
##      start end
## [1,]     7   7
## [2,]    13  13
```

As expected, we have a list.

Let's first examine the results for the Night Heron, which are stored at index 1 of this list.
We now know that to pluck out these results we must use `[[`.

```
dash_positions[[1]]
```

```
##      start end
## [1,]     6   6
## [2,]    20  20
## [3,]    26  26
```

Congrats – we're out of the list! What we have now is three vectors sandwiched together into one, one for each occurrence of the dash. We are only interested in the last one.

`tail` is a handy function for extracting the last value of a vector.

```
tail(dash_positions[[1]], n=1) # why n = 1? why not 2?
```

```
##      start end
## [3,]    26  26
```

Technically, we only care about the ending position of the dash (at index 2) but either value will do.

```
last_dash <- tail(dash_positions[[1]], n=1)[1]
last_dash
```

```
## [1] 26
```

We're almost done: we now know that the common name occurs between positions 1 and 25 (why not 26?). We can use `str_sub` to extract the substring.

```
str_sub(birds[1], end = last_dash - 1)
```

```
## [1] "Black-crowned Night-Heron"
```

Here is everything together:

```
dash_positions <- str_locate_all(birds, "-")
last_dash <- tail(dash_positions[[1]], n=1)[1]
str_sub(birds[1], end = last_dash - 1)
```

```
## [1] "Black-crowned Night-Heron"
```

What would the answer be for extracting `Little-egret`? Try it out!

1.7 Exercises

Be sure to install and load the following packages into your R environment before beginning this exercise set.

```
library(tidyverse)
library(edsdata)
library(gapminder)
```

Question 1. Let us explore the limit of the double data type. According to the textbook, the largest double value is greater than or equal to `1.79e+308`. Can you find the largest digit d (d must be one of $\{0..9\}$) such that `1.79de+308` is not `Inf`? Let us do this with a simple trial. Type `1.79de+308` with the `d` substituted with the values $0, 1, 2, 3, \ldots$ in this order, and stop when the value presented after pressing the return key becomes `Inf`.

After examining the values, store in a variable `the_digit` the value of `d` you have found.

Question 2. Let us learn how to convert data types. Recall `as.character`, `as.double`, and `as.integer` convert a given value to a string, a double, and an integer, respectively. We can also test to see if a value is a certain data type by replacing `as` with `is`. Recall also that to specify a string with its character sequence we can use either a matching pair of double quotation marks or a matching pair of single quotation marks sandwiching the sequence.

Following is a string called `a_happy_string` and a double named `double_trouble`:

```
a_happy_string <- "-4.5"
double_trouble <- 81.9
```

Here are some findings based on the above functions. Which of the following statements are TRUE?

1. `as.double(a_happy_string)` produces an error because we cannot covert a string to a double.
2. `as.integer(a_happy_string)` returns the value -4.5.
3. `as.character(double_trouble)` returns the value `"81.9"`.
4. `is.integer(1)` returns TRUE.

Question 3. Let us extract some information from strings. Following are three strings stored in three separate names.

```
str1 <- "State"
str2 <- "Department"
str3 <- "Office"
```

Do the following three tasks with these three strings:

- Inquire the length of each string, and store the three length values to `l1`, `l2`, and `l3`, respectively.
- Connect the three strings in the order of `str1`, `str2`, and `str3` using `str_c` and `str_c` with `sep = " "`, and store the two results in `join1` and `join2`.
- Obtain the substrings of `join1` between position pairs $(4, 10)$, $(-7, -1)$, and $(8, 8)$. Store the substrings in `sub1`, `sub2`, and `sub3`, respectively.

Question 4. Let us execute some operations to generate Boolean values. Use two variables `val1` and `val2` and assign them to the values `12.3` and `45.6`, respectively. Then put the two values in the six comparisons "equals", "not equals", "greater than", "greater than or equal to", "less than", and "less than or equal to". Store the six results in the variables `test1`, `test2`, `test3`, `test4`, `test5`, and `test6`, respectively.

Question 5. Let us play with Boolean operations. Let `str4` be a string variable. Assign the value of `hammerhead` to `str4`. Then create three Boolean variables `check1`, `check2`, and `check3` and assign, to these variables, the results of testing if the length of `str4` is equal to 10, if the length of `str4` is less than 5, and if the substring of `str4` from positions 10 to 10 is equal to `"a"`. Print the three Boolean value, and then compute the "or" of the three values and the "and" of the three values, and store these results in the names `the_OR` and `the_AND`, respectively.

Question 6. Suppose we have three Boolean names `tf1`, `tf2`, and `tf3`. Suppose we assign the following eight value pairs to them.

- FALSE, FALSE, and FALSE
- FALSE, FALSE, and TRUE
- FALSE, TRUE, and FALSE
- FALSE, TRUE, and TRUE
- TRUE, FALSE, and FALSE
- TRUE, FALSE, and TRUE
- TRUE, TRUE, and FALSE
- TRUE, TRUE, and TRUE

Answer the value of `tf1 | !tf2 & tf3` in the four combinations, and store them in eight names `bool1`, `bool2`, ..., `bool8`.

Question 7. Following are three Boolean expressions that evaluate to either TRUE or FALSE. Explain every step in the evaluation process before TRUE or FALSE is ultimately returned.

```
(2 - 1) == ((TRUE == TRUE) != FALSE)
(10 - (FALSE/2 + max(TRUE, FALSE))) >= (TRUE + 1)
(Inf > 5) == ((Inf > Inf) | (Inf >= Inf))
```

Question 8. A student is diligently studying Boolean data types and is stumped by the following:

I don't understand why `"Zoo"` > `"Napping"` is TRUE when "Napping" has more characters than "Zoo". But when I do `"ZZZ"` > `"ZZZping"`, it returns FALSE which makes sense because "ZZZping" has more characters than "ZZZ". So shouldn't the first expression evaluate to FALSE as well for the same reason 2 > 5 is FALSE?

What would you tell this student? Should we file a bug report to the R maintainers?

Question 9. Suppose we have four names `s1`, `s2`, `s3`, and `s4` that are defined as follows:

```
s1 <- "Fine"
s2 <- "Dine"
s3 <- "Sine"
s4 <- "Wine"
```

We then apply the following function and store the result in the name `snew`:

```
snew <- str_c(s1, s2, s3, s4, sep = " ")
snew
```

```
## [1] "Fine Dine Sine Wine"
```

- **Question 9.1** What is the length of `snew`? Do not compute this manually! Use an R function. Store your answer in the name `snewlen`.

- **Question 9.2** How many times does `"in"` appear in `snew`? You may find this manually or by using an R function. Assign your answer to the name `intimes`.

- **Question 9.3** Recall `str_sub( snew, a, b)` from `stringr` produces the substring of `snew` from position `a` to position `b`. Answer the combination of `a` and `b` that produces `"in"` such that `a` is the largest. This is also sometimes called a "right find" because the search is done from the right rather than the left. Store the combination of `a` and `b` you found in the names `a` and `b`.

Question 10: Vector manipulations. In this exercise we practice some basic construction and manipulation of vector data types.

- **Question 10.1.** Form a vector that contains the numbers 3, π, 4, 9, and e (Euler's constant), in that order. Your solution should reference the constants π and e without having to define them explicitly. Assign it to the name `vec1`.

- **Question 10.2.** Form a vector that contains the nine strings `"Dear"`, `"string"`, `","`, `"this"`, `"is"`, `"much"`, `"ado"`, `"about"`, and `" "`. Note that the last element is a string containing a single whitespace. Name this `vec2`.

- **Question 10.3.** Form a vector that *appends* `vec2` after `vec1`. Call it `vec3`. What is the data type of this new vector `vec3`? Use an R function to determine this and assign that answer to the name `vec3_type`. Then observe the difference in data types between `vec1`, `vec2`, and `vec3`.

Question 11: Sequences. This exercise explores the use of vectors to form various sequences using the `seq()` function. Each of these questions should require only one line of code to compute the answer.

- **Question 11.1** Form a vector that generates a sequence of numbers from 0 to 20 in steps of 2, e.g., 0, 2, 4, 6, 8, ..., 20. Assign this vector to the name `steps2`.

- **Question 11.2** Form a vector that generates a sequence containing multiples of 11 from 0 up to and including 1221. Assign this vector to the name `mult11`.

- **Question 11.3** Form a vector that generates a sequence containing the first 20 powers of 4, e.g., $4^0 = 1$, $4^1 = 4$, $4^2 = 16$, etc. Assign your answer to the name `powers4`.

Question 12: Element-wise operations. A benefit of working with vectors is that we can use vectors to accomplish *element-wise* operations, that is, some operation that is applied to every element of a vector. What makes element-wise operations attractive to use is that they can be applied with a single line of code. This exercise explores some element-wise operations.

- **Question 12.1** Let us compute the interest on several loans at once.

    - **Question 12.1.1** The vector `bank_loans` from the package `edsdata` contains 100,000 different loans. Assuming an annual simple interest rate of 4%, compute the amount each borrower would owe after one year if no payments were made during that time toward the loan. That means we can multiply the loan amount by 1.04 to get the amount owed after one year. Compute the amount owed for each loan in `bank_loans` after one year. Assign your answer to the name `amount_owed`.

    - **Question 12.1.2** What is the total amount of *interest* collected by the bank from all these loans after one year? Assign your answer to `total_interest_amount`.

- **Question 12.2** Suppose the population of Datatopia is growing steadily at 3% annually. On 1/1/2021, Datatopia had 1,821,411,277 people. Calculate the expected population of the country for the next 20 years.

  - **Question 12.2.1** Compute the expected population for the next 20 years as a 20-element vector, and store it in `p`. You can do this by first creating a sequence of 20 values `1..20`, powering 1.03 with the 20 values, and multiplying it by the initial population.

  - **Question 12.2.2** Set `pop_2025` and `pop_2040` to the population estimates for 2025 and 2040, respectively. Then set `pop_2025_to_2040` to a vector containing the estimates from 2025 to 2040 (the years 2025 and 2040 are both inclusive).

  - **Question 12.2.3** What happens when you try to access the following element in the vector `pop_2025_to_2040`?

    ```
    pop_2025_to_2040[17]
    ```

    Are you surprised by the result? Should you expect an error when trying this? Why or why not?

  - **Question 12.2.4** Set `pop_2025_to_2040_reverse` to a vector containing the populations in `pop_2025_to_2040`, but in reverse order. That is, starting first from 2040 and ending with 2025. To do this, use an index sequence.

  - **Question 12.2.5** Suppose that our population estimates for the years 2033 and 2034 are too unreliable to be useful and we would like to raise this error somehow. Flag this error in `pop_2025_to_2040` by setting the appropriate indices in `pop_2025_to_2040` to `NA`. `NA` is a special name that stands for "not a number".

  - **Question 12.2.6** The numbers in `pop_2025_to_2040` are large, which can sometimes be hard to interpret. Report the figures as *numbers in billions* instead. Accomplish this work using element-wise operation using a single line of R code and assign your answer to the name `pop_2025_to_2040_billions`.

  - **Question 12.2.7** How many years had a population exceeding 3 billion? Use your vector `pop_2025_to_2040_billions` to answer this. You will need to use another element-wise operation. Assign the number to the name `more_than_3_bill`. If your answer is coming out as `NA`, be sure to read the help and look at any arguments you can use.

Question 13: Exploring `stringr`. We have already learned how to do basic tasks with strings like inquiring about the number of characters in the string and finding substrings. But data science work often involves doing much more advanced tasks, e.g., splitting strings, replacing certain characters that match some pattern, etc. The `stringr` package from the tidyverse brings a suite of modern tools to use for working with strings in data science. Let's explore some of them in this second part. Be sure to bookmark its cheatsheet for reference[5].

`bands` is a character vector in the package `edsdata` that contains some band names.

- **Question 13.1** Obtain two copies of `bands`, one with every character in uppercase (e.g., LIKE THIS) and another with every character in lowercase (e.g., like this). Store the results in `upper` and `lower`, respectively. Use a `stringr` function.

---

[5]https://github.com/rstudio/cheatsheets/blob/master/strings.pdf

- **Question 13.2** Here is another character vector, `bird_in_a_hand` , that looks a bit different from bands. Even though both are seen as *character vectors* in R, what is the difference between `bird_in_a_hand` and bands?

```
bird_in_a_hand <- c("b","i","r","d",
                    "","i","n","","a","",
                    "h","a","n","d")
bird_in_a_hand
```

```
## [1] "b" "i" "r" "d" ""  "i" "n" ""  "a" ""  "h" "a" "n"
## [14] "d"
```

- **Question 13.3** Repeat **Question 13.1**, but for `bird_in_a_hand`. Assign your answers to the names `upper_bird` and `lower_bird`, respectively.

- **Question 13.4** The `stringr` function `str_remove_all()` can be used to remove matched patterns in a string. For instance, we could remove all lowercase a's and b's from the character vector `bird_in_a_hand` as follows:

```
str_remove_all(bird_in_a_hand, "[ab]")
```

Now take bands and modify it by removing each occurrence of a vowel in lowercase (`'a'`, `'e'`, `'i'`, `'o'`, `'u'`). Store the result in `devoweled`.

Question 14: Lists. This exercise gives practice with the `list` data type.

- **Question 14.1** Using the function `list` create a list `courses` whose elements are `"MTH118"`, `"GEG2490"`, `"CSC160"`, `"CSC353"`, `"ACC419"`, `"PSC356"`, and `"BIL155"`. Then:
  - Sort the list using the function `str_sort()` and store the result in `courses_sorted`.
  - Apply `str_sub` from position 4 to position 6 to the elements of `courses` and apply `str_sort` to the result. Store it in `numbers_sorted`.

- **Question 14.2** The name `fruits` is a character vector with some fruits.

```
fruits <- c(
  "apples and oranges and pears and bananas",
  "pineapples and mangos and guavas"
)
```

  - **Question 14.2.1** Accomplish the following tasks:
    * Create a list `fruits_list` from `fruits` using the `list` function.
    * Create a vector `fruits_unlisted` by applying `unlist` to `fruits_list`.
    * Try executing:

```
str_split(fruits, "[and]")
str_split(fruits_list, "[and]")
str_split(fruits_unlisted, "and")
```

    The middle one presents a warning message. Store the last result in `split_fruits`.

* Print `split_fruits`.

 - **Question 14.2.2** Let's recap what just happened:
    * What is the difference between `fruits`, `fruits_list`, `fruits_unlisted`, and `split_fruits`? Note the data type stored in each of these names (e.g., list, character vector, integer, etc.).
    * Why did you get a warning when trying `str_split(fruits_list, "[and]")`? Read the help for `str_split` to help you answer this.
    * Why does the word "and" no longer appear in the elements inside `split_fruits`?

Question 15: More `stringr`. Let's return to `stringr` again, this time looking at a few more functions that work with *regular expressions*. We actually saw one already when we worked on "devoweling" band names. Be sure you have read the textbook section on `stringr` before continuing with this part.

As a real-world example, the name `band_of_the_hour` is a string about the "Band of the Hour"[6] at the University of Miami. The string is available in the `edsdata` package.

* **Question 15.1** The function `str_match( s, p )` finds the first occurrence of the pattern `p` in the string `s`. Find the following patterns in `band_of_the_hour` using `str_match` and store the results in variables `m1`, …, `m3`:

 1. any uppercase alphabet `"[A-Z]"`,
 2. any series of lowercase alphabetical characters `"[a-z]+"`,
 3. any numeral `"[0-9]"`.

* **Question 15.2** Now use `str_match_all()` to find the occurrences of the following two patterns and store them in `ma1` and `ma2`: (a) any series of numerals and (b) any word that begins with an uppercase letter (e.g., "Fiesta", "Band", "University", etc.). Verify your answer by examining the contents of `ma1` and `ma2`.

* **Question 15.3** Now use `str_count()` to count the occurrences of `"Arrow"` and store in `arrow_count` and then use `str_split()` to split `band_of_the_hour` at each white space and store it in `band_of_the_hour_split`.

* **Question 15.4** The `stringr` function `str_subset()` keeps strings that match a pattern. For example:

```
str_subset(c("x", "y", "xx", "yy"), "x")
```

Now answer how you might obtain the elements in the list `band_of_the_hour_split` that have a `"0"` in it. Store the result in `zero_elements`.

Question 16. Suppose we have a vector of course IDs:

```
badly_formed_ids <- c("CSC_100", "BIO 111", "MTH161H",
                      "ECO--220", "MUS..160A")
badly_formed_ids
```

The school codes their IDs as follows:

* An ID starts with three uppercase letters.
* An ID has three numerals after the uppercase letters.

---

[6]https://en.wikipedia.org/wiki/Band_of_the_Hour

- Between the letters and the numerals, an ID may contain symbols that are neither letters nor numerals.
- An ID may have an additional alphabet character at the end.

Unfortunately, this format is not very good for data science. While it allows for a wide range of possible IDs, the flexibility is an obstacle to any analysis using R, e.g., how to tell which courses are offered through Computer Science (CSC)? How to find how many 200-level courses the school offers?

Let us "standardize" these course IDs into a uniform format consisting solely of the initial three letters followed by three numerals. After application of the function, the above IDs should look like:

```
c("CSC100", "BIO111", "MTH161", "ECO220", "MUS160")
```

Answer how you might obtain this with a single line of R code using a *single* stringr function.

2

Data Transformation

The work of data science begins with a dataset. These datasets can be so large that any manual inspection or review of them, say using editing software like TextEdit or Notepad++, becomes totally infeasible. To overcome this, data scientists rely on computational tools like R for working with datasets. Learning how to use these tools well lies at the heart of data science and what data scientists do daily at their desks.

A part of what makes these tools so powerful is that we often need to apply a series of actions to a dataset. Data scientists talk a lot about the importance of data cleaning, stating that without data cleaning no data analysis results are meaningful. Some go further to say that the most important step in the data science life cycle is data cleaning because, from their point of view, the analysis process following data cleaning is a routine to a great degree. As such, another important aspect of working with datasets is *transforming* data, i.e., rendering data suitable for analysis. When data is made into an analysis-ready form, we call such data *tidy data*. Transforming data to become tidy data is the focus of this chapter.

The tools we will cover in this chapter to accomplish this goal are also key members of the `tidyverse`. One is called `tibble`, which is a data structure for managing datasets, and another is called `dplyr`, which provides a grammar of data manipulation for acting upon datasets stored as tibbles. We will also learn about a third called `purrr` to help with the data manipulations, e.g., say when a column of data is in the wrong units.

2.1 Datasets and Tidy Data

Data scientists prefer working with data that is *tidy* because it facilitates data analysis. In this section we will introduce a vocabulary for working with datasets and describe what *tidy data* looks like.

2.1.1 Prerequisites

As before, let us load `tidyverse`.

```
library(tidyverse)
```

The `tidyverse` package comes with scores of datasets. By typing `data()` you can see a list of data sets available in the RStudio environment you are in. Quite a few data come with `tidyverse`. If your session has not yet loaded `tidyverse`, the list can be short.

DOI: 10.1201/9781003320845-2

2.1.2   A "hello world!" dataset

A dataset is a collection of *values*, which can be either a number or string. Let us begin by looking at our first dataset. We will examine the *Motor Trend Car Road Test* dataset which is made available through tidyverse. It was extracted from the 1974 *Motor Trend* US magazine, and contains data about fuel consumption and aspects of automobile design for 32 car models.

We can inspect it simply by typing its name.

```
mtcars
```

```
##                mpg cyl disp  hp drat    wt  qsec vs am
## Mazda RX4     21.0   6  160 110 3.90 2.620 16.46  0  1
## Mazda RX4 Wag 21.0   6  160 110 3.90 2.875 17.02  0  1
## Datsun 710    22.8   4  108  93 3.85 2.320 18.61  1  1
##               gear carb
## Mazda RX4        4    4
## Mazda RX4 Wag    4    4
## Datsun 710       4    1
```

Only the first few rows are shown here. You can pull up more information about the dataset by typing the name of it with a single question mark in front of it.

```
?mtcars
```

Datasets like these are often called *rectangular* tables. In a rectangular table, the rows have an identical number of cells and the columns have an identical number of cells, thus allowing access to any cell by specifying a row and a column together.

A conventional structure of rectangular data is as follows:

- The rows represent individual objects, whose information is available in the data set. We often call these *observations*.
- The columns represent properties of the observations. We often call these properties *variables* or *attributes*.
- The columns have unique names. We call them *variables names* or *attribute names*.
- Every value in the table belongs to some *observation* and some *variable*.

This dataset contains 352 values representing 11 variables and 32 observations. Note how it explicitly tells us the definition of an observation: a "car model" observation is defined as a combination of the variables that are present above, e.g., mpg, cyl, disp, etc.

2.1.3   In pursuit of tidy data

We are now ready to provide a definition of tidy data. We defer to Hadley Wickham (2014)[1] for a definition. We say that data is "tidy" when it satisfies four conditions:

1. Each *variable* forms a column.
2. Each *observation* forms a row.
3. Each value must have its own cell.
4. Each type of observational unit forms a table.

---

[1] https://vita.had.co.nz/papers/tidy-data.pdf

Data that exists in any other arrangement is, consequently, *messy*. A critical aspect in distinguishing between tidy and messy data forms is defining the **observational unit**. This can look different depending on the statistical question being asked. In fact, defining the observational unit is so important because data that is "tidy" in one application can be "messy" in another.

The goal of this chapter is to learn about methods for *transforming* "messy" data into "tidy" data, with some help from R and the tidyverse.

With respect to the mtcars dataset, we can glean the observational unit from its help page:

Fuel consumption and 10 aspects of automobile design and performance for 32 automobiles (1973–74 models).

Therefore, we expect each row to correspond to exactly one of the 32 different car models. With one small exception that we will return to later, the mtcars dataset fulfills the properties of tidy data. Let us look at other examples of datasets that fulfill or violate these properties.

2.1.4 Example: is it tidy?

Suppose that you are keeping track of weekly sales of three different kinds of cookies at a local Miami bakery in 2021. By instinct, you decide to keep track of the data in the following table.

bakery1

```
## # A tibble: 4 x 4
##    week gingerbread `chocolate peppermint` `macadamia nut`
##   <dbl>       <dbl>                  <dbl>           <dbl>
## 1     1          10                     23              12
## 2     2          16                     21              16
## 3     3          25                     20              24
## 4     4          12                     18              20
```

Alternatively, you may decide to encode the information as follows.

bakery2

```
## # A tibble: 3 x 5
##   week                   `1`   `2`   `3`   `4`
##   <chr>                <dbl> <dbl> <dbl> <dbl>
## 1 gingerbread             10    16    25    12
## 2 chocolate peppermint    23    21    20    18
## 3 macadamia nut           12    16    24    20
```

Do either of these tables fulfill the properties of tidy data?

First, we define the observational unit as follows:

A weekly sale for one of three different kinds of cookies sold at a Miami bakery in 2021. Three variables are measured per unit: the week it was sold, the kind of cookie, and the number of sales.

In `bakery1`, the variable we are trying to measure – *sales* – is actually split across three different columns and multiple observations appear in each row. In `bakery2`, the situation remains bad: both the *cookie type* and *sales* variables appear in each column and, still, multiple observations appear in each row. Therefore, neither of these datasets are *tidy*.

A tidy version of the dataset appears as follows. Compare this with the tables from `bakery1` and `bakery2`. Do not worry about the syntax and the functions used; we will learn about what these mean and how to use them in a later section.

```
bakery_tidy <- bakery1 |>
  pivot_longer(gingerbread:`macadamia nut`,
               names_to = "type", values_to = "sales")
bakery_tidy
```

```
## # A tibble: 12 x 3
##      week type                 sales
##     <dbl> <chr>                <dbl>
##  1      1 gingerbread             10
##  2      1 chocolate peppermint    23
##  3      1 macadamia nut           12
##  4      2 gingerbread             16
##  5      2 chocolate peppermint    21
##  6      2 macadamia nut           16
##  7      3 gingerbread             25
##  8      3 chocolate peppermint    20
##  9      3 macadamia nut           24
## 10      4 gingerbread             12
## 11      4 chocolate peppermint    18
## 12      4 macadamia nut           20
```

When a dataset is expressed in this manner, we say that it is in *long* format because the number of rows is comparatively larger compared to `bakery1` and `bakery2`. Admittedly, this form can make it harder to identify patterns or trends in the data by eye. However, tidy data opens the door to more efficient data science so that you can rely on existing tools to proceed with next steps. Without a standardized means of representing data, such tools would need to be developed from scratch each time you begin work on a new dataset.

Observe how this dataset fulfills the four properties of tidy data. The fourth property is fulfilled because the observational unit we are measuring is a weekly cookie sale, and we are measuring three variables – `week`, `type`, and `sales` – per observational unit. The detail of the observational unit description is important: these variables do not refer to measurements on some sale or bakery store; they refer specifically to measurements on a given weekly cookie sale for one of three kinds of cookies ("gingerbread", "chocolate peppermint", and "macadamia nut") sold at a local Miami bakery in 2021. If this dataset were to contain sales for a different year or cookie type not specified in our observational unit statement, then said observations would need to be sorted out into a different table.

A possible scenario in violation the third property might look like the following: the bakery decides to record sale *ranges* instead of a single estimate, e.g., in the case of making a forecast on future sales.

```
## # A tibble: 3 x 2
##    week forecast
##   <dbl> <chr>
## 1     1 200-300
## 2     2 300-400
## 3     3 200-500
```

In the next section we turn to the main data structures in R we will use for performing data transformations on datasets.

2.2 Working with Datasets

In this section we dive deeper into datasets and learn how to do basic tasks with datasets and query information from them.

2.2.1 Prerequisites

As before, let us load `tidyverse`.

```
library(tidyverse)
```

2.2.2 The data frame

Let us recall the `mtcars` dataset we visited in the last section.

```
mtcars
```

```
##                mpg cyl disp  hp drat    wt  qsec vs am
## Mazda RX4     21.0   6  160 110 3.90 2.620 16.46  0  1
## Mazda RX4 Wag 21.0   6  160 110 3.90 2.875 17.02  0  1
## Datsun 710    22.8   4  108  93 3.85 2.320 18.61  1  1
##               gear carb
## Mazda RX4        4    4
## Mazda RX4 Wag    4    4
## Datsun 710       4    1
```

Data frame is a term R uses to refer to data formats like the `mtcars` data set. In its simplest form, a data frame consists of vectors lined up together where each vector has a name.

How do we know how many rows and columns in the data as well as the names of the variables? The following functions answer those questions, respectively.

```
nrow(mtcars) # how many rows in the dataset?
```

```
## [1] 32
```

```
ncol(mtcars) # how many columns?
```

```
## [1] 11
```

```
colnames(mtcars) # what are the names of the columns?
```

```
##  [1] "mpg"  "cyl"  "disp" "hp"   "drat" "wt"   "qsec"
##  [8] "vs"   "am"   "gear" "carb"
```

We noted earlier that this dataset is tidy with one exception. Observe that the leftmost column in the table does not have the column header or the type designation. The strings appearing there are what we call *row names*; we learn of the existence of row names when we see that R prints the data without a column name for the row names.

The problem with row names is that a variable, here the name of the car model, is treated as a special attribute. The objective of tidy data is to store data consistently and this special treatment is, according to tidyverse, a violation of the principle.

2.2.3 Tibbles

An alternative to the data frame is the tibble which upholds best practices for working with data frames. It does not store row names as special columns like data frames do and the presentation of the table can be visually nicer to inspect than data frames when examining a dataset at the console.

To transform the `mtcars` data frame to a tibble is easy. We simply call the function `tibble`.

```
mtcars_tibble <- tibble(mtcars)
mtcars_tibble
```

```
## # A tibble: 32 x 11
##       mpg   cyl  disp    hp  drat    wt  qsec    vs    am
##     <dbl> <dbl> <dbl> <dbl> <dbl> <dbl> <dbl> <dbl> <dbl>
##  1  21       6   160   110  3.9   2.62  16.5     0     1
##  2  21       6   160   110  3.9   2.88  17.0     0     1
##  3  22.8     4   108    93  3.85  2.32  18.6     1     1
##  4  21.4     6   258   110  3.08  3.22  19.4     1     0
##  5  18.7     8   360   175  3.15  3.44  17.0     0     0
##  6  18.1     6   225   105  2.76  3.46  20.2     1     0
##  7  14.3     8   360   245  3.21  3.57  15.8     0     0
##  8  24.4     4   147.   62  3.69  3.19  20       1     0
##  9  22.8     4   141.   95  3.92  3.15  22.9     1     0
## 10  19.2     6   168.  123  3.92  3.44  18.3     1     0
## # ... with 22 more rows, and 2 more variables:
```

```
## #   gear <dbl>, carb <dbl>
```

The designation <dbl> appearing next to the columns indicates that the column has only double values. Observe that the names of the car models are no longer present. However, we may wish to keep the names of the models as it can bring useful information. tibble has thought of a solution to this problem for us: we can add a new column with the row name information. The required function is rownames_to_column.

```
mtcars_tibble <- tibble(rownames_to_column(mtcars, var = "model_name"))
mtcars_tibble
```

```
## # A tibble: 32 x 12
##    model~1  mpg   cyl  disp    hp  drat    wt  qsec    vs
##    <chr>   <dbl> <dbl> <dbl> <dbl> <dbl> <dbl> <dbl> <dbl>
##  1 Mazda ~  21      6  160   110  3.9   2.62  16.5     0
##  2 Mazda ~  21      6  160   110  3.9   2.88  17.0     0
##  3 Datsun~  22.8    4  108    93  3.85  2.32  18.6     1
##  4 Hornet~  21.4    6  258   110  3.08  3.22  19.4     1
##  5 Hornet~  18.7    8  360   175  3.15  3.44  17.0     0
##  6 Valiant  18.1    6  225   105  2.76  3.46  20.2     1
##  7 Duster~  14.3    8  360   245  3.21  3.57  15.8     0
##  8 Merc 2~  24.4    4  147.   62  3.69  3.19  20       1
##  9 Merc 2~  22.8    4  141.   95  3.92  3.15  22.9     1
## 10 Merc 2~  19.2    6  168.  123  3.92  3.44  18.3     1
## # ... with 22 more rows, 3 more variables: am <dbl>,
## #   gear <dbl>, carb <dbl>, and abbreviated variable name
## #   1: model_name
```

Throughout the text, we will store data using the tibble construct. However, because tibbles and data frames are close siblings, we may use the terms *tibble* and *data frame* interchangeably when talking about data that is stored in a rectangular format.

2.2.4   Accessing columns and rows

You can access an individual column in two ways: (1) by attaching the dollar sign to the name of the data frame and then the attribute name, and (2) using the function pull. We prefer to use the latter because of the |> operator which we will see later. Here are some example usages.

```
mtcars_tibble$cyl
```

```
##  [1] 6 6 4 6 8 6 8 4 4 6 6 8 8 8 8 8 8 4 4 4 4 8 8 8 8 4 4
## [28] 4 8 6 8 4
```

```
pull(mtcars_tibble, cyl)
```

```
##  [1] 6 6 4 6 8 6 8 4 4 6 6 8 8 8 8 8 8 4 4 4 4 8 8 8 8 4 4
## [28] 4 8 6 8 4
```

The result returned is the entire sequence for the column cyl.

If you know the position of a column in the dataset, you can use the function `select()` to get to the vector. The `cyl` is at position 3 of the data, so we obtain the following.

```
select(mtcars_tibble, 3)
```

```
## # A tibble: 32 x 1
##        cyl
##      <dbl>
##  1      6
##  2      6
##  3      4
##  4      6
##  5      8
##  6      6
##  7      8
##  8      4
##  9      4
## 10      6
## # ... with 22 more rows
```

Similarly, if we know the position of a row in the dataset, we can use `slice()`. The following will return all the associated information for the second row of the dataset.

```
slice(mtcars_tibble, 2)
```

```
## # A tibble: 1 x 12
##    model_~1   mpg   cyl  disp    hp  drat    wt  qsec    vs
##    <chr>    <dbl> <dbl> <dbl> <dbl> <dbl> <dbl> <dbl> <dbl>
## 1 Mazda R~    21     6   160   110   3.9  2.88  17.0     0
## # ... with 3 more variables: am <dbl>, gear <dbl>,
## #   carb <dbl>, and abbreviated variable name
## #   1: model_name
```

2.2.5   Extracting basic information from a tibble

You can use the function `unique` to obtain unique values in a column. Let us see the possible values for the number of cylinders.

```
unique(pull(mtcars_tibble, cyl))
```

```
## [1] 6 4 8
```

We find that there are three possibilities: 4, 6, and 8 cylinders.

We already know how to inquire about the maximum, minimum, and other properties of a vector. Let us check out the `mpg` attribute (miles per gallon) in terms of the maximum, the minimum, and sorting the values in the increasing order.

```
max(pull(mtcars_tibble, mpg))
```

```
## [1] 33.9
```

```
min(pull(mtcars_tibble, mpg))
```

```
## [1] 10.4
```

```
sort(pull(mtcars_tibble, mpg))
```

```
##  [1] 10.4 10.4 13.3 14.3 14.7 15.0 15.2 15.2 15.5 15.8
## [11] 16.4 17.3 17.8 18.1 18.7 19.2 19.2 19.7 21.0 21.0
## [21] 21.4 21.4 21.5 22.8 22.8 24.4 26.0 27.3 30.4 30.4
## [31] 32.4 33.9
```

2.2.6 Creating tibbles

Before moving on to dplyr, let us see how we can create a dataset. The package tibble offers some useful tools when you are creating data.

Suppose you have tests scores in Chemistry and Spanish for four students, Gail, Henry, Irwin, and Joan. You can create three vectors, names, Chemistry, and Spanish each representing the names, the scores in Chemistry, and the scores in Spanish.

```
students <- c("Gail", "Henry", "Irwin", "Joan")
chemistry <- c( 99, 98, 80, 92 )
spanish <- c(87, 85, 90, 88)
```

We can assemble them into a tibble using the function tibble. The function takes a series of columns, expressed as vectors, as arguments.

```
class <- tibble(students = students,
                chemistry_grades = chemistry,
                spanish_grades = spanish)
class
```

```
## # A tibble: 4 x 3
##     students chemistry_grades spanish_grades
##     <chr>               <dbl>          <dbl>
## 1 Gail                   99             87
## 2 Henry                  98             85
## 3 Irwin                  80             90
## 4 Joan                   92             88
```

The data type designation <chr> means "character" and so indicates that the column consists of strings.

Pop quiz: is the tibble class we just created an example of tidy data? Why or why not? If you are unsure, revisit the examples from the previous section and compare this tibble with those.

Let us see how we can query some basic information from this tibble.

```
pull(class, chemistry_grades) # all grades in chemistry
```

```
## [1] 99 98 80 92
```

```
min(pull(class, chemistry_grades))   # minimum chemistry score
```

```
## [1] 80
```

For small tables of data, we can also create a `tibble` using an easy row-by-row layout.

```
class <- tribble(~student,~chemistry_grades,~spanish_grades,
         "Gail",    99, 87,
         "Henry",   98, 85,
         "Irwin",   80, 90,
         "Joan",    92, 88)
class
```

```
## # A tibble: 4 x 3
##    student chemistry_grades spanish_grades
##    <chr>              <dbl>          <dbl>
## 1 Gail                  99             87
## 2 Henry                 98             85
## 3 Irwin                 80             90
## 4 Joan                  92             88
```

We can also form tibbles using sequences as follows.

```
tibble(x=1:5,
       y=x*x,
       z = 1.5*x - 0.2)
```

```
## # A tibble: 5 x 3
##        x     y     z
##    <int> <int> <dbl>
## 1     1     1   1.3
## 2     2     4   2.8
## 3     3     9   4.3
## 4     4    16   5.8
## 5     5    25   7.3
```

The seq that is native of R allows us to create a sequence. The syntax is `seq(START,END,GAP)`, where the sequence starts from `START` and then adds `GAP` to the sequence until the value exceeds `END`. We can create the sequence with the name "x", and then add three other columns based on the value of "x".

Here is another example.

```
tibble(x = seq(1,4,0.5),
       y = sin(x),
       z = cos(x),
       w = x^3 - 10*x^2 + x - 2)
```

```
## # A tibble: 7 x 4
##       x      y       z      w
##   <dbl>  <dbl>   <dbl>  <dbl>
## 1     1  0.841   0.540    -10
## 2   1.5  0.997  0.0707  -19.6
## 3     2  0.909  -0.416    -32
## 4   2.5  0.598  -0.801  -46.4
## 5     3  0.141  -0.990    -62
## 6   3.5 -0.351  -0.936  -78.1
## 7     4 -0.757  -0.654    -94
```

2.2.7 Loading data from an external source

Usually data scientists need to load data from files. The package readr of tidyverse offers ways for that. With the package readr you can read from, among others, comma-separated files (CSV files) and tab-separated files (TSV files).

To read files, we specify a string the location of the file and then use the function for reading the file, read_csv if it is a CSV file and read_tab if it is a TSV file. If you have a file that uses another delimiter, a more general read_delim function exists as well.

Here is an example of reading a CSV file from a URL available on the internet.

```
path <- str_c("https://data.bloomington.in.gov/",
              "dataset/117733fb-31cb-480a-8b30-fbf425a690cd/",
              "resource/2b2a4280-964c-4845-b397-3105e227a1ae/",
              "download/pedestrian-and-bicyclist-counts.csv")
bloom <- read_csv(path)
```

The data set shows the traffic in the city of Bloomington, the hometown of the Indiana University at Bloomington, Indiana.

We can inspect the first few rows of the tibble using the function slice_head.

```
slice_head(bloom, n = 3)
```

```
## # A tibble: 3 x 11
##    Date      7th a~1 7th u~2 7th u~3 7th u~4 Bline~5 Pedes~6
##    <chr>       <dbl>   <dbl>   <dbl>   <dbl>   <dbl>   <dbl>
## 1 Wed, Fe~      186     221     155      66     688     490
## 2 Thu, Fe~      194     166      98      68     676     450
## 3 Fri, Fe~      147     200     142      58     603     399
## # ... with 4 more variables: Cyclists <dbl>,
## #   `Jordan and 7th` <dbl>, `N College and RR` <dbl>,
## #   `S Walnut and Wylie` <dbl>, and abbreviated variable
## #   names 1: `7th and Park Campus`, 2: `7th underpass`,
## #   3: `7th underpass Pedestrians`,
## #   4: `7th underpass Cyclists`,
## #   5: `Bline Convention Cntr`, 6: Pedestrians
```

Note that some columns have spaces in them. To access the column corresponding to the attribute, we cannot simply type the column because of the white space. To access these columns, we surround the attribute with backticks (').

```
pull(bloom, `N College and RR`)
```

2.2.8 Writing results to a file

Saving a tibble to file is easy. You use `write_csv(DATA_NAME,PATH)` where `DATA_NAME` is the name of the data frame to save and `PATH` is the "path name" of the file.

Below, the action is to store the tibble `bloom` as "bloom.csv" in the current working directory.

```
write_csv(bloom, "bloom.csv")
```

2.3 `dplyr` Verbs

The past section showed two basic data structures – data frames and tibbles – that can be used for loading, creating, and saving datasets. We also saw how to query basic information from these structures. In this section we turn to the topic of data *transformation*, that is, actions we can apply to a dataset to transform it into a new, and hopefully more useful, dataset. Recall that data transformation is the essence of achieving tidy data.

The `dplyr` packages provides a suite of functions for providing such transformations. Put another way, `dplyr` provides a *grammar* of data manipulation where each function can be thought of as the *verbs* that act upon the subject, the dataset (in tibble form). In this section we study the main `dplyr` verbs.

2.3.1 Prerequisites

As before, let us load `tidyverse`.

```
library(tidyverse)
```

Let us load `mtcars` as before and call it `mtcars_tibble` and then, as before, convert the row names to a column. Call the new attribute "model_name".

```
mtcars_tibble <- tibble(rownames_to_column(mtcars, "model_name"))
mtcars_tibble
```

```
## # A tibble: 32 x 12
##     model~1   mpg   cyl  disp    hp  drat    wt  qsec    vs
##     <chr>   <dbl> <dbl> <dbl> <dbl> <dbl> <dbl> <dbl> <dbl>
## 1 Mazda ~    21      6   160   110   3.9  2.62  16.5     0
## 2 Mazda ~    21      6   160   110   3.9  2.88  17.0     0
```

```
##  3 Datsun~ 22.8     4 108      93 3.85 2.32 18.6    1
##  4 Hornet~ 21.4     6 258     110 3.08 3.22 19.4    1
##  5 Hornet~ 18.7     8 360     175 3.15 3.44 17.0    0
##  6 Valiant 18.1     6 225     105 2.76 3.46 20.2    1
##  7 Duster~ 14.3     8 360     245 3.21 3.57 15.8    0
##  8 Merc 2~ 24.4     4 147.     62 3.69 3.19 20      1
##  9 Merc 2~ 22.8     4 141.     95 3.92 3.15 22.9    1
## 10 Merc 2~ 19.2     6 168.    123 3.92 3.44 18.3    1
## # ... with 22 more rows, 3 more variables: am <dbl>,
## #   gear <dbl>, carb <dbl>, and abbreviated variable name
## #   1: model_name
```

2.3.2   A fast overview of the verbs

The main important verbs from dplyr that we will cover are shown in the following figure.

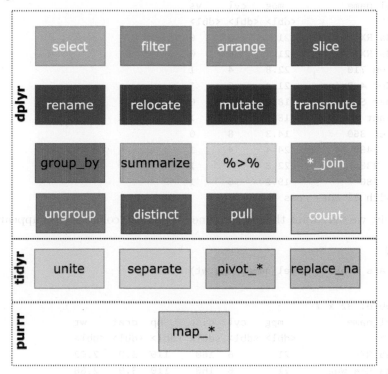

This section will cover the following:

- select, for selecting or deselecting columns
- filter, for filtering rows
- arrange, for reordering rows
- slice, for selecting rows with criteria or by row numbers
- rename, for renaming attributes
- relocate, for adjusting the order of the columns
- mutate and transmute, for adding new columns
- group_by and summarize, for grouping rows together and summarizing information about the group

We will also discuss the |> operator to coordinate multiple actions seamlessly.

Be sure to bookmark the `dplyr` cheatsheet[2] which will come in handy and useful for exploring more verbs available.

2.3.3 Selecting columns with `select`

The selection of attributes occurs when you want to focus on a subset of the attributes of a dataset at hand. The function `select` allows the selection in multiple possible ways.

In the simplest form of `select`, we list the attributes we wish to include in the data with a comma in between. For instance, we may only want to focus on the model name, miles per gallon, the number of cylinders, and the engine design.

```
select(mtcars_tibble, model_name, mpg, cyl, vs)
```

```
## # A tibble: 32 x 4
##    model_name           mpg   cyl    vs
##    <chr>              <dbl> <dbl> <dbl>
##  1 Mazda RX4             21     6     0
##  2 Mazda RX4 Wag         21     6     0
##  3 Datsun 710          22.8     4     1
##  4 Hornet 4 Drive      21.4     6     1
##  5 Hornet Sportabout   18.7     8     0
##  6 Valiant             18.1     6     1
##  7 Duster 360          14.3     8     0
##  8 Merc 240D           24.4     4     1
##  9 Merc 230            22.8     4     1
## 10 Merc 280            19.2     6     1
## # ... with 22 more rows
```

Alternatively, we may want the model name and all the columns that appear between `mpg` and `wt`.

```
select(mtcars_tibble, model_name, mpg:wt)
```

```
## # A tibble: 32 x 7
##    model_name           mpg   cyl  disp    hp  drat    wt
##    <chr>              <dbl> <dbl> <dbl> <dbl> <dbl> <dbl>
##  1 Mazda RX4             21     6   160   110   3.9  2.62
##  2 Mazda RX4 Wag         21     6   160   110   3.9  2.88
##  3 Datsun 710          22.8     4   108    93  3.85  2.32
##  4 Hornet 4 Drive      21.4     6   258   110  3.08  3.22
##  5 Hornet Sportabout   18.7     8   360   175  3.15  3.44
##  6 Valiant             18.1     6   225   105  2.76  3.46
##  7 Duster 360          14.3     8   360   245  3.21  3.57
##  8 Merc 240D           24.4     4   147.   62  3.69  3.19
##  9 Merc 230            22.8     4   141.   95  3.92  3.15
## 10 Merc 280            19.2     6   168.  123  3.92  3.44
## # ... with 22 more rows
```

---

[2]https://rstudio.com/wp-content/uploads/2015/02/data-wrangling-cheatsheet.pdf

We can also provide something more complex. select can receive attribute matching options like starts_with, ends_with, and contains. The following example demonstrates the use of some of these.

```
select(mtcars_tibble, cyl | !starts_with("m") & contains("a"))
```

```
## # A tibble: 32 x 5
##       cyl  drat    am  gear  carb
##     <dbl> <dbl> <dbl> <dbl> <dbl>
## 1      6  3.9      1     4     4
## 2      6  3.9      1     4     4
## 3      4  3.85     1     4     1
## 4      6  3.08     0     3     1
## 5      8  3.15     0     3     2
## 6      6  2.76     0     3     1
## 7      8  3.21     0     3     4
## 8      4  3.69     0     4     2
## 9      4  3.92     0     4     2
## 10     6  3.92     0     4     4
## # ... with 22 more rows
```

The criterion for selection here: in addition to mpg and cyl, any attribute whose name starts with some character other than "m" and contains "a" somewhere.

Going one step further, we can also supply a regular expression to do the matching. Recall that ^ and $ are the start and end of a string, respectively, and [a-z]{3,5} means any lowercase alphabet sequence having length between 3 and 5. Have a look at the following example.

```
select(mtcars_tibble, matches("^[a-z]{3,5}$"))
```

```
## # A tibble: 32 x 7
##      mpg   cyl  disp  drat  qsec  gear  carb
##    <dbl> <dbl> <dbl> <dbl> <dbl> <dbl> <dbl>
## 1   21      6  160   3.9   16.5    4     4
## 2   21      6  160   3.9   17.0    4     4
## 3   22.8    4  108   3.85  18.6    4     1
## 4   21.4    6  258   3.08  19.4    3     1
## 5   18.7    8  360   3.15  17.0    3     2
## 6   18.1    6  225   2.76  20.2    3     1
## 7   14.3    8  360   3.21  15.8    3     4
## 8   24.4    4  147.  3.69  20      4     2
## 9   22.8    4  141.  3.92  22.9    4     2
## 10  19.2    6  168.  3.92  18.3    4     4
## # ... with 22 more rows
```

The regular expression here means return any columns that have "lowercase name with length between 3 and 5".

2.3.4   Filtering rows with `filter`

Let us turn our attention now to the rows. The function `filter` allows us to select rows using some criteria. The syntax is to provide a Boolean expression for what should be included in the filtered dataset.

We can select all car models with 8 cylinders. Note how `cyl == 8` is an expression that evaluates to either TRUE or FALSE depending on whether the the attribute `cyl` of the row has a value of 8.

```
filter(mtcars_tibble, cyl == 8)
```

```
## # A tibble: 14 x 12
##    model~1    mpg   cyl  disp    hp  drat    wt  qsec    vs
##    <chr>    <dbl> <dbl> <dbl> <dbl> <dbl> <dbl> <dbl> <dbl>
##  1 Hornet~  18.7      8   360   175  3.15  3.44  17.0     0
##  2 Duster~  14.3      8   360   245  3.21  3.57  15.8     0
##  3 Merc 4~  16.4      8   276.  180  3.07  4.07  17.4     0
##  4 Merc 4~  17.3      8   276.  180  3.07  3.73  17.6     0
##  5 Merc 4~  15.2      8   276.  180  3.07  3.78  18       0
##  6 Cadill~  10.4      8   472   205  2.93  5.25  18.0     0
##  7 Lincol~  10.4      8   460   215  3     5.42  17.8     0
##  8 Chrysl~  14.7      8   440   230  3.23  5.34  17.4     0
##  9 Dodge ~  15.5      8   318   150  2.76  3.52  16.9     0
## 10 AMC Ja~  15.2      8   304   150  3.15  3.44  17.3     0
## 11 Camaro~  13.3      8   350   245  3.73  3.84  15.4     0
## 12 Pontia~  19.2      8   400   175  3.08  3.84  17.0     0
## 13 Ford P~  15.8      8   351   264  4.22  3.17  14.5     0
## 14 Masera~  15        8   301   335  3.54  3.57  14.6     0
## # ... with 3 more variables: am <dbl>, gear <dbl>,
## #   carb <dbl>, and abbreviated variable name
## #   1: model_name
```

We could be more picky and refine our search by including more attributes to filter by.

```
filter(mtcars_tibble, cyl == 8, am == 1, hp > 300)
```

```
## # A tibble: 1 x 12
##   model_~1   mpg   cyl  disp    hp  drat    wt  qsec    vs
##   <chr>    <dbl> <dbl> <dbl> <dbl> <dbl> <dbl> <dbl> <dbl>
## 1 Maserat~    15     8   301   335  3.54  3.57  14.6     0
## # ... with 3 more variables: am <dbl>, gear <dbl>,
## #   carb <dbl>, and abbreviated variable name
## #   1: model_name
```

Here, we requested a new tibble that contains rows with 8 cylinders, a manual transmission, and a gross horsepower over 300.

We may be interested in fetching a particular row in the dataset, say, the information associated with the car model "Datsun 710". We can also use `filter` to achieve this task.

```
filter(mtcars_tibble, model_name == "Datsun 710")
```

```
## # A tibble: 1 x 12
##   model_~1   mpg   cyl  disp    hp  drat    wt  qsec    vs
##   <chr>    <dbl> <dbl> <dbl> <dbl> <dbl> <dbl> <dbl> <dbl>
## 1 Datsun ~  22.8     4   108    93  3.85  2.32  18.6     1
## # ... with 3 more variables: am <dbl>, gear <dbl>,
## #   carb <dbl>, and abbreviated variable name
## #   1: model_name
```

2.3.5 Re-arranging rows with `arrange`

It may be necessary to rearrange the order of the rows to aid our understanding of the meaning of the dataset. The function `arrange` allows us to do just that.

To arrange rows, we state a list of attributes in the order we want to use for re-arranging. For instance, we can rearrange the rows by gross horsepower (`hp`).

```
arrange(mtcars_tibble, hp)
```

```
## # A tibble: 32 x 12
##    model~1   mpg   cyl  disp    hp  drat    wt  qsec    vs
##    <chr>   <dbl> <dbl> <dbl> <dbl> <dbl> <dbl> <dbl> <dbl>
##  1 Honda ~  30.4     4  75.7    52  4.93  1.62  18.5     1
##  2 Merc 2~  24.4     4 147.     62  3.69  3.19  20       1
##  3 Toyota~  33.9     4  71.1    65  4.22  1.84  19.9     1
##  4 Fiat 1~  32.4     4  78.7    66  4.08  2.2   19.5     1
##  5 Fiat X~  27.3     4  79      66  4.08  1.94  18.9     1
##  6 Porsch~  26       4 120.     91  4.43  2.14  16.7     0
##  7 Datsun~  22.8     4  108     93  3.85  2.32  18.6     1
##  8 Merc 2~  22.8     4 141.     95  3.92  3.15  22.9     1
##  9 Toyota~  21.5     4 120.     97  3.7   2.46  20.0     1
## 10 Valiant  18.1     6  225    105  2.76  3.46  20.2     1
## # ... with 22 more rows, 3 more variables: am <dbl>,
## #   gear <dbl>, carb <dbl>, and abbreviated variable name
## #   1: model_name
```

By default, `arrange` will reorder in ascending order. If we wish to reorder in descending order, we put the attribute in a `desc` function call. While we are at it, let us break ties in `hp` and order by miles per gallon (`mpg`).

```
arrange(mtcars_tibble, desc(hp), mpg)
```

```
## # A tibble: 32 x 12
##   model~1   mpg   cyl  disp    hp  drat    wt  qsec    vs
##   <chr>   <dbl> <dbl> <dbl> <dbl> <dbl> <dbl> <dbl> <dbl>
## 1 Masera~  15       8   301   335  3.54  3.57  14.6     0
## 2 Ford P~  15.8     8   351   264  4.22  3.17  14.5     0
## 3 Camaro~  13.3     8   350   245  3.73  3.84  15.4     0
## 4 Duster~  14.3     8   360   245  3.21  3.57  15.8     0
```

```
##  5 Chrysl~  14.7     8  440     230  3.23  5.34  17.4      0
##  6 Lincol~  10.4     8  460     215  3     5.42  17.8      0
##  7 Cadill~  10.4     8  472     205  2.93  5.25  18.0      0
##  8 Merc 4~  15.2     8  276.    180  3.07  3.78  18        0
##  9 Merc 4~  16.4     8  276.    180  3.07  4.07  17.4      0
## 10 Merc 4~  17.3     8  276.    180  3.07  3.73  17.6      0
## # ... with 22 more rows, 3 more variables: am <dbl>,
## #   gear <dbl>, carb <dbl>, and abbreviated variable name
## #   1: model_name
```

2.3.6   Selecting rows with `slice`

The function is for selecting rows by specifying the rows position. You can specify one row with its row number, a range of rows with a number pair A:B where you can have an expression involving the function n to specify the number of rows in the data.

The following use of slice() uses the range (n()-10):(n()-2) is the range starting from the tenth row from the last and ending at the second to last row.

```
slice(mtcars_tibble, (n()-10) : (n()-2))
```

```
## # A tibble: 9 x 12
##    model_~1   mpg   cyl  disp    hp  drat    wt  qsec    vs
##    <chr>    <dbl> <dbl> <dbl> <dbl> <dbl> <dbl> <dbl> <dbl>
## 1 Dodge C~  15.5     8  318    150  2.76  3.52  16.9      0
## 2 AMC Jav~  15.2     8  304    150  3.15  3.44  17.3      0
## 3 Camaro ~  13.3     8  350    245  3.73  3.84  15.4      0
## 4 Pontiac~  19.2     8  400    175  3.08  3.84  17.0      0
## 5 Fiat X1~  27.3     4   79     66  4.08  1.94  18.9      1
## 6 Porsche~  26       4  120.    91  4.43  2.14  16.7      0
## 7 Lotus E~  30.4     4   95.1  113  3.77  1.51  16.9      1
## 8 Ford Pa~  15.8     8  351    264  4.22  3.17  14.5      0
## 9 Ferrari~  19.7     6  145    175  3.62  2.77  15.5      0
## # ... with 3 more variables: am <dbl>, gear <dbl>,
## #   carb <dbl>, and abbreviated variable name
## #   1: model_name
```

We can also use `slice_head(n = NUMBER)` and `slice_tail(n = NUMBER)` to select the top NUMBER rows and the last NUMBER rows, respectively.

```
slice_head(mtcars_tibble, n = 2)
```

```
## # A tibble: 2 x 12
##    model_~1   mpg   cyl  disp    hp  drat    wt  qsec    vs
##    <chr>    <dbl> <dbl> <dbl> <dbl> <dbl> <dbl> <dbl> <dbl>
## 1 Mazda R~    21     6  160    110  3.9   2.62  16.5      0
## 2 Mazda R~    21     6  160    110  3.9   2.88  17.0      0
## # ... with 3 more variables: am <dbl>, gear <dbl>,
## #   carb <dbl>, and abbreviated variable name
## #   1: model_name
```

```
slice_tail(mtcars_tibble, n = 2)
```

```
## # A tibble: 2 x 12
##    model_~1   mpg   cyl  disp    hp  drat    wt  qsec    vs
##    <chr>     <dbl> <dbl> <dbl> <dbl> <dbl> <dbl> <dbl> <dbl>
## 1 Maserat~    15     8   301   335  3.54  3.57  14.6     0
## 2 Volvo 1~   21.4    4   121   109  4.11  2.78  18.6     1
## # ... with 3 more variables: am <dbl>, gear <dbl>,
## #   carb <dbl>, and abbreviated variable name
## #   1: model_name
```

If we are interested in some particular row, we can use slice for that as well.

```
slice(mtcars_tibble, 3)
```

```
## # A tibble: 1 x 12
##    model_~1   mpg   cyl  disp    hp  drat    wt  qsec    vs
##    <chr>     <dbl> <dbl> <dbl> <dbl> <dbl> <dbl> <dbl> <dbl>
## 1 Datsun ~   22.8    4   108    93  3.85  2.32  18.6     1
## # ... with 3 more variables: am <dbl>, gear <dbl>,
## #   carb <dbl>, and abbreviated variable name
## #   1: model_name
```

We can also select a *random* row by using slice_sample. In this example, each row has an equal chance of being selected.

```
slice_sample(mtcars_tibble)
```

```
## # A tibble: 1 x 12
##    model_~1   mpg   cyl  disp    hp  drat    wt  qsec    vs
##    <chr>     <dbl> <dbl> <dbl> <dbl> <dbl> <dbl> <dbl> <dbl>
## 1 AMC Jav~   15.2    8   304   150  3.15  3.44  17.3     0
## # ... with 3 more variables: am <dbl>, gear <dbl>,
## #   carb <dbl>, and abbreviated variable name
## #   1: model_name
```

2.3.7 Renaming columns with rename

This function allows you to rename a specific column. The syntax is NEW_NAME = OLD_NAME. Below, we replace the name wt with weight amd cyl with cylinder.

```
rename(mtcars_tibble, weight = wt, cylinder = cyl)
```

```
## # A tibble: 32 x 12
##    model_name   mpg cylin~1  disp    hp  drat weight  qsec
##    <chr>       <dbl>  <dbl> <dbl> <dbl> <dbl>  <dbl> <dbl>
## 1 Mazda RX4     21      6   160   110   3.9    2.62  16.5
## 2 Mazda RX4~    21      6   160   110   3.9    2.88  17.0
## 3 Datsun 710   22.8     4   108    93  3.85    2.32  18.6
## 4 Hornet 4 ~   21.4     6   258   110  3.08    3.22  19.4
```

```
##  5 Hornet Sp~   18.7      8   360   175  3.15   3.44  17.0
##  6 Valiant      18.1      6   225   105  2.76   3.46  20.2
##  7 Duster 360   14.3      8   360   245  3.21   3.57  15.8
##  8 Merc 240D    24.4      4   147.   62  3.69   3.19  20
##  9 Merc 230     22.8      4   141.   95  3.92   3.15  22.9
## 10 Merc 280     19.2      6   168.  123  3.92   3.44  18.3
## # ... with 22 more rows, 4 more variables: vs <dbl>,
## #   am <dbl>, gear <dbl>, carb <dbl>, and abbreviated
## #   variable name 1: cylinder
```

2.3.8   Relocating column positions with `relocate`

Sometimes you may want to change the order of columns by moving a column from the present location to another. We can relocate a column using the `relocate` function by specifying which column should go where. The syntax is `relocate(DATA_NAME,ATTRIBUTE,NEW_LOCATION)`.

The specification for the new location is either by `.before=NAME` or by `.after=NAME`, where `NAME` is the name of a column.

```
relocate(mtcars_tibble, am, .before = mpg)
```

```
## # A tibble: 32 x 12
##     model~1    am   mpg   cyl  disp    hp  drat    wt  qsec
##     <chr>   <dbl> <dbl> <dbl> <dbl> <dbl> <dbl> <dbl> <dbl>
##  1 Mazda ~     1  21       6   160   110  3.9   2.62  16.5
##  2 Mazda ~     1  21       6   160   110  3.9   2.88  17.0
##  3 Datsun~     1  22.8     4   108    93  3.85  2.32  18.6
##  4 Hornet~     0  21.4     6   258   110  3.08  3.22  19.4
##  5 Hornet~     0  18.7     8   360   175  3.15  3.44  17.0
##  6 Valiant     0  18.1     6   225   105  2.76  3.46  20.2
##  7 Duster~     0  14.3     8   360   245  3.21  3.57  15.8
##  8 Merc 2~     0  24.4     4   147.   62  3.69  3.19  20
##  9 Merc 2~     0  22.8     4   141.   95  3.92  3.15  22.9
## 10 Merc 2~     0  19.2     6   168.  123  3.92  3.44  18.3
## # ... with 22 more rows, 3 more variables: vs <dbl>,
## #   gear <dbl>, carb <dbl>, and abbreviated variable name
## #   1: model_name
```

Here we moved the column am to the front, just before mpg.

2.3.9   Adding new columns using `mutate`

The function `mutate` can be used for modification or creation of a new column using some function of the values of existing columns. Let us see an example before getting into the details.

Suppose we are interested in calculating the ratio between the numbers of cylinders and forward gears for each car model. We can do this by appending a new column with the calculated ratios using `mutate`.

```
mtcars_with_ratio <- mutate(mtcars_tibble,
                            cyl_gear_ratio = cyl / gear)
mtcars_with_ratio
```

```
## # A tibble: 32 x 13
##    model~1  mpg   cyl  disp    hp  drat    wt  qsec    vs
##    <chr>   <dbl> <dbl> <dbl> <dbl> <dbl> <dbl> <dbl> <dbl>
##  1 Mazda ~  21      6   160   110  3.9   2.62  16.5     0
##  2 Mazda ~  21      6   160   110  3.9   2.88  17.0     0
##  3 Datsun~  22.8    4   108    93  3.85  2.32  18.6     1
##  4 Hornet~  21.4    6   258   110  3.08  3.22  19.4     1
##  5 Hornet~  18.7    8   360   175  3.15  3.44  17.0     0
##  6 Valiant  18.1    6   225   105  2.76  3.46  20.2     1
##  7 Duster~  14.3    8   360   245  3.21  3.57  15.8     0
##  8 Merc 2~  24.4    4   147.   62  3.69  3.19  20       1
##  9 Merc 2~  22.8    4   141.   95  3.92  3.15  22.9     1
## 10 Merc 2~  19.2    6   168.  123  3.92  3.44  18.3     1
## # ... with 22 more rows, 4 more variables: am <dbl>,
## #   gear <dbl>, carb <dbl>, cyl_gear_ratio <dbl>, and
## #   abbreviated variable name 1: model_name
```

Unfortunately, the new column appears at the very end which may not be desirable. We can fix this with the following adjustment.

```
mtcars_with_ratio <- mutate(mtcars_tibble,
                            cyl_gear_ratio = cyl / gear,
                            .before = mpg)
mtcars_with_ratio
```

```
## # A tibble: 32 x 13
##    model_name  cyl_g~1   mpg   cyl  disp    hp  drat    wt
##    <chr>        <dbl>  <dbl> <dbl> <dbl> <dbl> <dbl> <dbl>
##  1 Mazda RX4    1.5     21      6   160   110  3.9   2.62
##  2 Mazda RX4 ~  1.5     21      6   160   110  3.9   2.88
##  3 Datsun 710   1       22.8    4   108    93  3.85  2.32
##  4 Hornet 4 D~  2       21.4    6   258   110  3.08  3.22
##  5 Hornet Spo~  2.67    18.7    8   360   175  3.15  3.44
##  6 Valiant      2       18.1    6   225   105  2.76  3.46
##  7 Duster 360   2.67    14.3    8   360   245  3.21  3.57
##  8 Merc 240D    1       24.4    4   147.   62  3.69  3.19
##  9 Merc 230     1       22.8    4   141.   95  3.92  3.15
## 10 Merc 280     1.5     19.2    6   168.  123  3.92  3.44
## # ... with 22 more rows, 5 more variables: qsec <dbl>,
## #   vs <dbl>, am <dbl>, gear <dbl>, carb <dbl>, and
## #   abbreviated variable name 1: cyl_gear_ratio
```

By specifying an additional .before argument with the value mpg, we inform dplyr that the new column cyl_gear_ratio should appear before the column mpg, which is the first column in the dataset.

Generally speaking, the syntax for mutate is:

```
mutate(DATA_SET_NAME, NEW_NAME = EXPRESSION, OPTION)
```

---

where:

- The NEW_NAME = EXPRESSION specifies the name of the new attribute and how to compute it, and OPTION is an option to specify the location of the new attribute relative to the existing attributes.
- The position option is either of the form .before=VALUE or of the form .after=VALUE with VALUE specifying the name of the column where the new column will appear before or after; it can also receive a number indicating the position for the newly inserted column.

- The EXPRESSION can be either a mathematical expression or a function call.

Let us see another example. In addition to calculating the ratio from before, we will create another column containing the make of the car. We will do this by extracting the first word from model_name using a regular expression. Let us amend our mutate code from before to include the changes.

```
mtcars_mutated <- mutate(mtcars_tibble,
                  cyl_gear_ratio = cyl / gear,
                  make = str_replace(model_name, " .*", ""),
                  .before = mpg)
mtcars_mutated
```

```
## # A tibble: 32 x 14
##     model_name  cyl_g~1 make    mpg   cyl disp    hp drat
##     <chr>        <dbl> <chr> <dbl> <dbl> <dbl> <dbl> <dbl>
##  1 Mazda RX4      1.5  Mazda 21      6  160    110  3.9
##  2 Mazda RX4 ~   1.5  Mazda 21      6  160    110  3.9
##  3 Datsun 710    1    Dats~ 22.8    4  108     93  3.85
##  4 Hornet 4 D~   2    Horn~ 21.4    6  258    110  3.08
##  5 Hornet Spo~   2.67 Horn~ 18.7    8  360    175  3.15
##  6 Valiant       2    Vali~ 18.1    6  225    105  2.76
##  7 Duster 360    2.67 Dust~ 14.3    8  360    245  3.21
##  8 Merc 240D     1    Merc  24.4    4  147.    62  3.69
##  9 Merc 230      1    Merc  22.8    4  141.    95  3.92
## 10 Merc 280      1.5  Merc  19.2    6  168.   123  3.92
## # ... with 22 more rows, 6 more variables: wt <dbl>,
## #   qsec <dbl>, vs <dbl>, am <dbl>, gear <dbl>,
## #   carb <dbl>, and abbreviated variable name
## #   1: cyl_gear_ratio
```

To form the make column, we use the function str_replace; we look for substrings that match the pattern " .*" (one white space and then any number of characters following it) and replace it with an empty string "", leaving only the first word, as desired. String operations are applicable to strings, which is what appears in the column make.

Note that the original dataset, before mutation, remains unchanged in `mtcars_tibble`.

```
mtcars_tibble
```

```
## # A tibble: 32 x 12
##    model~1   mpg   cyl  disp    hp  drat    wt  qsec    vs
##    <chr>   <dbl> <dbl> <dbl> <dbl> <dbl> <dbl> <dbl> <dbl>
##  1 Mazda ~  21       6   160   110  3.9   2.62  16.5     0
##  2 Mazda ~  21       6   160   110  3.9   2.88  17.0     0
##  3 Datsun~  22.8     4   108    93  3.85  2.32  18.6     1
##  4 Hornet~  21.4     6   258   110  3.08  3.22  19.4     1
##  5 Hornet~  18.7     8   360   175  3.15  3.44  17.0     0
##  6 Valiant  18.1     6   225   105  2.76  3.46  20.2     1
##  7 Duster~  14.3     8   360   245  3.21  3.57  15.8     0
##  8 Merc 2~  24.4     4   147.   62  3.69  3.19  20       1
##  9 Merc 2~  22.8     4   141.   95  3.92  3.15  22.9     1
## 10 Merc 2~  19.2     6   168.  123  3.92  3.44  18.3     1
## # ... with 22 more rows, 3 more variables: am <dbl>,
## #   gear <dbl>, carb <dbl>, and abbreviated variable name
## #   1: model_name
```

How many different makes are there? We can use `unique` for removing duplicates to find out.

```
unique(pull(mtcars_mutated, make))
```

```
##  [1] "Mazda"    "Datsun"   "Hornet"   "Valiant"
##  [5] "Duster"   "Merc"     "Cadillac" "Lincoln"
##  [9] "Chrysler" "Fiat"     "Honda"    "Toyota"
## [13] "Dodge"    "AMC"      "Camaro"   "Pontiac"
## [17] "Porsche"  "Lotus"    "Ford"     "Ferrari"
## [21] "Maserati" "Volvo"
```

When an existing column is given in the specification, no new column is created, and the existing column is modified instead. For instance, the following rounds `wt` to the nearest integer value.

```
mutate(mtcars_tibble, wt = round(wt))
```

```
## # A tibble: 32 x 12
##    model~1   mpg   cyl  disp    hp  drat    wt  qsec    vs
##    <chr>   <dbl> <dbl> <dbl> <dbl> <dbl> <dbl> <dbl> <dbl>
##  1 Mazda ~  21       6   160   110  3.9      3  16.5     0
##  2 Mazda ~  21       6   160   110  3.9      3  17.0     0
##  3 Datsun~  22.8     4   108    93  3.85     2  18.6     1
##  4 Hornet~  21.4     6   258   110  3.08     3  19.4     1
##  5 Hornet~  18.7     8   360   175  3.15     3  17.0     0
##  6 Valiant  18.1     6   225   105  2.76     3  20.2     1
##  7 Duster~  14.3     8   360   245  3.21     4  15.8     0
##  8 Merc 2~  24.4     4   147.   62  3.69     3  20       1
##  9 Merc 2~  22.8     4   141.   95  3.92     3  22.9     1
```

```
## 10 Merc 2~  19.2      6  168.    123  3.92      3  18.3      1
## # ... with 22 more rows, 3 more variables: am <dbl>,
## #   gear <dbl>, carb <dbl>, and abbreviated variable name
## #   1: model_name
```

We can also modify multiple columns in a single pass, say, `wt`, `mpg`, and `qsec` should all be rounded to the nearest integer. We can accomplish this using a combination of `mutate` with the helper `dplyr` verb `across`.

```
mutate(mtcars_tibble,
       across(c(mpg, wt, qsec), round))
```

```
## # A tibble: 32 x 12
##     model~1   mpg  cyl  disp    hp  drat    wt  qsec    vs
##     <chr>   <dbl> <dbl> <dbl> <dbl> <dbl> <dbl> <dbl> <dbl>
## 1 Mazda ~    21     6   160   110  3.9      3    16     0
## 2 Mazda ~    21     6   160   110  3.9      3    17     0
## 3 Datsun~    23     4   108    93  3.85     2    19     1
## 4 Hornet~    21     6   258   110  3.08     3    19     1
## 5 Hornet~    19     8   360   175  3.15     3    17     0
## 6 Valiant    18     6   225   105  2.76     3    20     1
## 7 Duster~    14     8   360   245  3.21     4    16     0
## 8 Merc 2~    24     4   147.   62  3.69     3    20     1
## 9 Merc 2~    23     4   141.   95  3.92     3    23     1
## 10 Merc 2~   19     6   168.  123  3.92     3    18     1
## # ... with 22 more rows, 3 more variables: am <dbl>,
## #   gear <dbl>, carb <dbl>, and abbreviated variable name
## #   1: model_name
```

2.3.10   The function `transmute`

The function `transmute` is a variant of `mutate` where we keep only the new columns generated.

```
only_the_new_stuff <- transmute(mtcars_tibble,
                        cyl_gear_ratio = cyl / gear,
                        make = str_replace(model_name, " .*", ""))
only_the_new_stuff
```

```
## # A tibble: 32 x 2
##    cyl_gear_ratio make
##             <dbl> <chr>
## 1            1.5  Mazda
## 2            1.5  Mazda
## 3            1    Datsun
## 4            2    Hornet
## 5            2.67 Hornet
## 6            2    Valiant
## 7            2.67 Duster
## 8            1    Merc
## 9            1    Merc
```

```
## 10          1.5  Merc
## # ... with 22 more rows
```

2.3.11   The pair `group_by` and `summarize`

Suppose you are interested in exploring the relationship between the number of cylinders in a car model and the miles per gallon it has. One way to examine this is to look at some summary statistic, say the *average*, of the miles per gallon for car models with 6 cylinders, car models with 7 cylinders, and car models with 8 cylinders.

When thinking about the problem in this way, we have effectively divided up all of the rows in the dataset into three groups, where the group a car model will belong to is determined by the number of cylinders it has.

dplyr accomplishes this using the function `group_by()`. The syntax for `group_by()` is simple: simply list the attributes with which you want to build groups. Let us give an example on how to use it.

```
grouped_by_cl <- group_by(mtcars_tibble, cyl)
slice_head(grouped_by_cl, n=2) # show 2 rows per group
```

```
## # A tibble: 6 x 12
## # Groups:   cyl [3]
##   model_~1   mpg   cyl  disp    hp  drat    wt  qsec    vs
##   <chr>    <dbl> <dbl> <dbl> <dbl> <dbl> <dbl> <dbl> <dbl>
## 1 Datsun ~  22.8     4   108    93  3.85  2.32  18.6     1
## 2 Merc 24~  24.4     4   147.   62  3.69  3.19  20       1
## 3 Mazda R~  21       6   160   110  3.9   2.62  16.5     0
## 4 Mazda R~  21       6   160   110  3.9   2.88  17.0     0
## 5 Hornet ~  18.7     8   360   175  3.15  3.44  17.0     0
## 6 Duster ~  14.3     8   360   245  3.21  3.57  15.8     0
## # ... with 3 more variables: am <dbl>, gear <dbl>,
## #   carb <dbl>, and abbreviated variable name
## #   1: model_name
```

We can spot two rows shown per each `cyl` group. `group_by()` alone is often not useful. To make something out of this, we need to *summarize* some piece of information using these groups, e.g. the average `mpg` per group as is needed for our task.

The summary function is called `summarize()`. Let us amend our above grouping code to include the summary.

```
grouped_by_cl <- group_by(mtcars_tibble, cyl)
summarized <- summarize(grouped_by_cl,
         count = n(),
         avg_mpg = mean(mpg))
summarized
```

```
## # A tibble: 3 x 3
##     cyl count avg_mpg
##   <dbl> <int>   <dbl>
## 1     4    11    26.7
```

```
## 2      6      7     19.7
## 3      8     14     15.1
```

This table looks more like what we would expect. Our `summary` calculates two summaries, each reflected in a column in the above table:

- `count`, the number of car models belonging to the group
- `avg_mpg`, the average miles per gallon of car models in the group

The summary results make sense. More cylinders translates to more power, but it also means more moving parts which can hurt efficiency. Therefore, it seems an association exists where the more cylinders a car has, the lower its miles per gallon.

These functions come handy when you want to examine data by grouping rows and summarize some information with respect to each group.

2.3.12   Coordinating multiple actions using |>

Let us revise a bit our previous study. Curious about the joint effect of the numbers of cylinders and the transmission of the car, you decide to group by both `cyl` and `am`. After summarizing the groups, you calculate the counts in each group and the average `mpg`. Finally, after the summary is done, you would like to remove any groups from the summary that have less than 2 cars.

Your analysis pipeline, then, would be composed of three steps:

1. Group rows by `cyl` and `am`, the number of cylinders.
2. Summarize to calculate the average miles per gallon per group.
3. Filter out rows that are below the average miles per gallon.

A first solution for this task might look like the following.

```
# step 1
grouped_by_cl <- group_by(mtcars_tibble, cyl, am)
# step 2
summarized <- summarize(grouped_by_cl,
          count = n(),
          avg_mpg = mean(mpg))
# step 3
avg_mpg_counts <- filter(summarized, count > 2)
avg_mpg_counts
```

```
## # A tibble: 5 x 4
## # Groups:    cyl [3]
##      cyl    am count avg_mpg
##    <dbl> <dbl> <int>   <dbl>
## 1      4     0     3    22.9
## 2      4     1     8    28.1
## 3      6     0     4    19.1
## 4      6     1     3    20.6
## 5      8     0    12    15.0
```

Observe how the code we have just written is quite cumbersome. It introduces several intermediate products that we do not need, namely, the names `grouped_by_cl` and `summarized`. It can also be difficult to come up with descriptive names.

Conveniently, there is a construct from base R called the "pipe" which allows us to pass the results from one function as input to another. The way to use piping is simple.

- You start by stating the initial dataset.
- For each operation to form, you append |> and then the operation, where you omit the dataset name part.
- If you need to save the result in a data set, you use the assignment operator <- at the beginning as usual.

Thus, we can rewrite our first solution as follows.

```
avg_mpg_counts <- mtcars |>
  group_by(cyl, am) |>
  summarize(count = n(),
            avg_mpg = mean(mpg)) |>
  filter(count > 2)
```

This solution is much easier to read than our first; we can clearly identify the transformations being performed on the data. It is good to read |> as "then".

Note that there is not much mystery with |>. All the pipe operator does is place an object into the first argument of a function. So, when we say `mtcars |> group_by(cyl, am)`, the pipe changes this to `group_by(mtcars, cyl, am)`. Or, more generally, if we have `x |> func(y)`, this is changed to `func(x, y)`.

Here is another use of the pipe, using our `mtcars_mutated` tibble from earlier.

```
mtcars_mutated |>
  pull(make) |>
  unique()
```

```
##  [1] "Mazda"    "Datsun"   "Hornet"   "Valiant"
##  [5] "Duster"   "Merc"     "Cadillac" "Lincoln"
##  [9] "Chrysler" "Fiat"     "Honda"    "Toyota"
## [13] "Dodge"    "AMC"      "Camaro"   "Pontiac"
## [17] "Porsche"  "Lotus"    "Ford"     "Ferrari"
## [21] "Maserati" "Volvo"
```

Neat! This one demonstrates some of the usefulness of `pull` over the traditional $ for accessing column data.

2.3.13 Practice makes perfect!

This section has covered a lot of `dplyr` functions for transforming datasets and, despite our best efforts, understanding what these functions are doing can quickly become overwhelming. The only way to truly understand these functions – and which ones should be used when confronted with a situation – is to practice using them.

Begin with the `mtcars` dataset from this section and run through each of the functions and the examples discussed here on your own. Observe what the dataset looks like before and after the transformation and try to understand what the transformation is.

Once you develop enough familiarity with these functions, try making small changes to our examples and coming up with your own transformations to apply. Be sure to include the |> operator whenever possible.

You may also wish to look at some of the datasets available to you when running the command data().

2.4   Tidy Transformations

In this section we turn to transformation techniques that are essential for achieving tidy data.

2.4.1   Prerequisites

As before, let us load tidyverse.

```
library(tidyverse)
```

2.4.2   Uniting and separating columns

The third tidy data guideline states that each value must have its own cell. Sometimes this value may be split across multiple columns or merged in a single column.

In the case of the Miami bakery example, we saw that when the bakery records sale forecasts, the lower and upper bounds of the range are fused in a single cell. This makes extraction and analysis of these values difficult, especially when R treats the forecast column as a character sequence.

```
forecast_sales <- tibble(
  week     = c(1, 2, 3),
  forecast = c("200-300", "300-400", "200-500")
)
forecast_sales
```

```
## # A tibble: 3 x 2
##    week forecast
##   <dbl> <chr>
## 1     1 200-300
## 2     2 300-400
## 3     3 200-500
```

A solution would be to *split* forecast into multiple columns, one giving the lower bound and the other the upper bound. The tidyr function separate accomplishes the work.

```
forecast_sales |>
  separate(forecast, c("low", "high"), "-", convert = TRUE)
```

```
## # A tibble: 3 x 3
##   week   low  high
##   <dbl> <int> <int>
## 1    1   200   300
## 2    2   300   400
## 3    3   200   500
```

We separate the columns based on the presence of the "-" character. The convert argument is set so that the lower and upper values can be treated as proper integers.

The tibble table5 displays the number of TB cases documented by the World Health Organization in Afghanistan, Brazil, and China between 1999 and 2000. The "year", however, is a single value that has been split across a century and year column.

table5

```
## # A tibble: 6 x 4
##   country     century year  rate
## * <chr>       <chr>   <chr> <chr>
## 1 Afghanistan 19      99    745/19987071
## 2 Afghanistan 20      00    2666/20595360
## 3 Brazil      19      99    37737/172006362
## 4 Brazil      20      00    80488/174504898
## 5 China       19      99    212258/1272915272
## 6 China       20      00    213766/1280428583
```

The unite function can be used to merge a split value. Its functionality is similar to separate.

```
table5 |>
  unite("year", century:year, sep="")
```

```
## # A tibble: 6 x 3
##   country     year  rate
##   <chr>       <chr> <chr>
## 1 Afghanistan 1999  745/19987071
## 2 Afghanistan 2000  2666/20595360
## 3 Brazil      1999  37737/172006362
## 4 Brazil      2000  80488/174504898
## 5 China       1999  212258/1272915272
## 6 China       2000  213766/1280428583
```

We specify an empty string ("") in the sep argument to indicate no character delimiter should be used when merging the values.

Note also that the rate column needs tidying. We leave the tidying of this column as an exercise for the reader.

2.4.3 Pulling data from multiple sources

The fourth principle of tidy data stated that an observational unit should form a table. However, often times the observational unit we are measuring is split across multiple tables.

Let us suppose we are measuring student assessments in a class. The data is given to us in the form of two tables, one for exams and the other for assignments. We load the scores into our R environment with tibble.

```
exams <- tibble(name = c("Adriana", "Beth", "Candy", "Emily"),
                midterm = c(90, 80, 95, 87),
                final = c(99, 50, 70, 78))
assignments <- tibble(name = c("Adriana", "Beth", "Candy", "Florence"),
                assign1 = c(80, 88, 93, 88),
                assign2 = c(91, 61, 73, 83))
exams
```

```
## # A tibble: 4 x 3
##   name    midterm final
##   <chr>     <dbl> <dbl>
## 1 Adriana      90    99
## 2 Beth         80    50
## 3 Candy        95    70
## 4 Emily        87    78
```

```
assignments
```

```
## # A tibble: 4 x 3
##   name    assign1 assign2
##   <chr>     <dbl>   <dbl>
## 1 Adriana      80      91
## 2 Beth         88      61
## 3 Candy        93      73
## 4 Florence     88      83
```

If the observational unit is an *assessment* result, then some assessments are in one table and some assessments are in another. Therefore, according to this definition, the current arrangement of the data is *not* tidy. The data should be kept together in a single table.

You can combine two tibbles using a common attribute as the key for combining; that is, finding values appearing in both tibbles and then connecting rows having the names in common. In general, if there are multiple matches between the two tibbles concerning the attribute, each possible row matches will appear.

The construct for stitching together two tibbles together in this manner is called the *join*. The general syntax is:

```
JOIN_METHOD_NAME(DATA1, DATA2, by="NAME").
```

Here DATA1 and DATA2 are the names of the tibbles and NAME is the name of the key attributes. There are four types of join functions. The differences among them are in how they treat non-matching values.

- left_join: Exclude any rows in DATA2 with no matching values in DATA1.
- right_join; Exclude any rows in DATA1 with no matching values in DATA2.
- inner_join: Exclude any rows in DATA2 and DATA1 with no matching values in the other data frame.
- full_join: No exclusions.

The example below shows the results of four join operations.

```
scores_left <- left_join(assignments, exams, by="name")
scores_left
```

```
## # A tibble: 4 x 5
##   name     assign1 assign2 midterm final
##   <chr>      <dbl>   <dbl>   <dbl> <dbl>
## 1 Adriana       80      91      90    99
## 2 Beth          88      61      80    50
## 3 Candy         93      73      95    70
## 4 Florence      88      83      NA    NA
```

```
scores_right <- right_join(assignments, exams, by="name")
scores_right
```

```
## # A tibble: 4 x 5
##   name    assign1 assign2 midterm final
##   <chr>     <dbl>   <dbl>   <dbl> <dbl>
## 1 Adriana      80      91      90    99
## 2 Beth         88      61      80    50
## 3 Candy        93      73      95    70
## 4 Emily        NA      NA      87    78
```

```
scores_inner <- inner_join(assignments, exams, by = "name")
scores_inner
```

```
## # A tibble: 3 x 5
##   name   assign1 assign2 midterm final
##   <chr>    <dbl>   <dbl>   <dbl> <dbl>
## 1 Adriana     80      91      90    99
## 2 Beth        88      61      80    50
## 3 Candy       93      73      95    70
```

```
scores_full <- full_join(assignments, exams, by="name")
scores_full
```

```
## # A tibble: 5 x 5
##   name     assign1 assign2 midterm final
##   <chr>      <dbl>   <dbl>   <dbl> <dbl>
## 1 Adriana       80      91      90    99
## 2 Beth          88      61      80    50
## 3 Candy         93      73      95    70
## 4 Florence      88      83      NA    NA
## 5 Emily         NA      NA      87    78
```

The results of the join reveal some anomalies in our data. Namely, we see that Emily does not have any assignment scores nor does Florence have any exam scores. Hence, in the left, right, and full joins, we see values labeled NA appear where they would have those values. We call these missing values, which can be thought of as "holes" in the data. We will return to missing values in a later section.

An alternative to the join is to stack up the rows using bind_rows.

```
bind_rows(assignments, exams)
```

```
## # A tibble: 8 x 5
##    name      assign1 assign2 midterm final
##    <chr>       <dbl>   <dbl>   <dbl> <dbl>
## 1 Adriana        80      91      NA    NA
## 2 Beth           88      61      NA    NA
## 3 Candy          93      73      NA    NA
## 4 Florence       88      83      NA    NA
## 5 Adriana        NA      NA      90    99
## 6 Beth           NA      NA      80    50
## 7 Candy          NA      NA      95    70
## 8 Emily          NA      NA      87    78
```

Observe how this one does not join values where possible, and so there is redundancy in the rows that appear, e.g., Adriana appears twice. As a result, many missing values appear in the resulting table.

2.4.4 Pivoting

Let us return to the resulting table after the inner join.

```
scores_inner
```

```
## # A tibble: 3 x 5
##    name     assign1 assign2 midterm final
##    <chr>      <dbl>   <dbl>   <dbl> <dbl>
## 1 Adriana       80      91      90    99
## 2 Beth          88      61      80    50
## 3 Candy         93      73      95    70
```

The third property of tidy data is fulfilled now that the observational unit forms a single table. However, the joined table is still messy. The grades are split across four different columns and, therefore, multiple observations occur at each row.

To remedy this, we use *pivot* and, in terms of R, the function pivot_longer from the tidyr package. The syntax for pivot_longer can be complex, and so we do not go over it in detail. Here is how we can use it.

```
scores_long <- scores_inner |>
  pivot_longer(c(assign1, assign2, midterm, final),
               names_to = "assessment", values_to = "score")
scores_long
```

```
## # A tibble: 12 x 3
##    name     assessment score
##    <chr>    <chr>      <dbl>
##  1 Adriana  assign1       80
##  2 Adriana  assign2       91
##  3 Adriana  midterm       90
##  4 Adriana  final         99
##  5 Beth     assign1       88
##  6 Beth     assign2       61
##  7 Beth     midterm       80
##  8 Beth     final         50
##  9 Candy    assign1       93
## 10 Candy    assign2       73
## 11 Candy    midterm       95
## 12 Candy    final         70
```

The usage above takes `scores_inner`, merges all the assessment columns, creates a new column with name `assessment`, and presents the corresponding values under the column `score`. Graphically, this is what a pivot longer transformation computes.

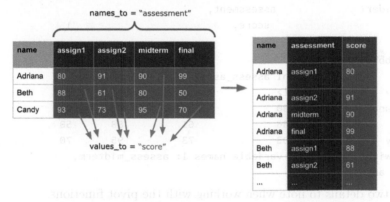

Observe how we can easily read off the three variables from this table: `name`, `assessment`, and `score`. We can be confident in knowing that this is tidy data.

If we wish to go in the other direction, we can use `pivot_wider`. The function `pivot_wider` grabs a pair of columns and spreads the pair into a series of columns. One column of the pair serves as the source for the new column names after spreading. For each value appearing in the source column, the function creates a new column by the name. The value appearing opposite to the source value appears as the value for the column corresponding to the source.

```
scores_long |>
  pivot_wider(names_from = assessment, values_from = score)
```

```
## # A tibble: 3 x 5
##   name    assign1 assign2 midterm final
##   <chr>     <dbl>   <dbl>   <dbl> <dbl>
## 1 Adriana      80      91      90    99
## 2 Beth         88      61      80    50
## 3 Candy        93      73      95    70
```

Here is a visual demonstrating the pivot wider transformation:

Note how this simply undoes what we have done, returning us to the original `scores_inner` table. We can also prefix each of the new columns with `assess_`.

```
scores_long |>
  pivot_wider(names_from = assessment,
              values_from = score, names_prefix = "assess_")
```

```
## # A tibble: 3 x 5
##   name    assess_assign1 assess_assign2 assess_m~1 asses~2
##   <chr>            <dbl>          <dbl>      <dbl>   <dbl>
## 1 Adriana             80             91         90      99
## 2 Beth                88             61         80      50
## 3 Candy               93             73         95      70
## # ... with abbreviated variable names 1: assess_midterm,
## #   2: assess_final
```

There are two details to note when working with the pivot functions.

- `pivot_wider` should not be thought of as an "undo" operation. Like `pivot_longer` its primary purpose is also to make data tidy. Consider the following table and observe how each observation is scattered across two rows. The appropriate means to bring this data into tidiness is through an application of `pivot_wider`.

```
slice_head(table2, n = 5)
```

```
## # A tibble: 5 x 4
##   country     year  type        count
##   <chr>      <int> <chr>        <int>
## 1 Afghanistan 1999 cases          745
## 2 Afghanistan 1999 population 19987071
## 3 Afghanistan 2000 cases         2666
## 4 Afghanistan 2000 population 20595360
## 5 Brazil      1999 cases        37737
```

- `pivot_longer` and `pivot_wider` are not perfectly symmetrical operations. That is, there are cases where applying `pivot_wider`, followed by `pivot_longer`, will not reproduce the exact same dataset. Consider such an application on the following dataset. Keep in mind the column names and how the column data types change at each pivot step.

```
sales <- tibble(
  year    = c(2020, 2021, 2020, 2021),
  quarter = c(1, 2, 1, 2),
  sale    = c(70, 80, 62, 100)
)
```

2.5 Applying Functions to Columns

Situations can arise where we need to apply some function to a column. In this section we learn how to apply functions to columns using a construct called the `map`.

2.5.1 Prerequisites

As before, let us load `tidyverse`.

```
library(tidyverse)
```

We will use the `mtcars` tibble again in this section so let us prepare the tibble by migrating the row names to a dedicated column. Note how the pipe operator can be used to help with the work.

```
mtcars_tibble <- mtcars |>
  rownames_to_column("model_name")
```

2.5.2 What is a `function` anyway?

We have used several times by now the word "function". Here are some basic rules about functions.

- *A function is a block of code with a name that allows execution from other codes.* This mean that you can take any part of a working (i.e., all parentheses and brackets in the part have matching counterparts in the same part) and specify it to be a function.
- *If the function is active in the present run of R, each time a code call the function, the code of the function runs.* This means that R suspends the execution of the present code and processes the execution of the code of the function. When it finishes running the code of the function, it returns to the execution of the one it has suspended.
- *A function may take upon the role of computing a value.* You can design a function so that it uses a special function `return` at the end so as to specify the value it has computed. Note that the use of `return` is **optional** and, by default, R returns the last line of computation performed in the function.
- *If a function has the role of returning a value, the call itself represents the value it computes.* So, you store the value the function computes in a variable using an assignment.

- *A function may require some number of values to use in its calculation.* We call them *arguments*. When using a function that requires arguments, the arguments must appear in the call.

2.5.3   A very simple function

Here is a very simple function, `one_to_ten`, which prints the sequence of integers from 1 to 10. The definition of the function takes the form `one_ten <- function() { ... }`.

```
one_to_ten <- function() {
  print(1:10)
}
```

Here is what happens when you call the function.

```
one_to_ten()
```

```
## [1]  1  2  3  4  5  6  7  8  9 10
```

Note that the call stands alone, i.e., you can use it without anything else but its name and a pair of parentheses. By replacing the code appearing inside the curly brackets, you can define a different function with the same name `one_to_ten`.

Let us reverse the order in which the numbers appear.

```
one_to_ten <- function() {
  print(10:1)
}
```

Here is what happens when you call the function.

```
one_to_ten()
```

```
## [1] 10  9  8  7  6  5  4  3  2  1
```

The new behavior of `one_to_ten` substitutes the old one, and you cannot replay the behavior of the previous version (until, of course, you modify the function again).

2.5.4   Functions that compute a value

To make a function compute a value, you add a line `return(VALUE)` at the end of the code in the brackets. The function `my_family` returns a list of names for persons.

Remember the `c` function? The function creates a vector with 19 names as strings and returns the vector.

```
my_family <- function() {
  c("Amy", "Billie", "Casey", "Debbie", "Eddie", "Freddie", "Gary",
    "Hary", "Ivy", "Jackie", "Lily", "Mikey", "Nellie", "Odie",
    "Paulie", "Quincy", "Ruby", "Stacey", "Tiffany")
}
```

The call for the function produces the list that the function returns.

```
a <- my_family()
a
```

```
## [1] "Amy"      "Billie"  "Casey"   "Debbie"  "Eddie"
## [6] "Freddie"  "Gary"    "Hary"    "Ivy"     "Jackie"
## [11] "Lily"    "Mikey"   "Nellie"  "Odie"    "Paulie"
## [16] "Quincy"  "Ruby"    "Stacey"  "Tiffany"
```

Now whenever you need the 19-name list, you can either call the function or refer to the variable a that holds the list.

2.5.5 Functions that take arguments

To write a function that takes arguments, you determine how many arguments you need and determine the names you want to use for the arguments during the execution of the code for the function.

The function declaration now has the names of the arguments. You put them in the order you want to use with a comma in between. Below, we define a function that computes the max between 100 and the argument received. The function returns the argument so long as it is larger than 100.

```
passes_100 <- function(x) {
  max(100, x)
}
```

Here is a demonstration of how the function works.

```
passes_100(50)   # a value smaller than 100
```

```
## [1] 100
```

```
passes_100(2021)   # a value larger than 100
```

```
## [1] 2021
```

2.5.6 Applying functions using mutate

Let us now return to the discussion of how we can *apply* functions to a column. The meaning of apply is particular. What we mean by this is that we wish to run some function (which can receive an argument and return a value) to each row of a column. This can be useful if, say, some column is given in the wrong units or if the values in a column should be "cut off" at some threshold point.

Recall the tidied tibble mtcars_tibble.

```
mtcars_tibble
```

```
## # A tibble: 32 x 11
##       mpg   cyl  disp    hp  drat    wt  qsec    vs    am
##     <dbl> <dbl> <dbl> <dbl> <dbl> <dbl> <dbl> <dbl> <dbl>
##  1  21       6   160   110  3.9   2.62  16.5     0     1
##  2  21       6   160   110  3.9   2.88  17.0     0     1
##  3  22.8     4   108    93  3.85  2.32  18.6     1     1
##  4  21.4     6   258   110  3.08  3.22  19.4     1     0
##  5  18.7     8   360   175  3.15  3.44  17.0     0     0
##  6  18.1     6   225   105  2.76  3.46  20.2     1     0
##  7  14.3     8   360   245  3.21  3.57  15.8     0     0
##  8  24.4     4   147.   62  3.69  3.19  20       1     0
##  9  22.8     4   141.   95  3.92  3.15  22.9     1     0
## 10  19.2     6   168.  123  3.92  3.44  18.3     1     0
## # ... with 22 more rows, and 2 more variables:
## #   gear <dbl>, carb <dbl>
```

We can spot two areas that require transformation:

- Convert the `wt` column from pounds to kilograms.
- Cut off the values in `displ` so that no car model has a value larger than `400`.

We can address the first one by writing a function that multiples each value in the argument received by the conversion factor for kilograms. Let us test it out first with a simple vector.

```
wt_conversion <- function(x) {
  x * 0.454
}


wt_conversion(100:105)
```

```
## [1] 45.400 45.854 46.308 46.762 47.216 47.670
```

To incorporate this into the tibble, we make a call to `mutate` using our function `wt_conversion`, which *modifies* the column `wt`.

```
mtcars_transformed <- mtcars_tibble |>
  mutate(wt = wt_conversion(wt))
mtcars_transformed
```

```
## # A tibble: 32 x 11
##       mpg   cyl  disp    hp  drat    wt  qsec    vs    am
##     <dbl> <dbl> <dbl> <dbl> <dbl> <dbl> <dbl> <dbl> <dbl>
##  1  21       6   160   110  3.9   1.19  16.5     0     1
##  2  21       6   160   110  3.9   1.31  17.0     0     1
##  3  22.8     4   108    93  3.85  1.05  18.6     1     1
##  4  21.4     6   258   110  3.08  1.46  19.4     1     0
##  5  18.7     8   360   175  3.15  1.56  17.0     0     0
##  6  18.1     6   225   105  2.76  1.57  20.2     1     0
##  7  14.3     8   360   245  3.21  1.62  15.8     0     0
##  8  24.4     4   147.   62  3.69  1.45  20       1     0
##  9  22.8     4   141.   95  3.92  1.43  22.9     1     0
## 10  19.2     6   168.  123  3.92  1.56  18.3     1     0
```

```
## # ... with 22 more rows, and 2 more variables:
## #   gear <dbl>, carb <dbl>
```

We have successfully applied a function we wrote to a column in a tibble!

The second task is peculiar. As with the first example, we can define a function that computes the minimum between the argument and the value `400`.

```
cutoff_400 <- function(x) {
  min(400, x)
}
```

We could then apply the function to the column `disp` using a similar approach.

```
mtcars_transformed <- mtcars_tibble |>
  mutate(disp = cutoff_400(disp))
mtcars_transformed
```

```
## # A tibble: 32 x 11
##       mpg   cyl  disp    hp  drat    wt  qsec    vs    am
##     <dbl> <dbl> <dbl> <dbl> <dbl> <dbl> <dbl> <dbl> <dbl>
## 1    21       6  71.1   110  3.9   2.62  16.5     0     1
## 2    21       6  71.1   110  3.9   2.88  17.0     0     1
## 3    22.8     4  71.1    93  3.85  2.32  18.6     1     1
## 4    21.4     6  71.1   110  3.08  3.22  19.4     1     0
## 5    18.7     8  71.1   175  3.15  3.44  17.0     0     0
## 6    18.1     6  71.1   105  2.76  3.46  20.2     1     0
## 7    14.3     8  71.1   245  3.21  3.57  15.8     0     0
## 8    24.4     4  71.1    62  3.69  3.19  20       1     0
## 9    22.8     4  71.1    95  3.92  3.15  22.9     1     0
## 10   19.2     6  71.1   123  3.92  3.44  18.3     1     0
## # ... with 22 more rows, and 2 more variables:
## #   gear <dbl>, carb <dbl>
```

That didn't work out so well. The new `disp` column contains the same value `71.1` for every row in the tibble! Did `dplyr` make a mistake? Is the function we wrote just totally wrong?

The error, actually, is not in anything we wrote *per se*, but in how R *processes* the function `cutoff_400` during the mutate call. We expect to pass one number to the function `cutoff_400` so that we can compare it against 400, but our function instead receives a *vector* of values when used inside a `mutate` verb. That is, the entire `disp` column of values is passed as an argument to the function `cutoff_400`.

While this was no problem for the `wt_conversion` function, `cutoff_400` is not capable of handling a *vector* as an argument *and* returning a *vector* back.

To clarify the point, compare the result of these two functions after receiving the vector `395:410`.

```
wt_conversion(399:405)
```

```
## [1] 181.146 181.600 182.054 182.508 182.962 183.416
## [7] 183.870
```

```
cutoff_400(398:405)
```

```
## [1] 398
```

wt_conversion performs an element-wise operation to each element of the sequence and, therefore, the first example works as intended. In the second, cutoff_400 computes the minimum of the vector (398) and returns the result of just that computation; no element-wise comparison is made.

To make cutoff_400 work as intended, we turn to a new programming construct called the *map*, prepared by the package purrr[3].

2.5.7 purrr maps

The main construct we will be using from purrr is called the *map*. A map applies a function, say the cutoff_400 function we just wrote, to **each element** of a vector or list.

purrr offers many flavors of map, depending on what the output vector should look like:

- map_lgl() outputs a logical vector.
- map_int() outputs an integer vector.
- map_dbl() outputs a double vector.
- map_chr() outputs a character vector.
- map() outputs a list.

Here are some more examples of using map. Let us apply the wt_conversion to an input vector containing a sequence of values from 95 to 105.

```
map_dbl(399:405, wt_conversion)
```

```
## [1] 181.146 181.600 182.054 182.508 182.962 183.416
## [7] 183.870
```

Observe how this resulting vector is the same one we obtained when applying wt_conversion without a map.

We can also define functions and pass it in on the spot. We call these *anonymous functions*. The following is an *identity function*: it simply outputs what it takes in.

```
map_int(1:5, function(x) x)
```

```
## [1] 1 2 3 4 5
```

---

[3]https://github.com/rstudio/cheatsheets/blob/master/purrr.pdf

A catch here is that the code after the comma, i.e., `function(x)` x specifies *in place* the function to apply to each element of the series preceding the comma `1:5`. The function in question `function(x)` x specifies that the function will receive a value named x and *returns* the value of x without modification. Thus, we call it an identity function. The external function `map_int` states that the result of applying the identify function thus specified with `function(x)` x to each element of the sequence `1:5` will be presented as an integer.

We could write the above anonymous function more compactly.

```
map_int(1:5, \(x) x)
```

```
## [1] 1 2 3 4 5
```

The next one is perhaps more useful than the identify function. It computes the square of each element, i.e., x^2.

```
map_dbl(1:5, \(x) x ** 2)
```

```
## [1]  1  4  9 16 25
```

Why use map_dbl() instead of map_int()? By default, R treats numbers as doubles. While `1:5` is a vector of integers, each element is subject to the expression x ** 2, where x is an integer and 2 is a double. To make this operation compatible, R will "promote" x to a double, making the output of this expression a double as well.

The next one will always return a vector of 5's, regardless of the input. Can you see why? Do you also see why there are six elements, unlike five elements in the previous examples?

```
map_dbl(1:6, \(x) 5)
```

```
## [1] 5 5 5 5 5 5
```

2.5.8 **purrr** with **mutate**

By now we have seen enough examples of how to use `map` with a vector. Let us return to the issue of applying the function `cutoff_400` to the `disp` variable.

To incorporate this into a tibble, we encase our `map` inside a call to `mutate`, which modifies the column `disp`.

```
mtcars_transformed <- mtcars_tibble |>
  mutate(disp = map_dbl(disp, cutoff_400))
mtcars_transformed
```

```
## # A tibble: 32 x 11
##       mpg   cyl  disp    hp  drat    wt  qsec    vs    am
##     <dbl> <dbl> <dbl> <dbl> <dbl> <dbl> <dbl> <dbl> <dbl>
## 1   21       6   160   110  3.9   2.62  16.5     0     1
## 2   21       6   160   110  3.9   2.88  17.0     0     1
## 3   22.8     4   108    93  3.85  2.32  18.6     1     1
## 4   21.4     6   258   110  3.08  3.22  19.4     1     0
## 5   18.7     8   360   175  3.15  3.44  17.0     0     0
## 6   18.1     6   225   105  2.76  3.46  20.2     1     0
## 7   14.3     8   360   245  3.21  3.57  15.8     0     0
## 8   24.4     4   147.   62  3.69  3.19  20       1     0
## 9   22.8     4   141.   95  3.92  3.15  22.9     1     0
## 10  19.2     6   168.  123  3.92  3.44  18.3     1     0
## # ... with 22 more rows, and 2 more variables:
## #   gear <dbl>, carb <dbl>
```

We can inspect visually to see if there are any repeating values in `disp` or if any of those values turn out larger than 400 – there shouldn't be!

The following graphic illustrates the effect of the `purrr` map inside the mutate call.

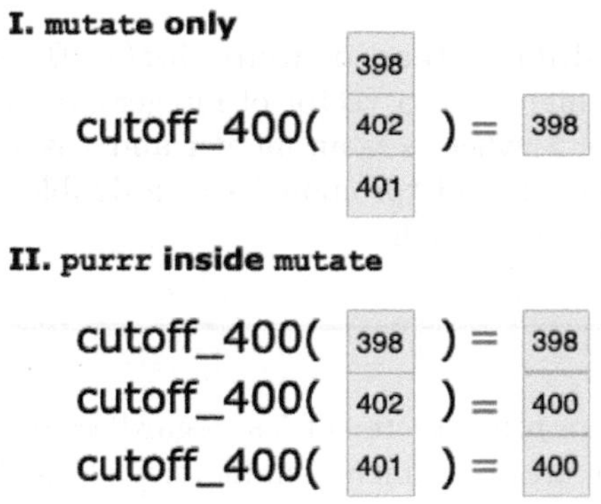

I. mutate only

$$\text{cutoff\_400(}\begin{array}{c}398\\402\\401\end{array}\text{)} = 398$$

II. purrr inside mutate

$$\text{cutoff\_400(}\;398\;\text{)} = 398$$
$$\text{cutoff\_400(}\;402\;\text{)} = 400$$
$$\text{cutoff\_400(}\;401\;\text{)} = 400$$

Note that the use of map allows a vector to be returned by the `cutoff_400` function, which can then be used as a column in the `mutate` call.

Pop quiz: In our two examples of applying a function to `wt` and `disp`, the new tibble (stored in `mtcars_transformed`) lost information about the values of `wt` and `disp` before the transformation. How could we amend our examples to still preserve the old information in case we would like to make comparisons between the before and after?

2.6 Handling Missing Values

In the section on joining tables together we saw a special value called NA crop up when rows did not align during the matching. We call these special quantities, as you might expect, *missing values* since they are "holes" in the data. In this section we dive more into missing values and how to address them in your datasets.

2.6.1 Prerequisites

As before, let us load the tidyverse.

```
library(tidyverse)
```

2.6.2 A dataset with missing values

The tibble trouble_temps contain temperatures from four cities across three consecutive weeks in the summer.

```
trouble_temps <- tibble(city = c("Miami", "Boston",
                                 "Seattle", "Arlington"),
                        week1 = c(89, 88, 87, NA),
                        week2 = c(91, NA, 86, 75),
                        week3 = c(88, 85, 88, NA))
trouble_temps
```

```
## # A tibble: 4 x 4
##   city      week1 week2 week3
##   <chr>     <dbl> <dbl> <dbl>
## 1 Miami        89    91    88
## 2 Boston       88    NA    85
## 3 Seattle      87    86    88
## 4 Arlington    NA    75    NA
```

As you might expect, this tibble contains missing values. We can see that Boston is missing a value from week2 and Arlington is missing values from both week1 and week3, possibly due to some faulty equipment.

2.6.3 Eliminating rows with missing values

If you need to get rid of all rows with NA, you can use drop_na which is part of dplyr.

```
temps_clean <- trouble_temps |>
  drop_na()
temps_clean
```

```
## # A tibble: 2 x 4
##   city    week1 week2 week3
##   <chr>   <dbl> <dbl> <dbl>
## 1 Miami      89    91    88
## 2 Seattle    87    86    88
```

2.6.4 Filling values by looking at neighbors

There is a way to fill missing values by dragging the non-NA value immediately below an NA to its position. In this manner, all NA's after the first non-NA will acquire a value. This works when the bottom row does not have an NA.

Let us fill the values using this setting.

```
trouble_temps |>
  fill(week1:week3, .direction = "up")
```

```
## # A tibble: 4 x 4
##   city     week1 week2 week3
##   <chr>    <dbl> <dbl> <dbl>
## 1 Miami       89    91    88
## 2 Boston      88    86    85
## 3 Seattle     87    86    88
## 4 Arlington   NA    75    NA
```

Note how the temperatures for Arlington remain unfilled.

In the case where the bottom row has an NA and the top row does not have an NA, you can drag the values upwards instead.

We can also combine the two actions in a bidirectional manner, either going down and then up or going up and then down.

```
trouble_temps |>
  fill(week1:week3, .direction = "updown")
```

```
## # A tibble: 4 x 4
##   city     week1 week2 week3
##   <chr>    <dbl> <dbl> <dbl>
## 1 Miami       89    91    88
## 2 Boston      88    86    85
## 3 Seattle     87    86    88
## 4 Arlington   87    75    88
```

The directional specifications are: "up", "down", "updown", and "downup". The default direction is "down", and so you do not have state it.

2.6.5 Filling values according to a global constant

If you want to make an across-the-board replacement of NA with a specific value, you can use the function replace_na from tidyr. For instance, the following replaces missing values in week1 with the value 70.

```
trouble_temps |>
  mutate(week1 = replace_na(week1, 70))
```

```
## # A tibble: 4 x 4
##   city      week1 week2 week3
##   <chr>     <dbl> <dbl> <dbl>
## 1 Miami        89    91    88
## 2 Boston       88    NA    85
## 3 Seattle      87    86    88
## 4 Arlington    70    75    NA
```

If you wish to apply this for all columns in the dataset, we can use provide `replace_na` as an anonymous function in a combination of `mutate` with `across`.

```
trouble_temps |>
  mutate(across(week1:week3, function(x) replace_na(x, 70)))
```

```
## # A tibble: 4 x 4
##   city      week1 week2 week3
##   <chr>     <dbl> <dbl> <dbl>
## 1 Miami        89    91    88
## 2 Boston       88    70    85
## 3 Seattle      87    86    88
## 4 Arlington    70    75    70
```

Note that if your dataset contains a mixture of strings and numbers, then a straightforward application like this will not work. Instead, you will need to split the process into two steps: first handling missing values in the strings columns and then, afterwards, taking care of the missing values in the numeric columns.

Alternatively, you can use `replace_na` to give an instruction on how to handle NA appearing in specific columns.

The syntax for the instruction is simple. For each attribute you make a placement, state its name, add an equal sign, and then add the value you want to use for replacement. The replacement instructions must appear in a list, even if there is only one replacement instruction.

Below, we fill any NA in `week1` with 89, in `week2` with 91, and in `week3` with 88.

```
trouble_temps |>
  replace_na(list(week1 = 89, week2 = 91, week3 = 88))
```

```
## # A tibble: 4 x 4
##   city      week1 week2 week3
##   <chr>     <dbl> <dbl> <dbl>
## 1 Miami        89    91    88
## 2 Boston       88    91    85
## 3 Seattle      87    86    88
## 4 Arlington    89    75    88
```

2.7 Exercises

Be sure to install and load the following packages into your R environment before beginning this exercise set.

Question 1 Recall from the textbook that data is *tidy* when it satisfies four conditions:

1. Each variable forms a column.
2. Each observation forms a row.
3. Each value must have its own cell.
4. Each type of observational unit forms a table.

```
is_it_tidy <- list(table5, table1, table3, table2)
```

is_it_tidy is a list containing 4 tibbles, with each dataset showing the same values of the four variables *country*, *year*, *population*, and *cases*, but each dataset organizing the values in a different way. All display the number of Tuberculosis (TB) cases documented by the World Health Organization in Afghanistan, Brazil, and China between 1999 and 2000.

Table 1

```
is_it_tidy[[1]]  # Table 1
```

Table 2

```
is_it_tidy[[2]] # Table 2
```

Table 3

```
is_it_tidy[[3]] # Table 3
```

Table 4

```
is_it_tidy[[4]] # Table 4
```

- **Question 1.1** Have a look at each of the four tibbles. What is the observational unit being measured?

- **Question 1.2** Using the observational unit you have defined, which of these tibbles, if any, fulfills the properties of *tidy data*? For this question, it is enough to state simply whether each tibble is tidy or not.

- **Question 1.3** Select one of the tibbles you found not to be tidy and explain which of the tidy data guidelines are violated.

Question 2 Gapminder is an independent educational non-profit project that identifies systematic misconceptions about important global trends. In this question we will explore an excerpt of the Gapminder data on life expectancy, GDP per capita, and population by country. This data is available in the tibble gapminder from the library gapminder.

Let us have a look at the data. We will make an explicit copy of the data called gap to prevent any worry of modifying the original data.

```
gap <- gapminder
gap
```

- **Question 2.1** Create a new variable called `gdp` that gives each country's GDP. This can be accomplished by multiplying the figures in population (`pop`) with GDP per capita (`gdpPercap`). Assign the resulting new tibble to the name `gap`.

- **Question 2.2** It can be helpful to report GDP per capita relative to some benchmark. Because the United States is the country where the authors reside, let us choose this as the reference country.

  Filter down `gap` to rows that pertain to United States. Extract the `gdpPercap` variable from the resulting tibble as a *vector* and assign it to a name called `usa_gdpPercap`.

- **Question 2.3** Obtain a tibble of unique country names that are in the variable `country`. We can accomplish this using the `dplyr` verb `distinct()`. Pipe your `gap` tibble into this function and store the resulting *tibble* into a name called `countries`.

- **Question 2.4** Replicate `usa_gdpPercap` once per each unique country in the dataset and store the resulting *vector* into a name called `usa_gdpPercap_rep`. Use the function `rep()`.

- **Question 2.5** Add a new column to `gap` called `gdpPercap_relative` which divides `gdpPercap` by this United States figure. Store the resulting tibble into the name `gap`.

- **Question 2.6** Relative to the United States, which country had the highest GDP per capita? And, in what year? Assign your answers to the names `highest_gdp_rel_to_us` and `year`, respectively. You should use a `dplyr` verb to help you answer this; do not attempt to find the answer manually.

- **Question 2.7** The last question made it seem that a majority of countries have a higher GDP per capita relative to the U.S. But that is just a tiny slice of the data and intuition may tell us otherwise. The *median* is a good measure for the central tendency of the data. Find the median of the variable `gdpPercap_relative` and assign your answer to the name `the_median`. Your answer should be a single double value.

- **Question 2.8** Think about the value of the median you just found and give an interpretation for it when compared to the bulk of the data. Is it true that the majority of countries have a higher GDP per capita compared to the United States?

  HINT: Remember that the median is the GDP per capita *relative to the United States*. If the median value was 1, what would that mean? If it was *greater* than 1? How about *less* than 1?

Question 3 In this question we will continue exploring the `gapminder` data to further practice `dplyr` verbs. As before, we will keep an explicit copy of the Gapminder data in a variable called `gap`.

```
gap <- gapminder
```

- **Question 3.1** How many observations are there **per continent**? Store the resulting tibble in a name called `continent_counts` with two variables: `continent` (the continents) and `n` (the counts).

- **Question 3.2** Let's have a look at the life expectancy in the continent Africa. What is the minimum, maximum, and average life expectancy in each year? You will need to use the pair `group_by()` and `summarize()` to answer this. Store the resulting tibble in a variable called `summarized_years`.

The first few rows of this tibble should look like:

year	min_life_exp	max_life_exp	avg_life_exp
1952	30.0	52.724	39.13550
1957	31.57	58.089	41.26635
...	...	...	...

- **Question 3.3** From `gap`, create a new variable named `amount_increase` which gives the amount life expectancy increased by when compared to 1952, **for each country.** Select only the variables `country`, `year`, `lifeExp`, and `life_exp_gain`. Store the resulting tibble into a name `from_1952`.

  HINT: Recall the *grouped mutate* construct discussed in the textbook: sometimes we wish to keep the groups after a `group_by()` and compute *within* them. Moreover, don't forget to `ungroup()` when you are done. Finally, the function `first()` can be used to extract the first value from something, e.g.,

  ```
  first(c(10, 4, 9, 42, -2))
  ```

- **Question 3.4** Which country had the *highest* life expectancy when compared to 1952 and in what year? Which country had the *lowest* and, similarly, what year did that occur? Use a `dplyr` verb to help you answer this. Assign your answers to the names `highest_country`, `highest_year`, `lowest_country`, and `lowest_year`.

Question 4 The Connecticut Department of Housing (DOH) publishes data about affordable housing. We've obtained data on affordable housing by town from 2011-2020 and collected this into a tibble named `affordable`, available in the `edsdata` package. Pull up the help for information about this dataset.

```
affordable
```

```
## # A tibble: 1,686 x 10
##    `Town Code` Town   Year 2010 ~1 Gover~2 Tenan~3 Singl~4
##          <dbl> <chr> <dbl>   <dbl>   <dbl>   <dbl>   <dbl>
## 1            1 Ando~  2020    1317      18       1      32
## 2            2 Anso~  2020    8148     349     764     147
## 3            3 Ashf~  2020    1903      32       0      36
## 4            4 Avon   2020    7389     244      16      44
## 5            5 Bark~  2020    1589       0       6      23
## 6            6 Beac~  2020    2509       0       4      46
## 7            7 Berl~  2020    8140     556      50     142
## 8            8 Beth~  2020    2044       0       2      13
## 9            9 Beth~  2020    7310     192      26     154
## 10          10 Beth~  2020    1575      24       0       9
## # ... with 1,676 more rows, 3 more variables:
## #   `Deed Restricted Units` <dbl>,
## #   `Total Assisted Units` <dbl>,
## #   `Percent Affordable` <dbl>, and abbreviated variable
```

```
## #    names 1: `2010 Census Units`,
## #          2: `Government Assisted`,
## #          3: `Tenant Rental Assistance`, ...
```

- **Question 4.1** Sort the data in *increasing* order by percent affordable, naming the sorted tibble `by_percent`. Create another tibble called `by_census` that is sorted in *decreasing* order by number of 2010 census units instead.

- **Question 4.2** Let us define "most affordable housing" as towns with housing affordability at least 30%. Create a tibble named `most_affordable` that gives the most affordable towns in the year 2020.

- **Question 4.3** Create a tibble named `affordable_by_year` that gives the number of towns with "most affordable housing" broken down by year. For instance, three towns had most affordable housing in the year 2015. This tibble should contain two variables named `Year` and `Number of Towns`.

- **Question 4.4** Based on this tibble, what would you say to the statement:

"It appears that the percent of most affordable housing in Connecticut towns, as defined as towns with housing affordability at least 30%, decreases over time when compared to 2011."

Is this a fair claim to make? Why or why not?

- **Question 4.5** It is usually a good idea to perform "sanity" checks on your data to make sure the data follows your intuition (or doesn't). For instance, we expect that by summing the variables `Government Assisted`, `Tenant Rental Assistance`, `Single Family CHFA/ USDA Mortgages`, and `Deed Restricted Units`, and then dividing this figure by the total number of 2010 census units, the percentage should equal the value in `Percent Affordable`.

Let us create two new columns in `affordable` that give:

  - Our own percent affordability variable named `my_affordable` that reports the above figure, *rounded to two decimal places*.

  - A variable named `equal_figures` that reports whether the two figures, `my_affordable` and `Percent Affordable` are equal.

Name the resulting tibble `with_my_affordable`.

- **Question 4.6** Do any of these figures differ? Form a tibble named `is_equal` using `with_my_affordable` that contains one row and one variable named `all_equal`. The single value in this tibble is a Boolean expressing whether there are any differences between the percent affordability figures.

Question 5 The U.S. Department of Agriculture (USDA) Economic Research Service[4] publishes data on unemployment rates in the USA. The data is available in `unemp_usda` from `edsdata`, and gives county-level socioeconomic indicators from 2000 to 2020. We will

---

[4]https://www.ers.usda.gov/data-products/county-level-data-sets/

use this dataset to examine the average yearly unemployment rate in each state in the USA during the recorded years.

- **Question 5.1** Select the state (`State`) and county (`Area_name`) columns and then only those columns that pertain to unemployment rate, that is, columns of the form `Un-employment_rate_X`, where `X` is some year. Store the resulting tibble in the name `un-emp_usda_relevant`.

- **Question 5.2** If our statistical question is about the average yearly unemployment rate in the USA from 2000 to 2020, does the data in `unemp_usda_relevant` fulfill the properties of tidy data? If so, why? If not, what tidy data principles are violated? Then, in English, describe what a tidy representation of the data would look like.

- **Question 5.3** Apply a pivot transformation to `unemp_usda_relevant` so that the four variables appear in the transformed table: `State`, `Area_name` (the county), `year`, and `unemployment_rate`. Store the resulting tibble in the name `unemp_usda_tidy`.

- **Question 5.4** The current form of the `year` variable in `unemp_usda_tidy` is awkward because we expect "year" to be a number, but "year" is prefixed by some string; this may be surprising to potential customers of this tibble. Tidy the column `year` by extracting only the year, e.g., `"2008"` from `"Unemployment_rate_2008"`. You will need to combine a function from `stringr` with a `dplyr` verb to accomplish this. Then convert `year` to a numerical column using `as.double()`. Store the resulting tibble in `unemp_usda_tidy`.

  HINT: A prerequisite to answering this question is to first write `stringr` code that can extract the string "2009" from the string "Unemployment_rate_2009". Once you have figured this sub-problem, then incorporate your `stringr` work into a `dplyr` verb.

- **Question 5.5** Form a tibble named `top_unemp_by_state` that gives the year with the highest unemployment rate for each state that appears in `unemp_usda_tidy`. This tibble should contain three columns (the state, the average unemployment rate, and the year where that unemployment rate occurred) and a single observation for each state reporting the figure.

  HINT: You will need to aggregate the county-level figures in order to compute a state-level average unemployment rate. Moreover, if you find `NA` in your solution, be sure to filter any missing values before computing the mean. Check the documentation for `mean` for hints on how to accomplish this.

- **Question 5.6** Based on these figures, can you say which year(s) saw the highest unemployment rates? Use a `dplyr` verb to help you answer this.

Question 6. Let's practice how to write and use functions.

- **Question 6.1** Complete the function below that converts a proportion to a percentage. For example, the value of `to_percent(0.5)` should be 50, i.e., 50%.

```
to_percent <- function(prop) {
  scale <- 100

}
```

- **Question 6.2** Try referring to the value of `scale` (1) inside the function and (2) outside the function by printing its value. For each case, what value is shown? Is an error produced? Why or why not?

- **Question 6.3** Consider the vowels in the English language. These are the five characters "a", "e", "i", "o", and "u".

  - **Question 6.3.1** Define a function called `vowel_remover`. It should take a single string as its argument and return a copy of that string, but with all vowels removed. You should use a `stringr` function to help you accomplish this.

  - **Question 6.3.2** Write a function called `num_non_vowels`. It should take a string as its argument and return a number. The number should be the number of characters in the argument string that are not vowels. One way to do that is to remove all the vowels and count the size of the remaining string.

- **Question 6.4** Recall that an important use of functions is that we can use it in a `purrr` *map*. Suppose that we have the following vector of fruits:

```
fruit_basket <- c("lychee", "banana", "mango")
```

Using a call to a `purrr` map function with the vector `fruit_basket`, create a copy of the vector `fruit_basket`, but with all the characters that are vowels removed from each element. Assign your answer to the vector `fruit_basket_nonvowels`.

Question 7 Let us examine annual compensation data reported by New York Local Authorities, available in `nysalary` from `edsdata`. Public authorities are required to regularly report salary and compensation information. This data is published through Open Data NY[5]. We have subsetted the data to include salary information for employees where the fiscal year ended on December 31, 2020. Let us have a look at this data.

```
nysalary
```

```
## # A tibble: 1,676 x 19
##    Authority~1 Fisca~2 Last ~3 Middl~4 First~5 Title Group
##    <chr>       <chr>   <chr>   <chr>   <chr>   <chr> <chr>
##  1 Albany Cou~ 12/31/~ Adding~ L       Ellen   Seni~ Admi~
##  2 Albany Cou~ 12/31/~ Boyea   <NA>    Kelly   Conf~ Admi~
##  3 Albany Cou~ 12/31/~ Calder~ <NA>    Philip  Chie~ Exec~
##  4 Albany Cou~ 12/31/~ Cannon  <NA>    Matthew Gove~ Admi~
##  5 Albany Cou~ 12/31/~ Cerrone A       Rima    Budg~ Mana~
##  6 Albany Cou~ 12/31/~ Chadde~ M       Helen   Mark~ Mana~
##  7 Albany Cou~ 12/31/~ Charla~ M       Elizab~ Dire~ Mana~
##  8 Albany Cou~ 12/31/~ Dickson C       Sara    Acco~ Admi~
##  9 Albany Cou~ 12/31/~ Finnig~ <NA>    James   Oper~ Admi~
## 10 Albany Cou~ 12/31/~ Greenw~ <NA>    Kathryn Dire~ Mana~
## # ... with 1,666 more rows, 12 more variables:
## #   Department <chr>, `Pay Type` <chr>,
## #   `Exempt Indicator` <chr>,
## #   `Base Annualized Salary` <chr>,
## #   `Actual Salary Paid` <chr>, `Overtime Paid` <chr>,
## #   `Performance Bonus` <chr>, `Extra Pay` <chr>,
## #   `Other Compensation` <chr>, ...
```

---

[5]https://data.ny.gov/Transparency/Salary-Information-for-Local-Authorities/fx93-cifz

- **Question 7.1** We tried to compute the average annual compensation like this:

```
nysalary |>
    summarize(avg_compensation = mean(`Total Compensation`))
```

Explain why this does not work. It may be helpful to inspect some values in the Total Compensation column.

- **Question 7.2** Extract the first value in the "Total Compensation" variable corresponding to Ellen Addington's annual compensation in the 2020 fiscal year. Call it addington_string.

- **Question 7.3** Convert addington_string to a number in *tens of thousands*. The stringr function str_remove_all() will be useful for removing non-numerical characters. For example, the value of str_remove_all("$100", "[$]") is the string "100". You will also need the function as.double(), which converts a string that looks like a number to an actual number. Assign the result to a name addington_number.

  To compute the average annual compensation, we would need to do this work for every employee in the dataset. This would be incredibly tedious to complete for 1,676 different employees! Instead, we can use functionals and the *map* construct to do the work for us.

- **Question 7.4** Define a function string_to_number that converts pay strings to pay numbers in *tens of thousands*. Your function should convert a pay string like "$137,000.00 to a number of dollars in tens of thousands, i.e., 13.7.

- **Question 7.5** Now apply the function string_to_number to every row in the tibble nysalary. Using a *map* and a dplyr verb, make a new tibble that is a copy of nysalary with one more variable called "Total Compensation ($)". It should be the result of applying string_to_number to the "Total Compensation" variable. Call this new tibble nysalary_cleaned.

- **Question 7.6** Try again to compute the average annual compensation using the cleaned dataset. Assign your answer to the name average_annual_comp.

Question 8 In 2017, the Australian Bureau of Statistics (ABS)[6] published the results of the Australian Marriage Law Postal Survey[7] in response to the question: *should the law be changed to allow same-sex couples to marry?* The majority of participating Australians voted in favor of same-sex couples. The ABS released data on responses and participation broken down by various criteria. This exercise will focus on the latter, and examine participation by state and territory, broken down by age. Following is a snapshot of a subset of the data:

Columns without names		18-19 years	20-24 years	25-29 years	30-34 years	35-39 years	40-44 years	45-49 years	50-54 years
				Empty row					
South Australia	Total participants	22,176	64,507	63,840	65,469	63,798	67,950	79,076	81,183
Merged cells	Eligible participants	28,119	89,604	89,625	91,868	88,373	91,494	102,579	101,167
	Participation rate	78.9	72.0	71.2	71.3	72.2	74.3	77.1	80.2
				Empty row					
Western Australia	Total participants	32,997	89,913	93,376	101,077	98,411	99,542	112,138	109,749
Merged cells	Eligible participants	42,104	126,399	132,880	142,301	136,468	134,620	145,403	137,096
	Participation rate	78.4	71.1	70.3	71.0	72.1	73.9	77.1	80.1

Unfortunately, as can be seen by the annotations we made, these data are not tidy; we show three different issues with the data. This exercise will practice how to bring this dataset into a tidy format so that it can be subject to analysis. The relevant data is available in the tibble abs_partp2017 from the edsdata package.

---

[6] https://www.abs.gov.au/

[7] https://www.abs.gov.au/ausstats/abs@.nsf/mf/1800.0

```
abs_partp2017
```

```
## # A tibble: 31 x 17
##    ...1       ...2    18-19~1 20-24~2 25-29~3 30-34~4 35-39~5
##    <chr>      <chr>     <dbl>   <dbl>   <dbl>   <dbl>   <dbl>
##  1 New Sout~  Tota~   1.09e5  2.97e5  2.95e5  3.07e5  3.12e5
##  2 <NA>       Elig~   1.42e5  4.07e5  4.09e5  4.26e5  4.29e5
##  3 <NA>       Part~   7.73e1  7.29e1  7.22e1  7.21e1  7.27e1
##  4 <NA>       <NA>    NA      NA      NA      NA      NA
##  5 Victoria   Tota~   8.32e4  2.50e5  2.56e5  2.66e5  2.64e5
##  6 <NA>       Elig~   1.01e5  3.25e5  3.37e5  3.50e5  3.43e5
##  7 <NA>       Part~   8.21e1  7.69e1  7.61e1  7.59e1  7.69e1
##  8 <NA>       <NA>    NA      NA      NA      NA      NA
##  9 Queensla~  Tota~   6.19e4  1.75e5  1.75e5  1.83e5  1.84e5
## 10 <NA>       Elig~   8.24e4  2.59e5  2.58e5  2.60e5  2.57e5
## # ... with 21 more rows, 10 more variables:
## #   `40-44 years` <dbl>, `45-49 years` <dbl>,
## #   `50-54 years` <dbl>, `55-59 years` <dbl>,
## #   `60-64 years` <dbl>, `65-69 years` <dbl>,
## #   `70-74 years` <dbl>, `75-79 years` <dbl>,
## #   `80-84 years` <dbl>, `85 years and over` <dbl>, and
## #   abbreviated variable names 1: `18-19 years`, ...
```

- **Question 8.1** If the observational unit is the 2017 participation of an Australian age bracket in a territory and we collect 5 measurements per this unit ("Territory/State", "age group", "total participants", "eligible participants", and "participation rate"), cite at least 2 more violations of the tidy data guidelines. Your answer should note violations other than the missing values caused by the issues raised in the above figure.

- **Question 8.2** Let us first deal with the missing values. These steps can be followed in order:

  - The unnamed columns (...1 and ...2) should be relabeled to "Territory/State" and "Participation Type", respectively.
  - For merged cells, missing values should be filled by looking at the first non-NA neighbor above, e.g., the second row should take on the value "New South Wales".
  - Missing rows should be discarded. This is a reasonable strategy based on what we know about the structure of the data.

  The resulting filled-in tibble should be assigned to a name abs_partp_filled.

- **Question 8.3** Apply pivot transformation(s) to bring abs_partp_filled into tidy format; the resulting tibble after this step should fulfill all tidy data guidelines. Assign this tibble to the name abs_partp_tidy.

- **Question 8.4** What proportion of results had a participation rate less than 60%?

- **Question 8.5** Which territory/state had the third smallest eligible voting population in the 18-19 age bracket?

- **Question 8.6** In the different territories surveyed, what is/are the most frequent age bracket(s) with the lowest participation rates in the survey? Your answer should be expressed as a tibble with two variables named Age group and n.

Question 9 This question is a continuation of **Question 8**. We will now analyze the 2017 Australian Marriage Law Postal Survey[8] another way by looking at the response data. To enrich the analysis, we will overlay the responses with educational qualification data from the 2016 Australian census of population and housing[9], also released through ABS. We have prepared these data for you, available in the tibbles abs_resp2017 and abs_census2016 from the edsdata package.

abs_resp2017

```
## # A tibble: 8 x 6
##    `Territory/State`     Yes Yes (~1     No No (%~2  Total
##    <chr>               <dbl>  <dbl>  <dbl>   <dbl>  <dbl>
## 1 New South Wales     2.37e6   57.8 1.74e6    42.2 4.11e6
## 2 Victoria            2.15e6   64.9 1.16e6    35.1 3.31e6
## 3 Queensland          1.49e6   60.7 9.61e5    39.3 2.45e6
## 4 South Australia     5.93e5   62.5 3.56e5    37.5 9.49e5
## 5 Western Australia   8.02e5   63.7 4.56e5    36.3 1.26e6
## 6 Tasmania            1.92e5   63.6 1.10e5    36.4 3.02e5
## 7 Northern Territory  4.87e4   60.6 3.17e4    39.4 8.04e4
## 8 Australian Capital~ 1.75e5   74   6.15e4    26   2.37e5
## # ... with abbreviated variable names 1: `Yes (%)`,
## #    2: `No (%)`
```

abs_census2016

```
## # A tibble: 64 x 4
##    `Education Qualification`     Terri~1  Count Perce~2
##    <chr>                         <chr>    <dbl>   <dbl>
##  1 Postgraduate Degree Level     New So~ 344490    5.65
##  2 Postgraduate Degree Level     Victor~ 260039    5.37
##  3 Postgraduate Degree Level     Queens~ 134242    3.54
##  4 Postgraduate Degree Level     South ~  50993    3.69
##  5 Postgraduate Degree Level     Wester~  76660    3.84
##  6 Postgraduate Degree Level     Tasman~  13408    3.19
##  7 Postgraduate Degree Level     Northe~   6298    3.51
##  8 Postgraduate Degree Level     Austra~  34819   10.8
##  9 Graduate Diploma and Graduate C~ New So~ 103340    1.70
## 10 Graduate Diploma and Graduate C~ Victor~ 119226    2.46
## # ... with 54 more rows, and abbreviated variable names
## #    1: `Territory/State`, 2: `Percent (%)`
```

Note that these data are at the Territory/State level, while the participation data in **Question 8** was broken down further into age brackets.

- **Question 9.1** Let us explore the relationship between education level and survey response. Using the census data, form a tibble that gives the percentage of Australians that hold at least a bachelor's degree, i.e., a qualification level that is either "Bachelor Degree Level", "Graduate Diploma and Graduate Certificate Level", or "Postgraduate

---

[8]https://www.abs.gov.au/ausstats/abs@.nsf/mf/1800.0
[9]https://www.abs.gov.au/AUSSTATS/abs@.nsf/Lookup/2071.0Main+Features100012016?OpenDocument

Degree Level." These designations are based on the Australian Standard Classification of Education (ASCED)[10]. The resulting tibble should have two variables, `Territory/State` and `At least Bachelor (%)`, and be assigned to a name `bachelor_by_territory`.

- **Question 9.2** Annotate `bachelor_by_territory` with the survey response data by joining `bachelor_by_territory` with `abs_resp2017`. Assign the resulting joined tibble to the name `with_response`.

- **Question 9.3** Note briefly the reason for selecting the join function you used. For instance, if you used `inner_join()`, why not `left_join()` or `right_join()`?

- **Question 9.4** Form a subset of `with_response` that has two rows giving the territory with the highest and lowest support for same-sex couples. Assign this tibble to the name `highest_lowest_support`.

- **Question 9.5** According to your findings, does there appear to be an association between survey response and territories with larger percentages of advanced degree holders? Why or why not?

Question 10 Consider the tibbles `election` and `unemp_usda` from the `edsdata` package.

These datasets give county-level results for presidential elections in the USA and the population and unemployment rate of all counties in the US, respectively. The data in `election` was made available by the MIT Election Data and Science Lab (MEDSL)[11] and contains county-level returns for presidential elections from 2000 to 2020. The data in `unemp_usda` was prepared by USDA, Economic Research Service[12] and gives county-level socioeconomic indicators for unemployment rates.

An important variable in both datasets is the FIPS code. FIPS codes are numbers which uniquely identify geographic areas. Every county has a unique five-digit FIPS code. For instance, `12086` is the FIPS code that identifies Miami-Dade, Florida.

- **Question 10.1** Select the relevant unemployment and voting returns data specifically for 2008. The resulting unemployment tibble should contain three columns: FIPS code, state, and the unemployment rate as of 2008. Store these tibbles in the names `election2008` and `unemp2008`.

- **Question 10.2** Some observations in `election2008` contain a missing FIPS code. Why might that be?

- **Question 10.3** Locate these rows and then filter them from your `election2008`. Assign the resulting tibble back to `election2008`.

- **Question 10.4** Suppose that we want to create a new tibble that contains **both** the election results and the unemployment data. More specifically, we would like to add unemployment information to the election data by *joining* `election2008` with `unemp2008`. Assign the resulting tibble to the name `election_unemp2008`.

  HINT: What is the key we can use to join these two tables? Note that the column names may be different for the key in each table. For example: we would like to join on the key `student_id` but one table has a column `studentID` and the other `student_id`. In the join function we use, we can say `???_join(tibble_a, tibble_b, by = c("studentID" = "student_id"))`.

---

[10]https://www.abs.gov.au/ausstats/abs@.nsf/0/F148CC2C8F5EA951CA256AAF001FCA39?opendocument
[11]https://doi.org/10.7910/DVN/VOQCHQ
[12]https://www.ers.usda.gov/data-products/county-level-data-sets/

- **Question 10.5** Explain why the join function you selected (e.g., right join, left join, etc.) is appropriate for this problem. Why not choose another join function instead?

  Let us explore the relationship between candidate votes and unemployment rate for each state.

- **Question 10.6** Create a tibble from `election_unemp2008` that contains, **for each state**, only the candidate that received the most number of votes. Assign the resulting tibble to the name `state_candidate_winner2008`. It should contain three variables: `state`, `candidate`, and `votes`. Here is what the first few rows of `state_candidate_winner2008` looks like:

state	candidate	votes
ALABAMA	JOHN MCCAIN	1266546
ALASKA	JOHN MCCAIN	193841
ARIZONA	JOHN MCCAIN	1230111
...	...	...

- **Question 10.7** The following tibble `unemp_by_state2008` gives an average unemployment rate for each state by averaging the unemployment rate over the respective counties.

```
unemp_by_state2008 <- election_unemp2008 |>
  group_by(state) |>
  summarize(avg_unemp_rate = mean(Unemployment_rate_2008,
                                  na.rm = TRUE))
unemp_by_state2008
```

  Create a new tibble that contains **both** the candidate winner voting data and the state-level average unemployment data. More specifically, we would like to add the state-level average unemployment data *to* the winner voting data by joining `state_candidate_winner2008` with `unemp_by_state2008`. Assign the resulting tibble to the name `state_candidate_winner_unemp2008`.

- **Question 10.8** Using `state_candidate_winner_unemp2008`, generate a tibble that gives the top 10 states with the highest average unemployment rate. Assign this tibble to the name `top_10`.

Question 11 At the College of Pluto, the six most popular majors are Astronomy, Biology, Chemistry, Data Science, Economics, and Finances. The applicants to the college specify their preference for a major, and the college selects the student with some criteria. The tibble `pluto` in the `edsdata` package gives the selection result from one year.

```
pluto
```

```
## # A tibble: 12 x 4
##     Major       Gender  Applied  Accepted
##     <chr>       <chr>     <dbl>     <dbl>
## 1 Astronomy    Male        825       511
## 2 Astronomy    Female      168       148
## 3 Biology      Male        560       352
## 4 Biology      Female       25        17
```

```
##  5 Chemistry     Male     325      120
##  6 Chemistry     Female   593      352
##  7 Data Science  Male     417      139
##  8 Data Science  Female   375      298
##  9 Economics     Male     191       53
## 10 Economics     Female   393      240
## 11 Finances      Male     373       22
## 12 Finances      Female   641      563
```

- **Question 11.1** Add a new variable `Proportion` that, for each gender, gives the *proportion* of accepted applicants to some major. Assign the resulting tibble to the name `pluto_with_prop`.

- **Question 11.2** Which major saw the highest proportion of accepted *male* applicants? How about accepted *woman* applicants? Use a `dplyr` verb to answer this. Your resulting tibble should have two rows, one for each gender, that gives the corresponding major with the *largest* proportion of accepted applicants.

- **Question 11.3** Using `pluto_with_prop`, write `dplyr` code that gives the top *two* majors with the largest *gap* in acceptance percentage between men and women. The resulting tibble should have two variables: the major and the quantity of the difference.

  HINT: The function `diff()` may be helpful for computing the difference within a group.

Note: The following exercises correspond to material that appears only in the accompanying website, at: https://ds4world.cs.miami.edu/.

Question 12: Examining racial breakdown in the College Scorecard. The chapter presented a case study of how to tidy the College Scorecard dataset. Let us play some more with the dataset. The table is available in the name `scorecard_fl` from the `edsdata` package.

```
scorecard_fl
```

We will be using the variables appearing on `relevant_cols`.

```
relevant_cols <- c("INSTNM", "CITY", "ZIP", "UGDS",
                   "NPT4_PUB", "NPT4_PRIV",
                   "UGDS_WHITE", "UGDS_BLACK",
                   "UGDS_HISP", "UGDS_ASIAN", "UGDS_AIAN")
```

- **Question 12.1** First, collect the variables appearing only in `mylist` and store the data in `with_race`. For this action, you can use the dplyr helper function `all_of` together with `select`.

- **Question 12.2** Of the variables we have selected, `UGDS` represents the total number of enrolled students (as a string). We already know what `NPT4_PUB` and `NPT4_PRIV` represent. What do `UGDS_WHITE` and `UGDS_AIAN` refer to? Have a look at the glossary[13] and data dictionary[14] to determine what these variables mean.

- **Question 12.3** As in the textbook, we will remove the four-digit route number in `ZIP` by replacing the part with the empty string. Call the new variable `ZIP5` and insert it after the `CITY` variable. Store the new data frame back in the name `with_race`.

---

[13] https://collegescorecard.ed.gov/data/glossary/
[14] https://collegescorecard.ed.gov/data/documentation/

- **Question 12.4** In the textbook, we looked at generating a Boolean column representing whether or not a college is a private or public institution. We also looked at generating from a string-valued column representing a number to a new column representing a number using `as.double`.

  Let's perform the following steps:

  - Create a new Boolean variable called `public` that indicates whether or not the college is a public institution, to be added before `ZIP5`.
  - The variables `UGDS`, `UGDS_WHITE`, `UGDS_BLACK`, `UGDS_HISP`, `UGDS_ASIAN`, and `UGDS_AIAN` are currently expressed as strings. Convert these columns to proper numeric columns using `as.double`. The operation can be performed in one step by using `across` within the `mutate` call.

  Store the resulting data frame in `with_race`.

- **Question 12.5** By multiplying `UGDS` by each of the five ratios, you can calculate the number of students in each of the categories. Call the number `n_XYZ` where `XYZ` represents the category and add the five numbers you can calculate from them after `UGDS`. Call the new tibble `student_counts`.

- **Question 12.6** You may observe that the five categories do not cover the entire racial composition. Let's create a new variable `n_others` by subtracting the five numbers from the total (`UGDS`). Add it after `n_aian`. Call the new tibble `student_counts_others`.

- **Question 12.7** Based on what you have calculated, find out which institution has the largest number of ...

  - Black students?
  - Hispanic students?
  - Asian students?

  You can find the answer by reordering the rows in the descending order of the ethnicity.

- **Question 12.8** Let us see which 5-digit ZIP code corresponds to the institutions with the largest number of White students. Group by ZIP code and compute the total count of `n_white` as `total`. Then form a single row that contains the ZIP code with the largest number of White students, along with the corresponding count.

- **Question 12.9** Which institution(s) correspond to the ZIP code that you found? Use a `dplyr` verb to help you answer this.

- **Question 12.10** Let us write a function `examine_by_zip` that accomplishes the task of finding the schools with the highest number of a student group broken down by some ethnicity (e.g., `n_white`) with the ZIP code aggregation.

  This function:

  - Receives a parameter representing a variable in `student_counts_others` (e.g., `n_white`), generates a summarized table, computes the total, and arranges the rows in the descending order of the total, in the same manner as **Question 12.8**.

  - The function then examines the first element of the `ZIP5` variable and uses it to obtain the schools whose ZIP matches the ZIP code, in the same manner as **Question 12.9**.

  After writing the function, run it with `examine_by_zip(n_white)` to ensure that the result matches the answer you obtained in the previous question.

Note: Referencing the variable `column_label` requires a double curly-bracketing when used within the function. This is an advanced `dplyr` usage that we will learn more about later. Here is an example usage of the incantation for the purpose of this exercise:

```
embraced <- function(column_label) {
  student_counts_others |>
    summarize(mean_num =
                mean({{ column_label }}, na.rm=TRUE))
}
embraced(n_white)
embraced(n_others)
```

```
examine_by_zip <- function(column_label) {


}

examine_by_zip(n_white)
```

3

Data Visualization

As you develop familiarity with processing data, you learn how to develop intuition from the data at hand by glancing at its values. Unfortunately, there is only so much you can do with glancing at values. There is a substantial limitation to what you can obtain when the data at hand is so large.

Visualization is a powerful tool in such cases. In this chapter we introduce another key member of the tidyverse, the `ggplot2` package, for visualization.

3.1 Introduction to `ggplot2`

R provides many facilities for creating visualizations. The most sophisticated of them, and perhaps the most elegant, is `ggplot2`. In this section we introduce generating visualizations using `ggplot2`.

3.1.1 Prerequisites

We will make use of the tidyverse in this chapter, so let's load it in as usual.

```
library(tidyverse)
```

3.1.2 The layered grammar of graphics

The structure of visualization with `ggplot2` is by way of something called the *layered grammar of graphics* – a name that will certainly impress your friends!
The name of the package `ggplot2` is a bit of a misnomer as the main function we call to visualize the data is `ggplot`. As with `dplyr` and `stringr`, the `ggplot2` cheatsheet[1] is quite helpful for quick referencing.

Each visualization with `ggplot` consists of some building blocks. We call these *layers*. There are three types of layers:

- the **base layer**, which consists of the background and the coordinate system,
- the **geom layers**, which consist of individual geoms, and
- the **ornament layers**, which consists of titles, legends, labels, etc.

---

[1] https://github.com/rstudio/cheatsheets/blob/main/data-visualization.pdf

DOI: 10.1201/9781003320845-3

We call the plot layers geom layers because each plot layer requires a call to a function with the name starting with geom_. There are many geoms available in ggplot and you can think of these as the buildings blocks that compose many of the diagrams you are already familiar with. For instance, *point* geoms are used to create scatter plots, *line* geoms for line graphs, *bar* geoms for bar charts, and *histogram* geoms for histograms – check out the cheat sheet for many more! We will explore the main geoms in this chapter.

To specify the base layer, we use the function ggplot(). Using the function alone is rather unimpressive.

```
ggplot()
```

All ggplot2 has done so far is set up a blank canvas. To make this plot more interesting, we need to specify a dataset and a coordinate system to use. To build up the discussion, let us turn to our first geom: the *point* geom.

3.2   Point Geoms

Let us begin our exploration with the *point* geom. As noted earlier, point geoms are useful in that they can be used to construct a *scatter plot*.

3.2.1   Prerequisites

We will make use of the tidyverse in this chapter, so let us load it in as usual.

```
library(tidyverse)
```

3.2.2 The `mpg` tibble

We will use the `mpg` dataset as our source for this section. This dataset is collected by the US Environmental Protection Agency and shows information about 38 models of car between 1999 and 2008. We have visited this data in earlier sections. Use `?mpg` to open its help page for more information.

The table `mpg` has 234 rows and 11 columns. Use Let us have a look at a snapshot of the data again.

```
mpg
```

```
## # A tibble: 234 x 11
##    manuf~1 model displ  year   cyl trans drv     cty   hwy
##    <chr>   <chr> <dbl> <int> <int> <chr> <chr> <int> <int>
##  1 audi    a4      1.8  1999     4 auto~ f        18    29
##  2 audi    a4      1.8  1999     4 manu~ f        21    29
##  3 audi    a4      2    2008     4 manu~ f        20    31
##  4 audi    a4      2    2008     4 auto~ f        21    30
##  5 audi    a4      2.8  1999     6 auto~ f        16    26
##  6 audi    a4      2.8  1999     6 manu~ f        18    26
##  7 audi    a4      3.1  2008     6 auto~ f        18    27
##  8 audi    a4 q~   1.8  1999     4 manu~ 4        18    26
##  9 audi    a4 q~   1.8  1999     4 auto~ 4        16    25
## 10 audi    a4 q~   2    2008     4 manu~ 4        20    28
## # ... with 224 more rows, 2 more variables: fl <chr>,
## #   class <chr>, and abbreviated variable name
## #   1: manufacturer
```

Another way to preview the data is using `glimpse`.

```
glimpse(mpg)
```

3.2.3 Your first visualization

We first specify the base layer. Unlike before, this time we specify our intention to use the `mpg` dataset.

```
ggplot(data = mpg)
```

We are still presented with a profoundly useless plot. Let us amend our code a bit.

```
ggplot(data = mpg) +
  geom_point(mapping = aes(x = displ, y = hwy))
```

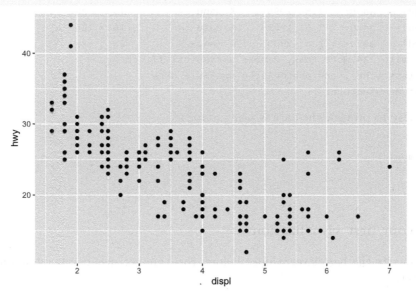

Ta-da, our first visualization! Let us unpack what we just did.

The first line of this code specifies the **base layer** with the argument `data =`. The second line describes the **geom layer** where a point geom is to be used along with some *mapping* from graphical elements in a plot to variables in a dataset. More specifically, we provide the specification for a point geom by calling the function `geom_point`. This geom is passed as an argument a mapping (the value that follows `mapping =`) from the Cartesian x and y coordinate locations to the variables `displ` and `hwy`. This is materialized by the `aes` function.

Aesthetics are visual characteristics of observations in a plot. Examples include coordinate positions, shape, size, or color. For instance, a plot will often use Cartesian coordinates[2]

---

[2]https://en.wikipedia.org/wiki/Cartesian_coordinate_system

where each axis represents a variable of a tibble and these variables are *mapped* on to the x and y axes, respectively. In the case of this plot, we map displ to the x-axis and hwy to the y-axis.

To round up the discussion, here are the key points from the code we have just written:

- There is one ggplot and one geom_point.
- The ggplot call preceded the geom_point call.
- The plus sign + connects the two calls.
- A data specification appears in the ggplot call.
- A mapping specification appears in the geom_point call.

The semantics of the code is as follows:

- Instruct ggplot to get ready for creating plots using mpg as the data.
- Instruct geom_point to create a plot where displ is *mapped* to the x-axis and hwy is *mapped* to the y-axis, and displ and hwy are two variables from the tibble mpg.

3.2.4   Scatter plots

Our first visualization is an example of a scatter plot. A *scatter plot* is a plot that presents the relation between two numerical variables.

In other words, a scatter plot of variables A and B draws data from a collection of pairs (a,b), where each pair comes from a single observation in the data set. The number of pairs you plot can be one or more. There is no restriction on the frequencies we observe the same pair, the same a, and the same b.

We can use a scatter plot to visualize the relationship between the highway fuel efficiency (hwy) and the displacement of its engine (displ).

```
ggplot(data = mpg) +
  geom_point(mapping = aes(x = displ, y = hwy))
```

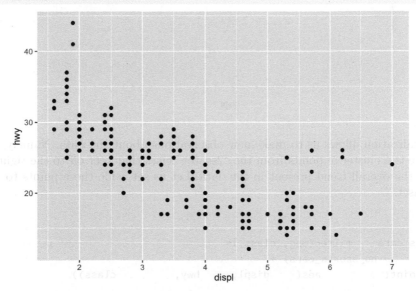

Each point on the plot is the pair of values of one car model in the dataset. Note that there are quite a few groups of points that align horizontally and quite a few groups of points that align vertically. The former are groups that share the same (close to the same) hwy values with each other, the latter are groups that share the same (close to the same) displ values with each other. We can observe a graceful trend downward in the plot – lower engine displacement is associated with more highway miles per gallon.

3.2.5   Adding color to your geoms

It appears that the points are following some downward trend. Let us examine this more closely by using ggplot to map colors to its points.

You can specify the attribute for ggplot to use to determine the colors, for instance, the class attribute. We make the specification in the aes of geom_point.

```
ggplot(data = mpg) +
  geom_point(mapping = aes(x = displ, y = hwy, color = class))
```

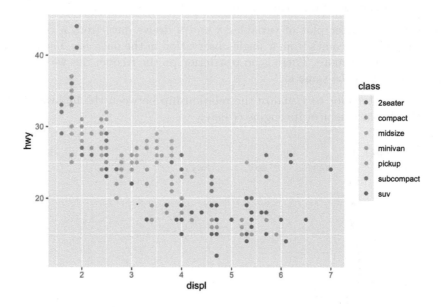

This visualization allows us to make new observations about the data. Namely, it appears that there is a cluster of points from the "2seater" class that veer off to the right and seem to break the overall trend present in the data. Let us set aside these points to compose a new dataset.

```
no_sports_cars <- filter(mpg, class != "2seater")
ggplot(data = no_sports_cars) +
  geom_point(mapping = aes(x = displ, y = hwy, color = class))
```

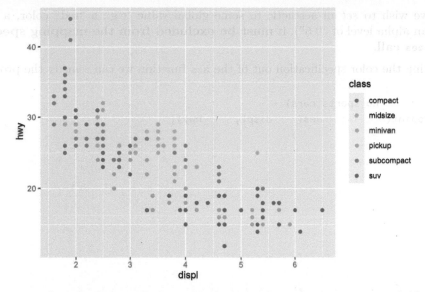

After the removal of the "2seater" class, the downward trend appears more vivid.

3.2.6 Mapping versus setting

Perhaps the most frequent mistake newcomers to `ggplot` make is conflating aesthetic *mapping* with aesthetic *setting*.

For instance, you may wish to set all points to a single color instead of coloring points according to the "type" of car. So, you devise the following `ggplot` code to color all points in the scatter blue.

```
ggplot(data = no_sports_cars) +
  geom_point(mapping = aes(x = displ, y = hwy, color = "blue"))
```

This code colors all points *red*, not blue! Can you spot the mistake in the above code?

When we wish to *set* an aesthetic to some global value (e.g., a "red" color, a "triangle" shape, an alpha level of "0.5"), it **must be excluded from the mapping specification in the aes call.**

By moving the color specification out of the aes function we can remedy the problem.

```
ggplot(data = no_sports_cars) +
  geom_point(mapping = aes(x = displ, y = hwy),
             color = "blue")
```

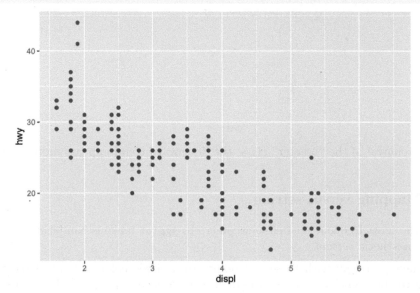

Observe how the aesthetics x and y are *mapped* to the variables displ and hwy, respectively, while the aesthetic color is *set* to a global constant value "blue".

When an aesthetic is *set*, we determine a visual property of a point that is *not* based on the values of a variable in the dataset. These will typically be provided as arguments to your geom_* function call. When an aesthetic is *mapped*, the visual property is determined based on the values of some variable in the dataset. This must be given within the aes function call.

3.2.7 Categorical variables

Coloring points according to some attribute is useful when dealing with *categorical variables*, that is, variables whose values come from a fixed set of categories. For instance, a variable named ice_cream_flavor may have values that are from the categories *chocolate, vanilla,* or *strawberry*; in terms of the mpg data, class can have values that are from the categories *compact, midsize, pickup, subcompact,* or *suv.*

Can we develop more insights from categorical variables?

Using the `mutate` function, we can create a new categorical variable `japanese_make`, which is either `TRUE` or `FALSE` depending on whether the manufacturer is one of Honda, Nissan, Subaru, or Toyota. We create a new dataset with the addition of this new variable.

```
no_sports_cars <- no_sports_cars |>
  mutate(japanese_make =
          manufacturer %in% c("honda", "nissan", "subaru", "toyota"))
```

Let us create a new plot using `japanese_make` as a coloring strategy.

```
ggplot(data = no_sports_cars) +
  geom_point(mapping = aes(x = displ, y = hwy, color = japanese_make))
```

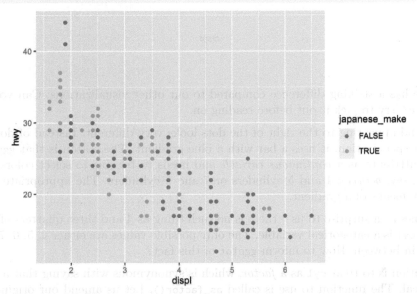

We can observe a downward trend in the data for cars with a Japanese manufacturer.

Let us try a second categorical variable: `cyl`.

```
ggplot(data = no_sports_cars) +
  geom_point(mapping = aes(x = displ, y = hwy, color = cyl))
```

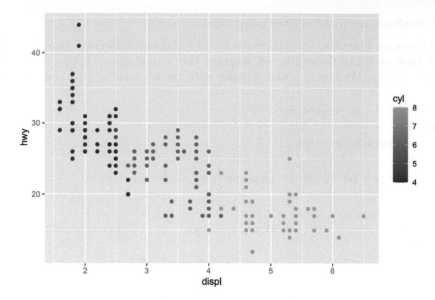

This one has a striking difference compared to our other visualizations. Can you spot the difference? Try to pick it out before reading on.

The legend appearing to the right of the dots looks very different. Instead of dots showing the color specification, it uses a bar with a blue gradient. The reason is that `ggplot` treats the `cyl` attribute as a *continuous variable* and needs to be able to select colors for values that are, say, *between* 4 and 5 cylinders or 7 and 8 cylinders. The appropriate way to do this is by means of a gradient.

This comes as a surprise to us – there is no such thing as 4 and three quarters of a cylinder because `cyl` is a categorical variable. The only possible values are either 4, 5, 6, 7, or 8, and nothing in between. How to inform `ggplot` of this fact?

The solution is to treat `cyl` as a *factor*, which is synonymous with saying that a variable is categorical. The function to use is called `as_factor()`. Let us amend our original attempt to include the call.

```
ggplot(data = no_sports_cars) +
  geom_point(mapping = aes(x = displ, y = hwy, color = as_factor(cyl)))
```

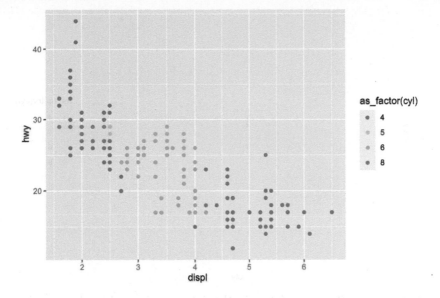

This one bears a resemblance that we are already familiar with.

Run your eyes from left to right along the horizontal axis. Observe how for a given hwy value, say points around hwy = 20, there is a clear transition from cars with 4 cylinders (in red) to cars with 6 cylinders (in cyan) and finally to cars with 8 cylinders (in purple). In contrast, if we look at points at, say around disp = 2, and run our eyes along the vertical axis, we do not observe such a transition in color – all the points still correspond to cars with 4 cylinders (in red).

This tells us that there is a stronger association between the continuous variable displ and the categorical variable cyl than between hwy and cyl.

Here is one more plot. Let us plot hwy against cty when coloring according to cyl. We naturally assume that the higher a car model's highway miles per gallon is (hwy), the higher its city miles per gallon (cty) is as well, and vice versa.

```
ggplot(no_sports_cars) +
  geom_point(aes(x = cty, y = hwy, color = as_factor(cyl)))
```

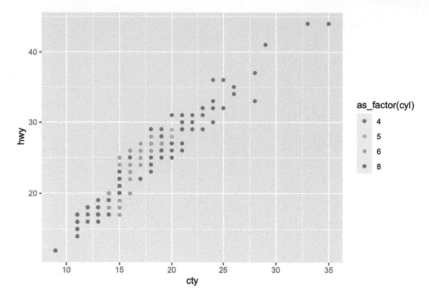

This one reveals a *positive association* in the data, as opposed to the *negative association* that we observed in the downward trend in the plot of `hwy` against `displ`. We can also see a greater transition in color as we move left to right in the plot, suggesting a stronger relationship between `cty`and `cyl`.

Observe that there are two points that seem to be very far off to the right and one point off to the left. We may call such data points *outliers* meaning that they do not appear to conform to the associations that other observations are following. Let us isolate these points using `filter`.

```
no_sports_cars |>
  filter(cty < 10 | cty > 32.5) |>
  relocate(cty, .after = year) |>
  relocate(hwy, .before = cyl)
```

```
## # A tibble: 7 x 12
##   manufa~1 model displ  year   cty   hwy   cyl trans drv
##   <chr>    <chr> <dbl> <int> <int> <int> <int> <chr> <chr>
## 1 dodge    dako~   4.7  2008     9    12     8 auto~ 4
## 2 dodge    dura~   4.7  2008     9    12     8 auto~ 4
## 3 dodge    ram ~   4.7  2008     9    12     8 auto~ 4
## 4 dodge    ram ~   4.7  2008     9    12     8 manu~ 4
## 5 jeep     gran~   4.7  2008     9    12     8 auto~ 4
## 6 volkswa~ jetta   1.9  1999    33    44     4 manu~ f
## 7 volkswa~ new ~   1.9  1999    35    44     4 manu~ f
## # ... with 3 more variables: fl <chr>, class <chr>,
## #   japanese_make <lgl>, and abbreviated variable name
## #   1: manufacturer
```

We see that the four Dodges and one Jeep are the far-left points and two Volkswagens are the far right points.

3.2.8 Continuous variables

Given what we learned when experimenting with `cyl`, we might be curious as to what can be gleaned when we intend on coloring points according to a *continuous variable*. Let us try it with the attribute `cty`, which represents the city fuel efficiency.

```
ggplot(data = no_sports_cars) +
  geom_point(mapping = aes(x = displ, y = hwy, color = cty))
```

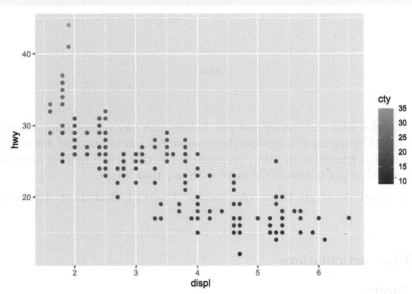

The blue gradient makes it hard to see changes in the color. Can we use a different gradient other than blue?

Yes! The solution is to add a layer `scale_color_gradient` at the end with two colors names of our choice. In the style of art deco, we pick two colors, `yellow3` and `blue`.

```
ggplot(data = no_sports_cars) +
  geom_point(mapping = aes(x = displ, y = hwy, color = cty)) +
  scale_color_gradient(low = "yellow3", high="blue")
```

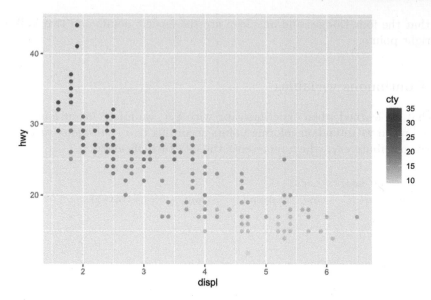

Contrast this plot with the one we saw just before with `hwy` versus `displ` when coloring according to `cyl`. The situation is reversed here: as we run our eye up and down for some value of `displ`, we see a transition in color; the same is not true when moving along horizontally. Thus, it seems that `cty` is associated more with `hwy` than `displ`.

By the way, where do we get those color names? There's a cheatsheet[3] for that!

3.2.9 Other articulations

3.2.9.1 Facets

Instead of using color to annotate points, we can also use something called *facets* which splits the plots into several subplots, one for each category of the categorical variable. Let us use faceting for our last visualization.

```
ggplot(data = no_sports_cars) +
  geom_point(mapping = aes(x = displ, y = hwy)) +
  facet_wrap(~ as_factor(cyl), nrow = 2)
```

---

[3]https://www.nceas.ucsb.edu/sites/default/files/2020-04/colorPaletteCheatsheet.pdf

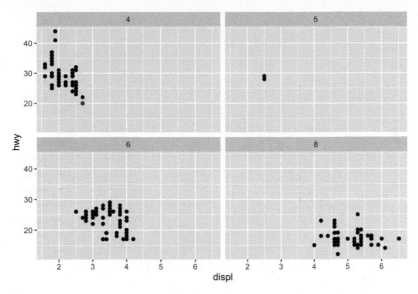

This one makes it very clear that hardly any car models in the dataset have 5 cylinders!

3.2.9.2 Shapes

You can use shapes and fill strengths to differentiate between points. The argument for the fill strength is `alpha = X` where X is the attribute.

```
ggplot(no_sports_cars) +
  geom_point(aes(x = cty, y = hwy, alpha = class))
```

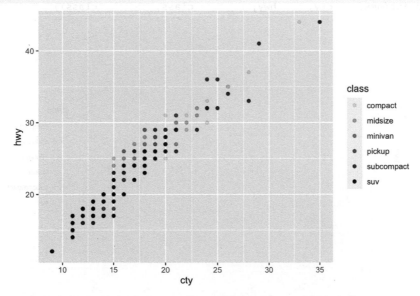

You can specify both `alpha` and `color`, even using different variables. See an example below.

```
ggplot(no_sports_cars) +
  geom_point(aes(x = cty, y = hwy, alpha = displ, shape = class))
```

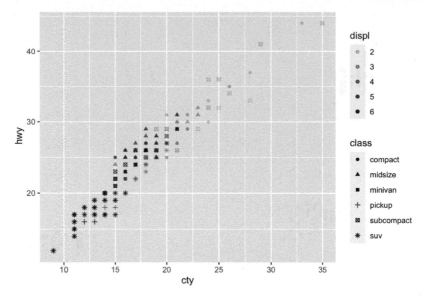

Welcome to your first 4-D visualization! Note how when we move up and down vertically for some fixed value of cty, the shapes do not grow more transparent; this is only observed as we move left and right at some fixed value of hwy, suggesting a stronger relationship between cty and displ. Put another way, we can say that displ varies more with cty than it does with hwy. We do not observe a strong effect with respect to the shapes in class.

Note that by mapping both the shape and color aesthetics to the same attribute, say, class, the legends are collapsed into one.

```
ggplot(no_sports_cars) +
  geom_point(aes(x = cty, y = hwy, shape = class, color = class))
```

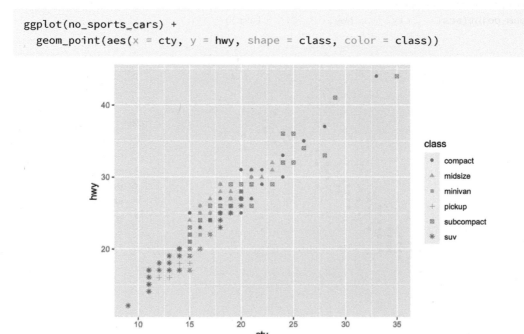

Can you say whether this visualization contains more or less information than our last one? How many dimensions are displayed here?

3.2.10   Jittering

You may notice that even though there are 234 entries in the data set, much fewer points (only 109 to be exact) appears in the initial (before filtering) scatter plot of mpg.

This is because many points collide on the plot. We call the phenomenon **overplotting**, meaning one point appearing over another. It is possible to nudge points in a random direction in a small quantity. By making all directions possible, we can make complete overplotting a rare event. We call the random nudging arrangement *jittering*.

To jitter, we add a positional argument position = "jitter" to geom_point. Note that this is not a part of the aesthetic specification (so is not inside aes).

```
ggplot(no_sports_cars) +
  geom_point(mapping = aes(x = cty, y = hwy, color = class),
             position = "jitter")
```

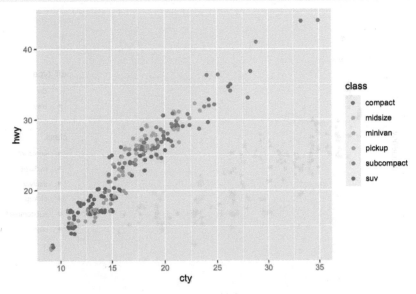

3.2.11   One more scatter plot

In our first visualization of this section, we plotted hwy against displ. In that plot, by substituting cty for hwy, we obtain a similar plot of cty against displ.

What if we want to see now both hwy and cty against displ? Is it possible to merge the two plots into one?

Yes, we can do the merge easily using pivoting. Recall that pivot_longer combines multiple columns into one. We can create a new data frame that combines the values from hwy and cty under the name efficiency while specifying whether the value is from hwy or from cty under the name eff_type.

```
no_sports_cars_pivot <- no_sports_cars |>
  pivot_longer(cols = c(hwy,cty),
               names_to = "eff_type",
               values_to = "efficiency")
```

Then we can plot `efficency` against `displ` by showing `eff_type` using the shape and `class` using the color.

```
ggplot(no_sports_cars_pivot) +
  geom_point(aes(x = displ, y = efficiency,
                 alpha = eff_type, shape = class),
             position = "jitter")
```

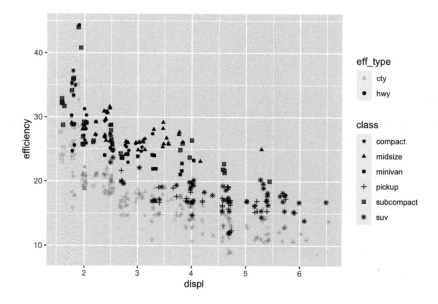

The "jitter" option makes the visualization quite busy. Let us take it away.

```
ggplot(no_sports_cars_pivot) +
  geom_point(aes(x = displ, y = efficiency,
                 alpha = eff_type, shape = class))
```

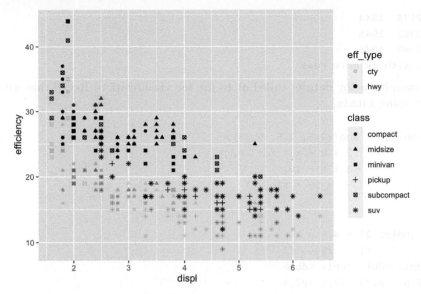

Note the pattern in the fill strength of points as dictated by `eff_type` – the upper region of the plot is shaded more boldly and the lower region very lightly. We leave it as an exercise to the reader to come up with some explanations as to why such a visible pattern emerges.

3.3 Line and Smooth Geoms

In the last section we introduced scatter plots and how to interpret them using `ggplot` with the point geom. In this section we introduce two new geoms, the point and smooth geoms, to build another (hopefully familiar) visualization: the line chart.

3.3.1 Prerequisites

We will make use of the tidyverse in this chapter, so let us load it in as usual.

```
library(tidyverse)
```

We will study a tibble called `airmiles`, which contains data about passenger miles on commercial US airlines between 1937 and 1960.

```
## # A tibble: 24 x 2
##    miles  date
##    <dbl> <dbl>
## 1    412  1937
## 2    480  1938
## 3    683  1939
## 4   1052  1940
## 5   1385  1941
## 6   1418  1942
## 7   1634  1943
```

```
##  8  2178  1944
##  9  3362  1945
## 10  5948  1946
## # ... with 14 more rows
```

We will examine a toy dataset called df to use for visualization. Recall that we can create a dataset using tibble.

```
df <- tibble(x = seq(-2.5, 2.5, 0.25),
             f1 = 2 * x - x * x + 20,
             f2 = 3 * x - 10,
             f3 = -x + 50 * sin(x))
df
```

```
## # A tibble: 21 x 4
##        x      f1     f2     f3
##    <dbl>  <dbl>  <dbl>  <dbl>
##  1 -2.5    8.75 -17.5  -27.4
##  2 -2.25  10.4  -16.8  -36.7
##  3 -2     12    -16    -43.5
##  4 -1.75  13.4  -15.2  -47.4
##  5 -1.5   14.8  -14.5  -48.4
##  6 -1.25  15.9  -13.8  -46.2
##  7 -1     17    -13    -41.1
##  8 -0.75  17.9  -12.2  -33.3
##  9 -0.5   18.8  -11.5  -23.5
## 10 -0.25  19.4  -10.8  -12.1
## # ... with 11 more rows
```

3.3.2   A toy data frame

df has four variables, x, y, z, and w. The range of x is [-2.5,2.5] with 0.25 as a step width. The functions for y, z, and w are $2x - x^2 + 20$, $3x - 10$, and $-x + \sin(x)$, respectively.

To visualize these three functions, we first need to *pivot* the data so that it becomes a long table. The reason for this step should become evident in a moment.

```
df_long <- df |>
  pivot_longer(c(f1, f2, f3), names_to = "type", values_to = "y") |>
  select(x, y, type)
df_long
```

```
## # A tibble: 63 x 3
##        x      y type
##    <dbl>  <dbl> <chr>
##  1 -2.5    8.75 f1
##  2 -2.5  -17.5  f2
##  3 -2.5  -27.4  f3
##  4 -2.25  10.4  f1
##  5 -2.25 -16.8  f2
```

```
##   6 -2.25 -36.7   f3
##   7 -2      12    f1
##   8 -2     -16    f2
##   9 -2     -43.5  f3
## 10 -1.75   13.4   f1
## # ... with 53 more rows
```

Note how we have two variables present, x and y, annotated by a third variable, type, designating which function the (x, y) pair belongs to.

3.3.3   The line geom

We start with visualizing y against x for each of the three functions. We can use the same strategy as geom_point by simply substituting geom_line for geom_point. However, we will pass an additional argument group to the aesthetic to inform ggplot which function a point comes from.

```
ggplot(data = df_long) +
  geom_line(mapping = aes(x = x, y = y, group = type))
```

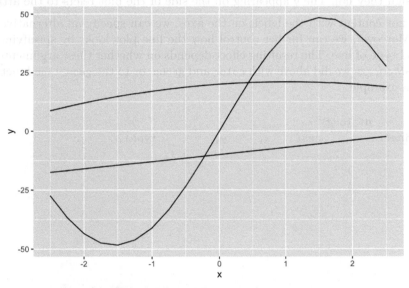

This plot is quite dull-looking and it can be hard to tell the lines apart from each other. How about we annotate each line with a color? To do this, we substitute the group argument for color.

```
ggplot(data = df_long) +
  geom_line(mapping = aes(x = x, y = y, color = type))
```

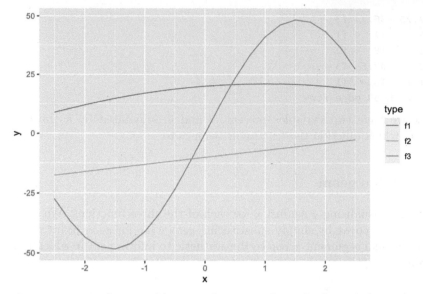

A curious phenomenon is the variables x and y coincide with the argument names x and y inside the aes. So the meaning of x and y are different depending on which side of the equality sign they fall. The y appearing on the side of the plot refers to the attribute.

If we are not content with the labels on the axes, we can specify an alternative using xlab or ylab. Moreover, we can further control how the line plot looks by specifying the shape, width, and type of line. The resulting effect depends on whether these arguments are passed to the *aesthetic*, as we did above with color and group, or to the geom_line function directly. Here is an example.

```
ggplot(data = df_long) +
  geom_line(mapping = aes(x = x, y = y, color = type),
            size = 2, linetype = "longdash") +
  xlab("x values") +
  ylab("y values")
```

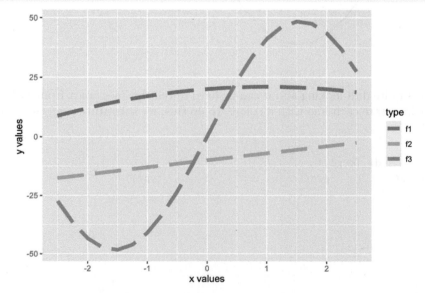

Observe how the color is varied for each of the functions, but the size and type of the line is the same across all of them. Can you tell why? If you think you got it, here is a follow-up question: what would you need to change to make *both* the color and line type different for each of the lines?

By the way, there are many line types offered by `ggplot`. Available line types are "twodash", "solid", "longdash", "dotted", "dotdash", "dashed", and "blank".

3.3.4 Combining `ggplot` calls with `dplyr`

Let us turn our attention to the function `f3` and set aside the functions `f1` and `f2` for now. We know how to do this using `filter` from `dplyr`.

```
only_f3 <- df_long |>
  filter(type == "f3")
```

The object `only_f3` keeps only those points corresponding to the function `f3`. We could then generate the line plot as follows.

```
ggplot(data = only_f3) +
  geom_line(mapping = aes(x = x, y = y))
```

However, we have discussed before how naming objects, and keeping track of them, can be cumbersome. Moreover, `only_f3` is only useful as input for the visualization; for anything else, it is a useless object sitting in memory.

We have learned that the pipe operator (`|>`) is useful for eliminating redundancy with `dplyr` operations. We can use the pipe again here, this time to "pipe in" a filtered data frame to use as a data source for visualization. Here is the re-worked code.

```
df_long |>
  filter(type == "f3") |>
  ggplot() +
  geom_line(mapping = aes(x = x, y = y))
```

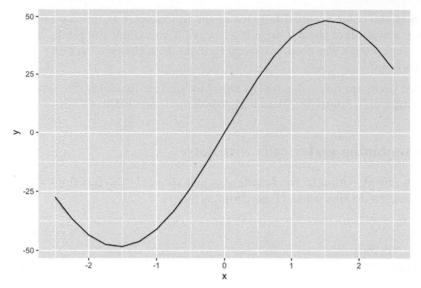

There is something unfortunate about this code: the pipe operator cannot be used when specifying the `ggplot` layers, so we have a motley mix of |> and + symbols in the code. Keep this in mind to keep the two straight in your head: use |> when working with `dplyr` and use + when working with `ggplot`.

This curve bears the shape of the famous sinusoidal wave true to trigonometry. However, upon closer inspection, you may notice that the curve is actually a concatenation of many straight-line pieces stitched together. Here is another example using `df_airmiles`.

```
ggplot(data = df_airmiles) +
  geom_line(mapping = aes(x = date, y = miles))
```

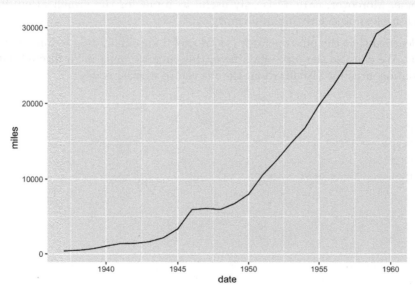

Is it possible to draw something smoother? For this, we turn to our next geom: the *smooth* geom.

3.3.5   Smoothers

We observed in our last plots that while line geoms can be used to plot a line chart, the result may not be as smooth as we would like. An alternative to a line geom is the *smooth* geom, which can be used to generate a *smooth line plot*.

The way to use it is pretty much the same as geom_line. Here is an example using the toy data frame, where the only change made is substituting the geom.

The argument se = FALSE we pass to geom_smooth is to disable a feature that displays confidence ribbons around the line. While these are certainly useful, we will not study them in this text.

```
ggplot(data = df_long) +
  geom_smooth(aes(x = x, y = y, color = type),
              se = FALSE)
```

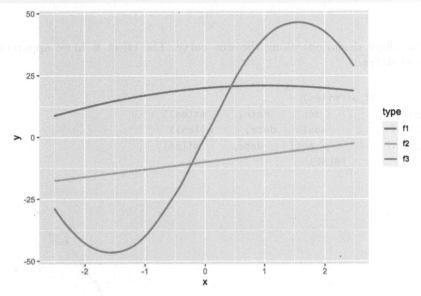

Observe that the piece-wise straight line of the sine function now looks like a proper curved line.

It is possible to mix line and smooth geoms together in a single plot.

```
ggplot(data = df_long) +
  geom_smooth(mapping = aes(x = x, y = y, color = type), se = FALSE) +
  geom_line(mapping = aes(x = x, y = y, color = type))
```

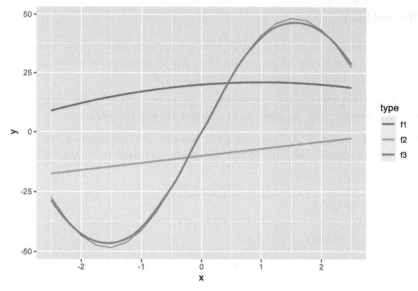

Notice the slight deviations along the sine curve. The effect is more apparent when we visualize `airmiles`.

```
ggplot(data = df_airmiles) +
  geom_point(mapping = aes(x = date, y = miles)) +
  geom_line(mapping = aes(x = date, y = miles)) +
  geom_smooth(mapping = aes(x = date, y = miles),
              se = FALSE)
```

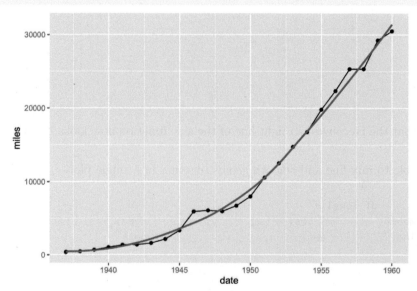

The line geom creates the familiar line graph we typically think of while the smooth geom "smooths" the line to aid the eye in seeing overall patterns; the line geom, in contrast, is much more "ridgy".

geom_smooth uses statistical methods to determine the smoother. One of the methods that can be used is *linear regression*, which is a topic we will see study in detail towards the end of the text. Here is an example.

```
ggplot(data = df_airmiles) +
  geom_point(mapping = aes(x = date, y = miles)) +
  geom_line(mapping = aes(x = date, y = miles)) +
  geom_smooth(mapping = aes(x = date, y = miles),
              method = "lm", se = FALSE)
```

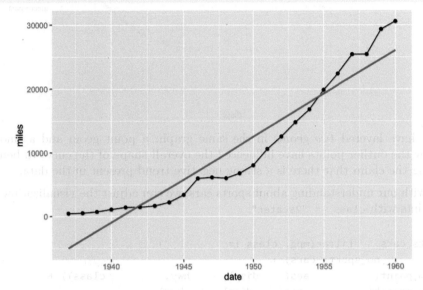

Let us close our discussion of line and smooth geoms using one more example of the smooth geom.

3.3.6   Observing a negative trend

The smooth geom can be useful to confirm the negative trend we have observed in the mpg data frame.

```
ggplot(data = mpg) +
  geom_point(mapping = aes(x = displ, y = hwy, color = class)) +
  geom_smooth(mapping = aes(x = displ, y = hwy),
              se = FALSE)
```

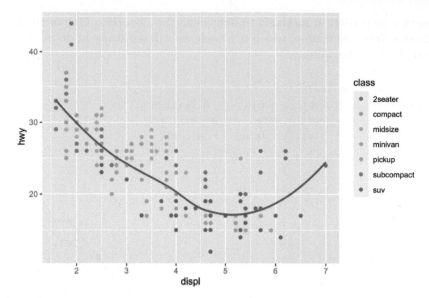

Here we have layered *two* geoms in the same graph: a point geom and a smooth geom. Note how the outlier points have influenced the overall shape of the curve to bend upward, muddying the claim that there is a strong negative trend present in the data.

Armed with our understanding about sports cars, we can adjust the visualization by setting aside points with class == "2seater".

```
no_sports_cars <- filter(mpg, class != "2seater")
ggplot(data = no_sports_cars) +
    geom_point(mapping = aes(x = displ, y = hwy, color = class)) +
    geom_smooth(mapping = aes(x = displ, y = hwy),
                se = FALSE)
```

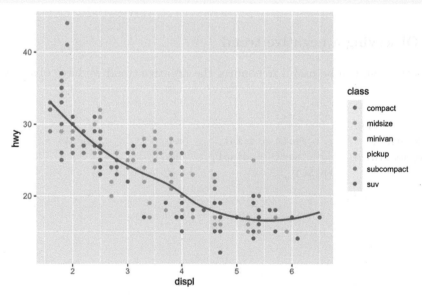

This plot shows a much more graceful trend downward.

3.3.7 Working with multiple geoms

Before moving to the next topic, we point out some technical concerns when working with multiple geoms. First, in the above code we have written, observe the same mapping for x and y was defined in two different places.

This could cause some unexpected surprises when writing code: imagine if we wanted to change the y-axis to cty instead of hwy, but we forgot to change both occurrences of hwy. This can be amended by moving the mapping into the ggplot function call.

```
no_sports_cars <- filter(mpg, class != "2seater")
ggplot(data = no_sports_cars,
       mapping = aes(x = displ, y = hwy)) +
    geom_point(mapping = aes(color = class)) +
    geom_smooth(se = FALSE)
```

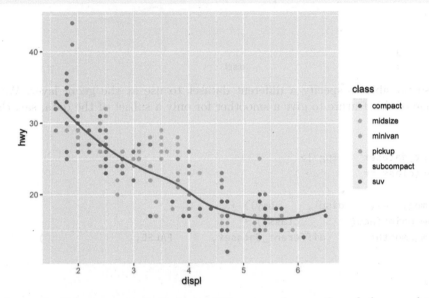

We can also write this more concisely by omitting some keywords as below, and the result would be the same.

```
no_sports_cars <- filter(mpg, class != "2seater")
ggplot(no_sports_cars,
       aes(x = displ, y = hwy)) +
    geom_point(aes(color = class)) +
    geom_smooth(se = FALSE)
```

What if we wanted a smoother for *each* type of car? The color aesthetic can also be moved into the ggplot function call. While we are it, we set the smoother to use linear regression for the smoothing.

```
ggplot(no_sports_cars,
       aes(x = displ, y = hwy, color = class)) +
    geom_point() +
    geom_smooth(se = FALSE, method = "lm")
```

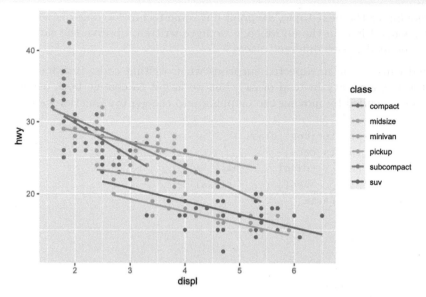

It is also possible to specify a different dataset to use at the geom layer. We can take advantage of this feature to give a smoother for only a subset of the data, say, the midsize cars.

```
different_dataset <- mpg |>
  filter(class == "midsize")

ggplot(mpg, aes(x = displ, y = hwy)) +
    geom_point(aes(color = class)) +
    geom_smooth(data = different_dataset, se = FALSE, method = "lm")
```

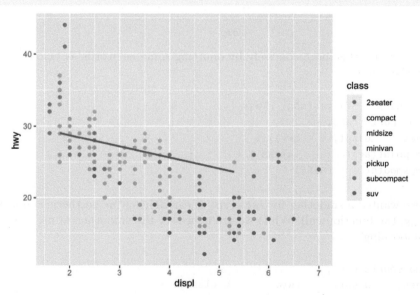

3.4   Categorical Variables

The point, line, and smooth plots are for viewing relations among numerical variables. As you are well aware, numerical variables are not the only type of variables in a data set. There are variables representing categories, and we call them *categorical variables*.

A categorical attribute has a fixed, finite number of possible values, which we call *categories*. The categories of a categorical attribute are distinct from each other.

A special categorical attribute is a *binary category*, where there are exactly two values. A binary category that we are probably the most familiar with is the *Boolean category*, which has "true" and "false" as its values. Because of the familiarity, we often identify a binary category as a Boolean category.

In datasets, categories in a categorical attribute are sometimes called *levels* and we refer to such a categorical attribute as a *factor*. Sometimes, categories are whole numbers 1, 2, ..., representing indexes. Such cases may require some attention when processing with R, because R may think of the variables as numbers.

We have seen examples of categorical variables before. The `class` attribute in the `mpg` data set is one.

3.4.1   Prerequisites

As before, let us load `tidyverse`. Moreover, we will make use of datasets that are not available in `tidyverse` but are available in the package `faraway`. So, we load the package too.

```
library(tidyverse)
library(faraway)
happy <- tibble(happy)
```

3.4.2   The `happy` and `diamonds` data frames

The table `happy` contains data on 39 students in a University of Chicago MBA class.

```
happy
```

```
## # A tibble: 39 x 5
##    happy money   sex  love  work
##    <dbl> <dbl> <dbl> <dbl> <dbl>
## 1     10    36     0     3     4
## 2      8    47     1     3     1
## 3      8    53     0     3     5
## 4      8    35     1     3     3
## 5      4    88     1     1     2
## 6      9   175     1     3     4
## 7      8   175     1     3     4
```

```
##  8      6      45      0      2      3
##  9      5      35      1      2      2
## 10      4      55      1      1      4
## # ... with 29 more rows
```

Armed with what we have learned about `ggplot2`, we can begin answering questions about this data set using data transformation and visualization techniques. For instance, which "happiness" scores are the most frequent among the students? Moreover, can we discover an association between feelings of belonging and higher scores of happiness? How about family income?

Before we inspect the `happy` data set any further, we will also consider another table `diamonds`, which contains data on almost 54,000 diamonds.

```
diamonds
```

```
## # A tibble: 53,940 x 10
##     carat cut    color clarity depth table price     x     y
##     <dbl> <ord>  <ord> <ord>   <dbl> <dbl> <int> <dbl> <dbl>
##  1   0.23 Ideal  E     SI2      61.5    55   326  3.95  3.98
##  2   0.21 Prem~  E     SI1      59.8    61   326  3.89  3.84
##  3   0.23 Good   E     VS1      56.9    65   327  4.05  4.07
##  4   0.29 Prem~  I     VS2      62.4    58   334  4.2   4.23
##  5   0.31 Good   J     SI2      63.3    58   335  4.34  4.35
##  6   0.24 Very~  J     VVS2     62.8    57   336  3.94  3.96
##  7   0.24 Very~  I     VVS1     62.3    57   336  3.95  3.98
##  8   0.26 Very~  H     SI1      61.9    55   337  4.07  4.11
##  9   0.22 Fair   E     VS2      65.1    61   337  3.87  3.78
## 10   0.23 Very~  H     VS1      59.4    61   338  4     4.05
## # ... with 53,930 more rows, and 1 more variable: z <dbl>
```

The *values* of the categorical variable *cut* are "fair", "good", "very good", "ideal", and "premium". We can look at how many diamonds are in each category by using `group_by()` and `summarize()`.

```
diamonds |>
  group_by(cut) |>
  summarize(count = n())
```

```
## # A tibble: 5 x 2
##   cut       count
##   <ord>     <int>
## 1 Fair       1610
## 2 Good       4906
## 3 Very Good 12082
## 4 Premium   13791
## 5 Ideal     21551
```

The table shows the number of diamonds of each cut. We call this a *distribution*. A distribution shows all the values of a variable, along with the frequency of each one. Recall that the summary does not persist on the `diamonds` data set and so the dataset remains the same after summarization.

3.4.3 Bar charts

The bar chart is a familiar way of visualizing categorical distributions. Each category of a categorical attribute has a number it has an association with, and a bar chart presents the numbers for the categories using bars, where the height of the bars represent the numbers.

Typically, the bars in a bar chart appear either all vertically or all horizontally with an equal space in between and with the same height but expanding horizontally (that is, the non-variable dimension of the bars).

```
ggplot(diamonds, aes(x = cut)) +
  geom_bar()
```

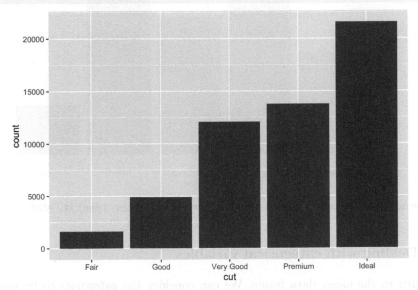

The x-axis displays the values for the *cut* attribute while the y-axis says "count". The label "count" is the result of geom_bar() generating bars. Since the number of observations is greater than there are categories, geom_bar() decides to count the occurrences of each category. The word "count" says that it is the result of counting.

We often call the inner working of the geom_bar() (and other geom functions) for number generation **stat**. Thus, ggplot2 transforms the raw table to a new dataset of categories with its corresponding counts. From this new table, the bar plot is constructed by mapping cut to the x-axis and count to the y-axis.

The default stat geom_bar() uses for counting is stat_count(), which counts the number of cases at each x position. If the counts are already present in the dataset and we would prefer to instead use these directly for the heights of the bars, we can set stat = "identity". For instance, consider this table about popular pies sold at a bakery. The "count" is already present in the sold variable.

```
store_pies <- tribble(
  ~pie,           ~sold,
  "Pecan",        906,
  "Key Lime",     620,
  "Pumpkin",      202,
  "Apple",        408,
```

```
    "Mississippi mud",   551
)
ggplot(data = store_pies) +
  geom_bar(mapping = aes(x = pie, y = sold),
           stat = "identity")
```

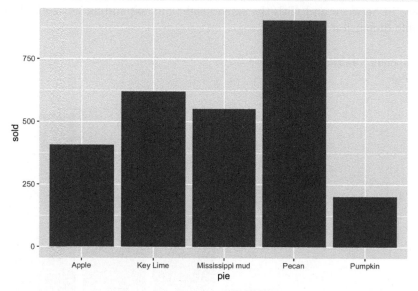

Note how we provide both x and y aesthetics when using the identity stat.

3.4.4   Dealing with categorical variables

Let us turn to the happy data frame. We can consider the *categories* to be points on the 10-point scale, and the *individuals* the students in each interval. Let us determine this distribution using group_by and sumarrize.

Below, we take the happy data set and execute grouping by happy. The category for the happy value is happiness in this new dataset happy_students. We summarize in terms of the counts n() and we state the count as an attribute number.

```
happy_students <- group_by(happy, happiness = happy) |>
  summarize(number = n())
happy_students
```

```
## # A tibble: 9 x 2
##    happiness number
##        <dbl>  <int>
## 1          2      1
## 2          3      1
## 3          4      4
## 4          5      5
## 5          6      2
```

```
## 6          7          8
## 7          8         14
## 8          9          3
## 9         10          1
```

We can now use this table, along with the graphing skills that we acquired above, to draw a bar chart that shows which scores are most frequent among the 39 students.

```
ggplot(happy_students) +
  geom_bar(aes(x = happiness, y = number),
           stat = "identity") +
  labs(x = "Happy score",
       y = "Count")
```

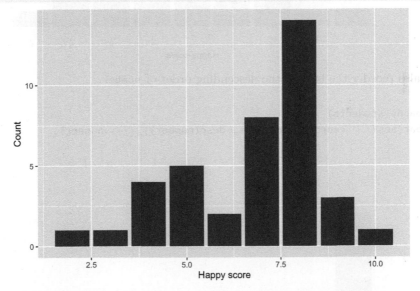

Here R treats the happy score as numerical values. That is the reason that we see 2.5, 5.0, 7.5, and 10.0 on the x-axis. Let us inform R that these are indeed categories by treating happiness as a factor.

```
ggplot(happy_students) +
  geom_bar(aes(x = as_factor(happiness), y = number),
           stat = "identity") +
  labs(x = "Happy score",
       y = "Count")
```

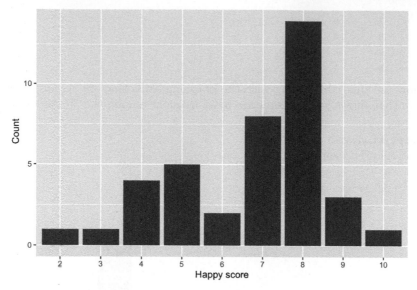

We can also reorder the bars in the descending order of number.

```
ggplot(happy_students) +
  geom_bar(aes(x = reorder(happiness, desc(number)), y = number),
           stat = "identity") +
  labs(x = "Happy score",
       y = "Count")
```

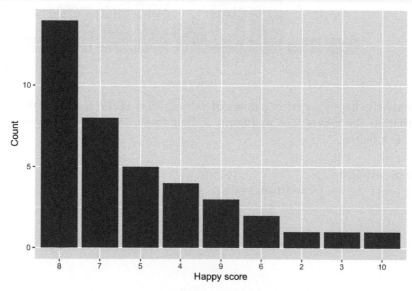

There is something unsettling about this chart. Though it does answer the question of which "happy" scores appear most frequently among the students, it doesn't list the scores in chronological order.

Let us return to the first plot.

```
ggplot(happy_students) +
  geom_bar(aes(x = as_factor(happiness), y = number),
           stat = "identity") +
  labs(x = "Happy score",
       y = "Count")
```

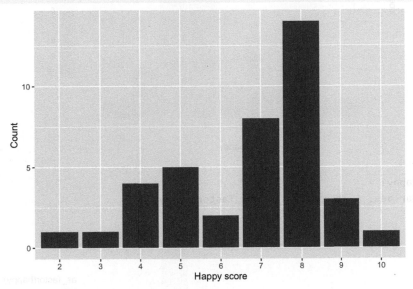

Now the scores are in increasing order.

We can attempt an answer to our second question: is there an association between feelings of belonging and "happy" scores? Put another way, what relationship, if any, exists between love and happy? For this, let us turn to positional adjustments in ggplot.

3.4.5   More on positional adjustments

With point geoms we saw the usefulness of the "jitter" position adjustment to overcome the problem of overplotting. Bar geoms similarly benefit from positional adjustments. For instance, we can set the color or fill of a bar plot.

```
ggplot(happy) +
  geom_bar(aes(x = happy, color = as_factor(happy)))
```

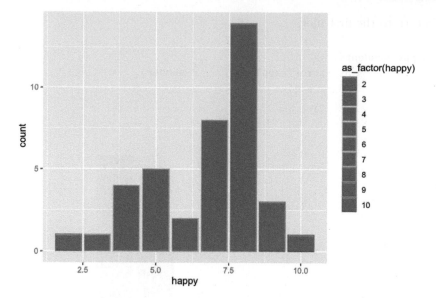

```
ggplot(happy) +
  geom_bar(aes(x = happy, fill = as_factor(happy)))
```

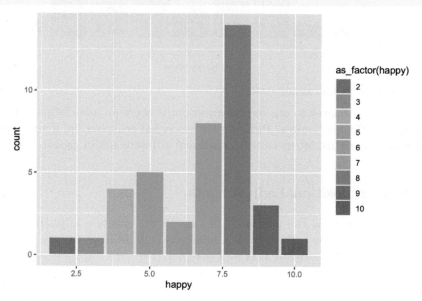

The legend that appears says "as_factor(happy)", because we used the value for coloring.
We can change the title with the use of labs(fill = ...) ornamentation.

```
ggplot(happy) +
  geom_bar(aes(x = happy, color = as_factor(happy))) +
  labs(color="happy")
```

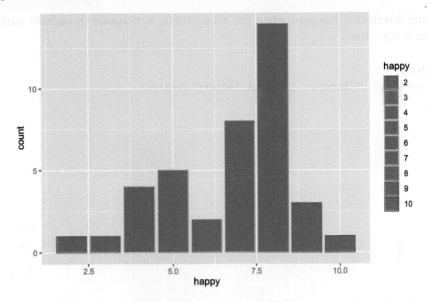

```
ggplot(happy) +
  geom_bar(aes(x = happy, fill = as_factor(happy))) +
  labs(fill="happy")
```

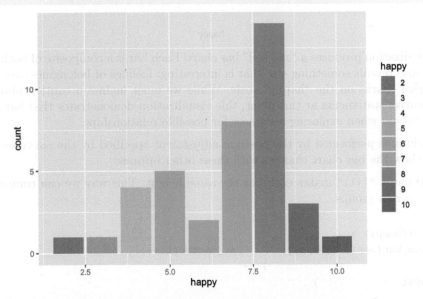

To make this code work, note how the parameter passed to the color and fill aesthetic is converted to a *factor*, i.e., a categorical variable, via as_factor(). As discussed, the happy variable is treated as numerical by R even though it is meant as a categorical variable in reality. ggplot2 can only color or fill a bar chart based on a categorical variable.

Something interesting happens when the fill aesthetic is mapped to another variable other than happy, e.g., love.

```
ggplot(happy) +
  geom_bar(aes(x = happy, fill = as_factor(love))) +
  labs(fill = "love")
```

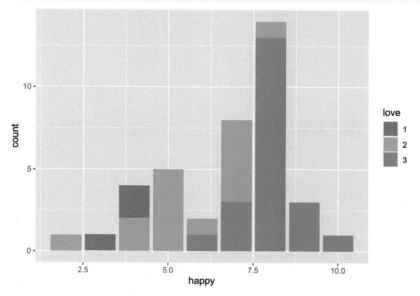

This visualization produces a "stacked" bar chart! Each bar is a composite of both happy and love. It also reveals something else that is interesting: feelings of belonginess are associated with higher marks on the "happy" scale. While we must maintain caution about making any causative statements at this point, this visualization demonstrates that bar charts can be a useful aid when exploring a dataset for possible relationships.

The stacking is performed by the position adjustment specified by the position argument. Observe how the bar chart changes with these other options:

- position = "fill" makes each bar the same height. This way we can compare proportions across groups.

```
ggplot(happy) +
  geom_bar(aes(x = happy, fill = as_factor(love)),
           position = "fill")  +
  labs(fill = "love")
```

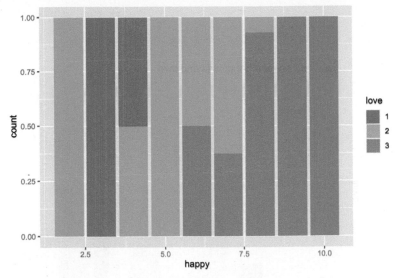

- `position = "dodge"` places the stacked bars directly *beside* one another. This makes it easier to compare individual values.

```
ggplot(happy) +
  geom_bar(aes(x = happy, fill = as_factor(love)),
           position = "dodge")+
  labs(fill = "love")
```

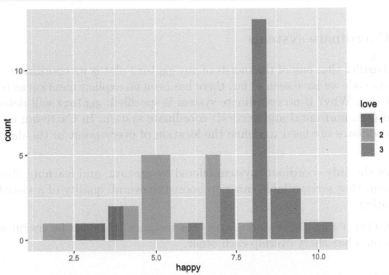

We can adjust the values of the x-axis to whole numbers using `as_factor` again. We can also the title "relation between happiness and love".

```
ggplot(happy) +
  geom_bar(aes(x = as_factor(happy), fill = as_factor(love)),
           position = "dodge") +
  labs(x = "Happy Score",
```

```
        fill = "love",
     title = "relation between happiness and love")
```

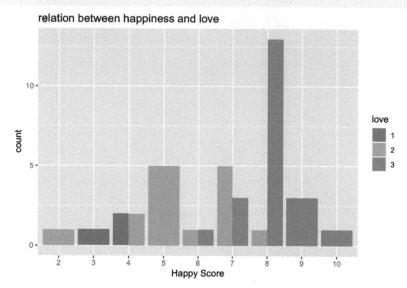

Bar charts are intended as visualizations of categorical variables. When the variable is numerical, the numerical relations between its values have to be taken into account when we create visualizations. That is the topic of the next section. Before ending the discussion here, we turn to one more important piece of `ggplot2` magic.

3.4.6   Coordinate systems

We noted earlier that one of the motifs of any `ggplot2` plot is its coordinate system. In all of the `ggplot2` code we have seen so far, there has been no explicit mention as to the coordinate system to use. Why? If no coordinate system is specified, `ggplot2` will default to using the Cartesian (i.e., horizontal and vertical) coordinate system. In Cartesian coordinates, the x and y coordinates are used to define the location of every point in the dataset, as we have just seen.

This is not the only coordinate system offered by `ggplot2`, and learning about other coordinate systems that are available can help boost the overall quality of a visualization and aid interpretation.

- `coord_flip()` flips the x and y axes. For instance, this can be useful when the x-axis labels on a bar chart overlap each other.

```
ggplot(happy) +
   geom_bar(aes(x = happy, fill = as_factor(love))) +
   coord_flip()
```

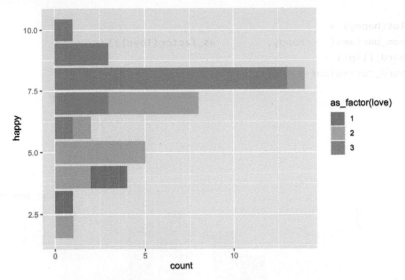

- `coord_polar()` uses polar coordinates. It is useful for plotting a Coxcomb chart. Note the connection between this and a bar chart.

```
ggplot(happy) +
  geom_bar(aes(x = happy, fill = as_factor(love))) +
  coord_polar()
```

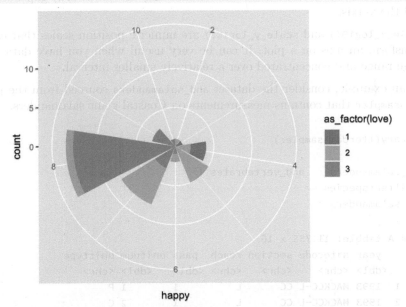

- `coord_cartesian(xlim, ylim)` can be passed arguments for "zooming in" the plot. For instance, we may want to limit the height of very tall bars (and, similarly, the effect of very small bars) in a bar chart by passing in a range of possible y-values to `ylim`.

```
ggplot(happy) +
  geom_bar(aes(x = happy, fill = as_factor(love))) +
  coord_flip() +
  coord_cartesian(ylim=c(1,10))
```

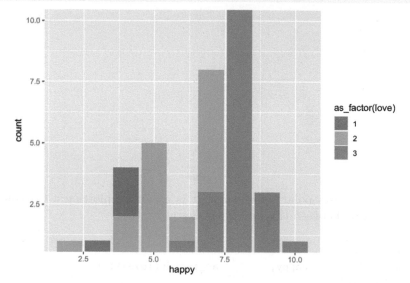

This can also be used as a trick for eliminating the (awkward) gap between the bars and the x-axis.

- scale_x_log10() and scale_y_log10() are numeric position scales that can be used to transform an axis on a plot. It can be very useful when you have data spread over a large range and concentrated over a relatively smaller interval.

As an example, consider the dataset and_salamanders sourced from the package lter-datasampler that contains measurements on Coastal giant salamanders.

```
library(lterdatasampler)

and_salamanders <- and_vertebrates |>
  filter(species == "Coastal giant salamander")
and_salamanders
```

```
## # A tibble: 11,758 x 16
##     year sitecode section reach  pass unitnum unittype
##    <dbl> <chr>    <chr>   <chr> <dbl>   <dbl> <chr>
## 1   1993 MACKCC-L CC      L         1       1 P
## 2   1993 MACKCC-L CC      L         1       2 C
## 3   1993 MACKCC-L CC      L         1       2 C
## 4   1993 MACKCC-L CC      L         1       2 C
## 5   1993 MACKCC-L CC      L         1       2 C
## 6   1993 MACKCC-L CC      L         1       2 C
## 7   1993 MACKCC-L CC      L         1       2 C
## 8   1993 MACKCC-L CC      L         1       2 C
```

```
##  9  1993 MACKCC-L CC        L           1        2 C
## 10  1993 MACKCC-L CC        L           1        2 C
## # ... with 11,748 more rows, and 9 more variables:
## #   vert_index <dbl>, pitnumber <dbl>, species <chr>,
## #   length_1_mm <dbl>, length_2_mm <dbl>, weight_g <dbl>,
## #   clip <chr>, sampledate <date>, notes <chr>
```

We can plot the relationship between salamander mass and length.

```
ggplot(and_salamanders) +
  geom_point(aes(x = length_1_mm, y = weight_g), alpha = 0.4)
```

```
## Warning: Removed 5429 rows containing missing values
## (`geom_point()`).
```

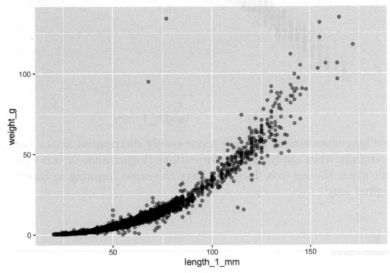

As illustrated by the alpha, we can observe most observations are concentrated where length is in the range [19, 100] and that observations are more dispersed in the range [100, 181]. A log transformation can help distribute the observations more evenly along the axis. We can apply this transformation to both axes as follows.

```
ggplot(and_salamanders) +
  geom_point(aes(x = length_1_mm, y = weight_g), alpha = 0.4) +
  scale_x_log10() +
  scale_y_log10()
```

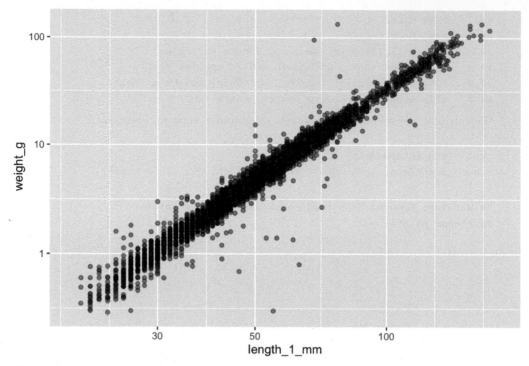

Observe how observations are now more evenly distributed across both axes. However, tread carefully when applying transformations (logarithmic or other) to an axis. Moving one unit along the log-axis does **not** have the same meaning as moving one unit along the original axis!

3.5   Numerical Variables

Many of the variables that data scientists study are *quantitative* or *numerical*, like the displacement and highway fuel efficiency, as we have seen before. Let us go back to the mpg data set and learn how to visualize its numerical values.

3.5.1   Prerequisites

As before, let us load tidyverse. We also make use of the patchwork library in this section to overlay multiple visualizations side-by-side.

```
library(tidyverse)
library(patchwork)
```

3.5.2 A slice of `mpg`

In this section we will draw graphs of the distribution of the numerical variable in the column hwy, which describes miles per gallon of car models on the highway. For simplicity, let us create a subset of the data frame that includes only the information we need.

```
mpg_sub <- select(mpg, manufacturer, model, hwy)
mpg_sub
```

```
## # A tibble: 234 x 3
##    manufacturer model       hwy
##    <chr>        <chr>     <int>
##  1 audi         a4           29
##  2 audi         a4           29
##  3 audi         a4           31
##  4 audi         a4           30
##  5 audi         a4           26
##  6 audi         a4           26
##  7 audi         a4           27
##  8 audi         a4 quattro   26
##  9 audi         a4 quattro   25
## 10 audi         a4 quattro   28
## # ... with 224 more rows
```

3.5.3 What is a histogram?

A *histogram* of a numerical variable looks very much like a bar chart, though it has some important differences that we will examine in this section. First, let us just draw a histogram of the highway miles per gallon.

The geom histogram generates a histogram of the values in a column. The histogram below shows the distribution of hwy.

```
ggplot(mpg_sub, aes(x = hwy)) +
  geom_histogram(fill = "darkcyan", color = "gray")
```

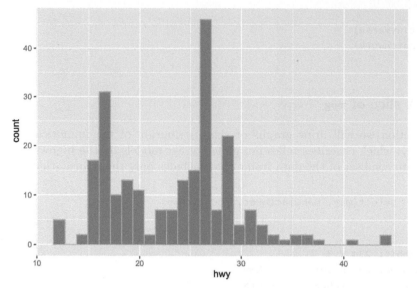

Note that, like bar charts, the mapping for the y-axis in the aesthetic is absent. This time, instead of internally computing a stat *count*, `ggplot` computes a stat *bin* called `stat_bin()`.

3.6   Histogram Shapes and Sizes

Histograms are useful for informing where the "bulk" of the data lies and can have different shapes. Let's have a look at some common shapes that occur in real data.

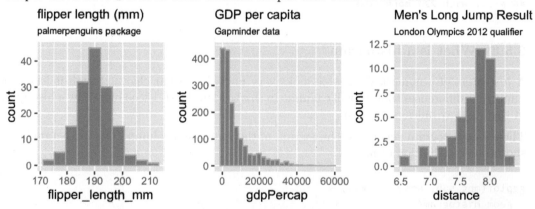

Plotting a histogram for penguin flipper lengths from the `palmerpenguins` package gives rise to a "bell-shaped" or symmetrical distribution that falls off evenly on both sides; the bulk of the distribution is clearly centered at the middle of the histogram, at around 190 mm.

Plotting GDP per capita data from the `gapminder` package gives rise to a distribution where the bulk of the countries have a GDP per capita less than $20,000 and countries with more are more extreme, especially at very high values of GDP per capita. Because of its distinctive appearance, distributions with this shape are called **right tailed** because the "mean" GDP per capita is "dragged" to the right in the direction of the tail.

The opposite is true for results of the men's long jump in the London Olympics 2012 qualifier. This histogram is generated from the `longjump` tibble in the `edsdata` package. In this case, we call such a distribution to be **left tailed** as the "mean" is pulled leftward.

3.6.1 The horizontal axis and bar width

In a histogram plot, we group the amounts into groups of contiguous (and thus, non-overlapping) intervals called bins. The histogram function of `ggplot` use the *left-out, right-in* convention in bin creation. What this means is that the interval between a value a and a value b, where $a < b$, includes b but not a. The convention does not apply the smallest bin, which has the left-end as well. Let us see an example.

Suppose we divide the interval from 0 to 100 into four bins of an equal size. The end points of the intervals are $0, 25, 50, 75,$ and 100. The following diagram demonstrates the situation:

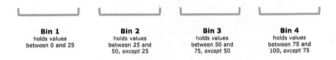

Bin 1 holds values between 0 and 25   **Bin 2** holds values between 25 and 50, *except 25*   **Bin 3** holds values between 50 and 75, *except 50*   **Bin 4** holds values between 75 and 100, *except 75*

We can write these intervals more concisely using open/close interval notion as: $[0, 25], (25, 50], (50, 75],$ and $(75, 100]$.

Suppose now that we are handed the following values:

$$\{0.0, 4.7, 5.5, 25.0, 25.5, 49.9, 50.0, 70.0, 72.2, 73.1, 74.4, 75.0, 99.0\}$$

If we filled in the bins with these values, the above diagram would look something like this:

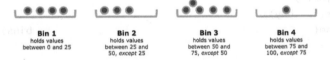

Bin 1 holds values between 0 and 25   **Bin 2** holds values between 25 and 50, *except 25*   **Bin 3** holds values between 50 and 75, *except 50*   **Bin 4** holds values between 75 and 100, *except 75*

Histograms are, in some sense, an extended version of the bar plot where the number of bins are adjustable. The heights represent *counts* (or *densities*, which we will see soon) and the widths are either the same or vary for each bar.

The end points of the bin-defining intervals are difficult to recognize just by looking at the chart. It is a little harder to see exactly where the ends of the bins are situated. For example, it is not easy to pinpoint exactly where the value 19 lies on the horizontal axis. Is it in the interval for the bin that stands on the line 20 on the x-axis or the one immediately to the left of it?

We can use, for a better visual assessment, a custom set of intervals as the bins. The specification of the bin set is by way of stating `breaks = BINS` as an argument in the call for `geom_histogram()`, where `BINS` is a sequence of breaking points.

Recall that we can define a numerical series from a number to another with fixed gap amount using function `seq()`. Below, we create a numerical series using `seq()` and then specify to use the sequence in the break points.

The sequence starts at 10 and ends at 50 with the gap of 1. Using the convention of end points in R, the sequence produces 40 intervals, from 10 to 11, from 11 to 12, ..., from 49 to 50, with the right end inclusive and the left end exclusive, except for the leftmost interval containing 10.

```
bins <- seq(10,50,1)
ggplot(mpg, aes(x = hwy)) +
  geom_histogram(fill = "darkcyan", color = "gray", breaks = bins)
```

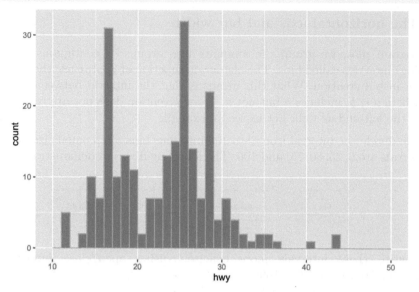

The tallest histogram bar is the one immediately to the right of the 25 white line on the x-axis, so it corresponds to the bin $(25, 26]$.

Let us try using a different step size, say 5.0.

```
bins <- seq(10,50,5)
ggplot(mpg, aes(x = hwy)) +
  geom_histogram(fill = "darkcyan", color = "gray", breaks = bins)
```

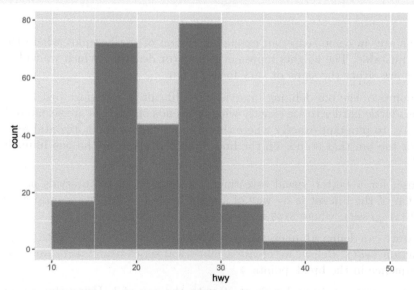

We observe two tall bars at $(15,20]$ and at $(25,30]$.

3.6.2   The counts in the bins

We can record the count calculation that geom_histogram carries out using the cut() function. The function takes an attribute and the bin intervals as its arguments and generates a table of counts.

The end point convention applies here, but in the data frame presentation, the description does not use the left square bracket for the lowest interval - it uses the left parenthesis like the other intervals.

Below, we create a new attribute bin that shows the bin name as its value using the cut(hwy, breaks = bins) call, then using count compute the frequency of each bin name in the bin attribute without dropping the empty bins in the counting result, and print the result on the screen.

```
bins <- seq(10,50,1)
binned <- mpg_sub |>
  mutate(bin = cut(hwy, breaks = bins)) |>
  count(bin, .drop = FALSE)
binned
```

```
## # A tibble: 40 x 2
##      bin          n
##      <fct>    <int>
##  1  (10,11]      0
##  2  (11,12]      5
##  3  (12,13]      0
##  4  (13,14]      2
##  5  (14,15]     10
##  6  (15,16]      7
##  7  (16,17]     31
##  8  (17,18]     10
##  9  (18,19]     13
## 10  (19,20]     11
## # ... with 30 more rows
```

Note that the label for 10 to 11 (appearing at the beginning) has the left parenthesis.

If you want to ignore the empty bins, change the option of .drop = FALSE to .drop = TRUE or take the option away.

```
bins <- seq(10,50,1)
binned <- mpg_sub |>
  mutate(bin = cut(hwy, breaks = bins)) |>
  count(bin, .drop = TRUE)
binned
```

```
## # A tibble: 27 x 2
##      bin        n
##      <fct>  <int>
##  1  (11,12]    5
##  2  (13,14]    2
##  3  (14,15]   10
```

```
##  4 (15,16]      7
##  5 (16,17]     31
##  6 (17,18]     10
##  7 (18,19]     13
##  8 (19,20]     11
##  9 (20,21]      2
## 10 (21,22]      7
## # ... with 17 more rows
```

We can try the alternate bin sequence, whose step size is 5.

```
bins2 <- seq(10,50,5)
binned2 <- mpg_sub |>
  mutate(bin = cut(hwy, breaks = bins2)) |>
  count(bin, .drop = TRUE)
binned2
```

```
## # A tibble: 7 x 2
##    bin          n
##    <fct>    <int>
## 1 (10,15]     17
## 2 (15,20]     72
## 3 (20,25]     44
## 4 (25,30]     79
## 5 (30,35]     16
## 6 (35,40]      3
## 7 (40,45]      3
```

3.6.3 Density scale

So far, the height of the bar has been the count, or the number of elements that are found in some bin. However, it can be useful to instead look at the *density* of points that are contained by some bin. When we plot a histogram in this manner, we say that it is in *density scale*.

In density scale, the height of each bar is the percent of elements that fall into the corresponding bin, *relative to the width of the bin*. Let us explain this using the following calculation.

Let the bin width be 5, which we will refer to as bin_width.

```
bin_width <- 5
```

The meaning of assigning 5 to the bin width is that each bin covers 5 consecutive units of the hwy value. Then we create a histogram with unit-size bins, divide the bars into consecutive groups of 5, and then even out the heights of the bars in each group.

More specifically, we execute the following steps.

- Using bin_width as a parameter, we create a sequence bins of bin boundaries from 10 to 50.

```
bins <- seq(10, 50, bin_width)
```

- Like before, when we "cut" the hwy values using bins as break points, we get the bins and their counts under variables bin and n, respectively.

```
binned <- mpg_sub |>
  mutate(bin = cut(hwy, breaks = bins)) |>
  count(bin, .drop = TRUE)
binned
```

```
## # A tibble: 7 x 2
##    bin          n
##    <fct>    <int>
## 1 (10,15]     17
## 2 (15,20]     72
## 3 (20,25]     44
## 4 (25,30]     79
## 5 (30,35]     16
## 6 (35,40]      3
## 7 (40,45]      3
```

- We obtain the proportions of the counts in the entire cars appearing in the tibble mpg_sub by dividing the counts by nrow(mpg_sub).

```
binned <- binned |>
  mutate(proportion = n/nrow(mpg_sub))
binned
```

```
## # A tibble: 7 x 3
##    bin          n proportion
##    <fct>    <int>      <dbl>
## 1 (10,15]     17     0.0726
## 2 (15,20]     72     0.308
## 3 (20,25]     44     0.188
## 4 (25,30]     79     0.338
## 5 (30,35]     16     0.0684
## 6 (35,40]      3     0.0128
## 7 (40,45]      3     0.0128
```

- For each bin, split the proportion in the bin among the 5 units the bin contains.

```
binned <- binned |>
  mutate(density = proportion/bin_width)
binned
```

```
## # A tibble: 7 x 4
##    bin          n proportion density
##    <fct>    <int>      <dbl>   <dbl>
## 1 (10,15]     17     0.0726  0.0145
```

```
## 2 (15,20]      72      0.308   0.0615
## 3 (20,25]      44      0.188   0.0376
## 4 (25,30]      79      0.338   0.0675
## 5 (30,35]      16      0.0684  0.0137
## 6 (35,40]       3      0.0128  0.00256
## 7 (40,45]       3      0.0128  0.00256
```

- Plot density against bin, and done!

```
ggplot(binned, aes(x = bin, y = density)) +
  geom_histogram(fill = "darkcyan", color = "gray",
                 stat = "identity")
```

```
## Warning in geom_histogram(fill = "darkcyan", color = "gray", stat = "identity"):
Ignoring unknown parameters: `binwidth`, `bins`, and
## `pad`
```

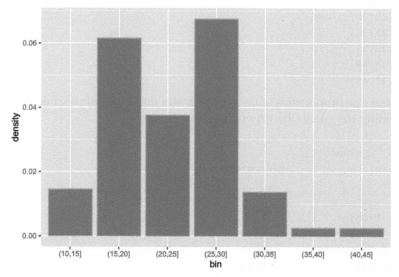

That was a lot of work and, fortunately, ggplot can take care of all that for us. The following code is a rewriting of the above where we omit the identity stat and instead map y onto density after being subject to the function after_stat.

```
ggplot(mpg, aes(x = hwy)) +
  geom_histogram(aes(y = after_stat(density)),
                 fill = "darkcyan", color = "gray", breaks = bins)
```

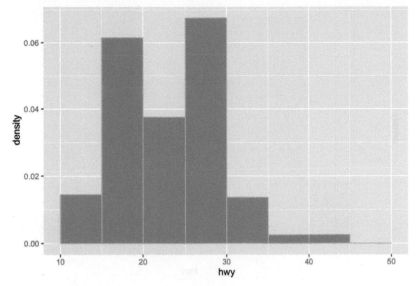

The function call to `after_stat` requires some explanation. In the above code, observe that we mapped y onto a variable `density`, but `density` is *not* a variable present in the `mpg` dataset! Histogram geoms, like bar geoms, internally compute new variables like `count` and `density`. By specifying `density` in the function call `after_stat`, we flag to `ggplot` that evaluation of this aesthetic mapping should be deferred until after the stat transformation has been computed.

Compare this with the density plot crafted by hand. Observe that the density values we have calculated match the height of the bars.

3.6.4   Why bother with density scale?

There are some discrepancies to note between a histogram in density scale and a histogram with count scale. While the count scale may be easier to digest visually than density scale, the count scale can be misleading when using bins with different widths. The problem: the height of each bar does *not* account for the difference in the widths of the bins.

Suppose the values in `hwy` are binned into three uneven categories.

```
uneven_bins <- c(10, 15, 30, 45)
ggplot(mpg, aes(x = hwy)) +
  geom_histogram(aes(y = after_stat(density)),
                 fill = "darkcyan", color = "grey",
                 breaks = uneven_bins, position = "identity")
```

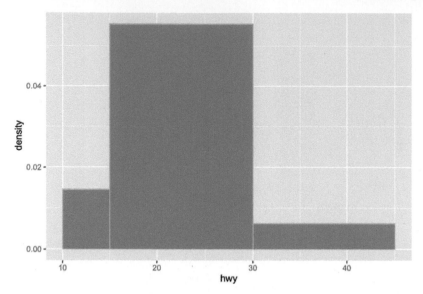

Here are the counts in the three bins. Observe that the first and last bins have roughly the same number of elements.

```
mpg_sub |>
  mutate(bin = cut(hwy, breaks = uneven_bins)) |>
  count(bin, .drop = FALSE)
```

```
## # A tibble: 3 x 2
##   bin         n
##   <fct>     <int>
## 1 (10,15]     17
## 2 (15,30]    195
## 3 (30,45]     22
```

Let us compare the following two histograms that use uneven_bins.

```
g1 <- ggplot(mpg, aes(x = hwy)) +
  geom_histogram(fill = "darkcyan", color = "grey",
                 breaks = uneven_bins, position = "identity")

uneven_bins <- c(10, 15, 30, 45)

g2 <- ggplot(mpg, aes(x = hwy)) +
  geom_histogram(aes(y = after_stat(density)),
                 fill = "darkcyan", color = "grey",
                 breaks = uneven_bins, position = "identity")

g1 + g2
```

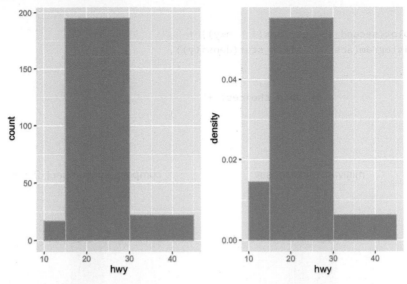

Note how the histogram in count scale *exaggerates* the height of the (30, 45] bar. The height shown is simply the number of car models in that bin with no regard to the width of the bin. While both the [10, 15] and (30, 45] bars may have the same number of car models in the bin, the *density* of the [10, 15] bar is greater because there are more elements contained by a smaller bin width. Put another way, the (30, 45] bar can provide more coverage because it is so "spread out", i.e., its bin width is much larger than that of the [10, 15] bar.

For this reason, we will prefer to plot our histograms in density scale rather than count scale.

3.6.5 Density scale makes direct comparisons possible

Histograms in density scale allow comparisons to be made between histograms generated from datasets of different sizes or make different bin choices. Consider the following two subsets of car models in mpg.

```
first_subset <- mpg |>
  filter(class %in% c("minivan", "midsize", "suv"))
second_subset <- mpg |>
  filter(class %in% c("compact", "subcompact", "2seater"))
```

We can generate a histogram in density scale for each subset. Note that both histograms apply the same bin choices given in the following bin_choices.

```
bin_choices <- c(10, 15, 20, 22, 27, 30, 45)

g1 <- ggplot(first_subset, aes(x = hwy)) +
  geom_histogram(aes(y = after_stat(density)),
                 fill = "darkcyan",
                 color = "grey",
                 breaks = bin_choices) +
  labs(title = "minivan, midsize, suv")
```

```
g2 <- ggplot(second_subset, aes(x = hwy)) +
  geom_histogram(aes(y = after_stat(density)),
                 fill = "darkcyan",
                 color = "grey",
                 breaks = bin_choices) +
  labs(title = "compact, subcompact, 2seater")


g1 + g2
```

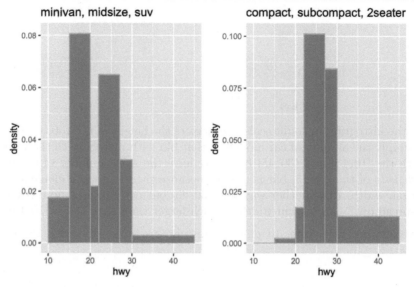

For instance, we can immediately tell that the density of observations in the (30, 45] bin is greater in the second subset than in the first. We can also glean that the "bulk" of car models in the first subset is concentrated more in the range [10, 30] than in the second subset.

We can confirm this by overlaying a "smoothed" curve atop each of the histograms. Let us add a new geom to our ggplot code using a *density* geom.

```
g1 <- ggplot(first_subset, aes(x = hwy)) +
  geom_histogram(aes(y = after_stat(density)),
                 fill = "white", color = "grey",
                 breaks = bin_choices) +
  geom_density(adjust = 3, fill = "purple", alpha = 0.3) +
  labs(title = "minivan, midsize, suv")

g2 <- ggplot(second_subset, aes(x = hwy)) +
  geom_histogram(aes(y = after_stat(density)),
                 fill = "white", color = "grey",
                 breaks = bin_choices) +
  geom_density(adjust = 3, fill = "purple", alpha = 0.3) +
  labs(title = "compact, subcompact, 2seater")


g1 + g2
```

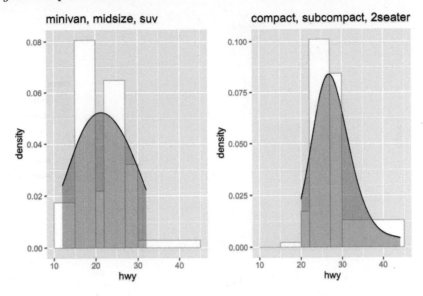

The area beneath the curve is a percentage and tells us where the "bulk" of the data is. We can see that the top of the curve is shifted to the left for the first subset when compared to the second subset. Overlaying a density geom over a histogram geom is possible only when the histograms are drawn in density scale.

3.6.6 Histograms and positional adjustments

As with bar charts and scatter plots, we can use positional adjustments with histograms.

Below, we use bins of width 5 and then color the portions of the bars according to the classes. To make the breakdown portions appear on top of each other, we use the `position = "stack"` adjustment.

```
bins <- seq(10,50,5)
ggplot(mpg, aes(x = hwy)) +
  geom_histogram(aes(fill = class), breaks = bins,
                 position = "stack")
```

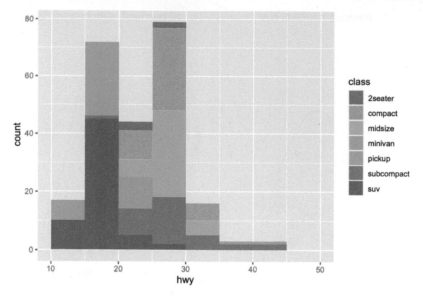

Instead of stacking, the bars can be made to appear side by side.

```
bins <- seq(10,50,5)
ggplot(mpg, aes(x = hwy)) +
  geom_histogram(aes(fill = class), breaks = bins,
                 position = "dodge")
```

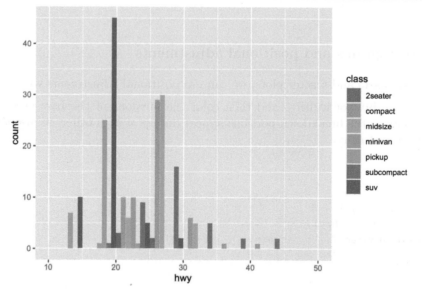

3.7   Drawing Maps

Sometimes our dataset contains information about geographical quantities, such as latitude, longitude, or a physical area that corresponds to a "landmark" or region of interest. This section explores methods from ggplot2 and a new package called tigris for downloading and working with *spatial* data.

3.7.1 Prerequisites

We will make use of several packages in this section.

```
library(tidyverse)
library(patchwork)
library(mapview)
library(tigris)
```

We also use tigris, patchwork, and mapview, so we load in these packages as well.

3.7.2 Simple maps with polygon geoms

ggplot2 provides functionality for drawing maps. For instance, the following dataset from ggplot2 contains longitude and latitude points that correspond to the mainland United States.

```
us <- map_data("state") |>
  as_tibble() |>
  select(long, lat, group, region)
us
```

```
## # A tibble: 15,537 x 4
##      long   lat group region
##     <dbl> <dbl> <dbl> <chr>
##  1 -87.5  30.4      1 alabama
##  2 -87.5  30.4      1 alabama
##  3 -87.5  30.4      1 alabama
##  4 -87.5  30.3      1 alabama
##  5 -87.6  30.3      1 alabama
##  6 -87.6  30.3      1 alabama
##  7 -87.6  30.3      1 alabama
##  8 -87.6  30.3      1 alabama
##  9 -87.7  30.3      1 alabama
## 10 -87.8  30.3      1 alabama
## # ... with 15,527 more rows
```

We can plot a map of the United States using a polygon geom where the x and y aesthetics are mapped to the longitude and latitude positions, respectively. We add a coordinate system layer to project this portion of the earth onto a 2D coordinate plane.

```
us_map <- ggplot(us) +
  geom_polygon(aes(x = long, y = lat, group = group),
               fill = "white", color = "black") +
  coord_quickmap()
us_map
```

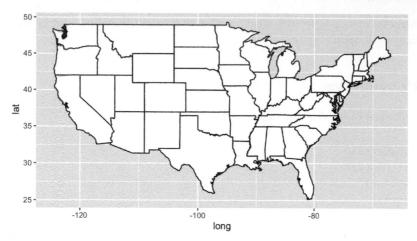

For instance, we can annotate the map with the position track of Atlantic storms in 2006 using the `storms` dataset provided by `dplyr`.

```
storms2006 <- storms |>
  filter(year == 2006)
storms2006
```

```
## # A tibble: 190 x 13
##      name  year month   day  hour   lat  long status
##     <chr> <dbl> <dbl> <int> <dbl> <dbl> <dbl> <chr>
##  1 Zeta   2006     1     1     0  25.6 -38.3 tropical sto~
##  2 Zeta   2006     1     1     6  25.4 -38.4 tropical sto~
##  3 Zeta   2006     1     1    12  25.2 -38.5 tropical sto~
##  4 Zeta   2006     1     1    18  25   -38.6 tropical sto~
##  5 Zeta   2006     1     2     0  24.6 -38.9 tropical sto~
##  6 Zeta   2006     1     2     6  24.3 -39.7 tropical sto~
##  7 Zeta   2006     1     2    12  23.8 -40.4 tropical sto~
##  8 Zeta   2006     1     2    18  23.6 -40.8 tropical sto~
##  9 Zeta   2006     1     3     0  23.4 -41   tropical sto~
## 10 Zeta   2006     1     3     6  23.3 -41.3 tropical sto~
## # ... with 180 more rows, and 5 more variables:
## #   category <ord>, wind <int>, pressure <int>,
## #   tropicalstorm_force_diameter <int>,
## #   hurricane_force_diameter <int>
```

Observe how the `storms2006` contains latitude (`lat`) and longitude (`long`) positions. We can generate an overlaid scatter plot using these positions. Note how this dataset is given a specification at the geom layer, which is different from the dataset used for creation of the ggplot object in `us_map`.

```
us_map +
  geom_point(data = storms2006,
             aes(x = long, y = lat, color = name))
```

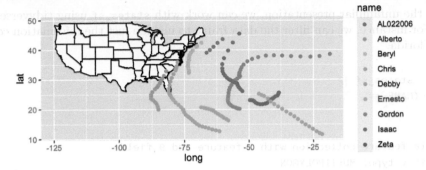

The visualization points out three storms (Alberto, Ernesto, and Beryl) that made landfall in the United States.

3.7.3 Shape data and simple feature geoms

tigris is a package that allows users to download TIGER/Line shape data directly from the US Census Bureau website. For instance, we can retrieve shape data pertaining to the United States.

```
states_sf <- states(cb = TRUE)
```

```
## Retrieving data for the year 2020
```

The function defaults to retrieving data from the most recent year available (currently, 2020). The result is a "simple feature collection" expressed as a data frame.

```
states_sf
```

```
## Simple feature collection with 2 features and 9 fields
## Geometry type: MULTIPOLYGON
## Dimension:     XY
## Bounding box:  xmin: -109.0502 ymin: 17.88328 xmax: -65.2207 ymax: 37.00023
## Geodetic CRS:  NAD83
##   STATEFP   STATENS    AFFGEOID GEOID STUSPS        NAME
## 1      35 00897535 0400000US35    35     NM  New Mexico
## 2      72 01779808 0400000US72    72     PR Puerto Rico
##   LSAD        ALAND      AWATER
## 1   00 314198560935   726482115
## 2   00   8868948653  4922329963
##                         geometry
## 1 MULTIPOLYGON (((-109.0502 3...
## 2 MULTIPOLYGON (((-65.23805 1...
```

The data frame gives one row per state/territory. The shape information in the variable geometry is a polygon corresponding to that geographical area. It is a representation of a shape by a clockwise enumeration of the corner location. By drawing a straight line between each pair of neighboring points in the enumeration between the last and the first, you can draw a shape and that shape is an approximation of the region of interest.

Despite the unfamiliar presentation, we can work with states_sf using tidyverse tools like dplyr. For instance, we can filter the data frame to include just the information corresponding to Florida.

```
fl_sf <- states_sf |>
  filter(NAME == "Florida")
fl_sf
```

```
## Simple feature collection with 1 feature and 9 fields
## Geometry type: MULTIPOLYGON
## Dimension:     XY
## Bounding box:  xmin: -87.63494 ymin: 24.5231 xmax: -80.03136 ymax: 31.00089
## Geodetic CRS:  NAD83
##   STATEFP  STATENS    AFFGEOID GEOID STUSPS    NAME LSAD
## 1      12 00294478 0400000US12    12     FL Florida   00
##          ALAND      AWATER                            geometry
## 1 138958484319 45975808217 MULTIPOLYGON (((-80.17628 2...
```

The shape data in the geometry variable can be visualized using a simple feature geom ("sf" for short). The sf geom can automatically detect the presence of a geometry variable stored in a data frame and draw a map accordingly.

```
fl_sf |>
  ggplot() +
  geom_sf() +
  theme_void()
```

We can refine the granularity of the shape data to the county level. Here, we collect shape data for all Florida counties.

```
fl_county_sf <- counties("FL", cb = TRUE)
```

The Census Bureau makes available cartographic boundary shape data, which often look better when doing thematic mapping. We retrieve these data by setting the cb flag to TRUE.

Here is a preview of the county-level data.

```
fl_county_sf
```

```
## Simple feature collection with 2 features and 12 fields
## Geometry type: MULTIPOLYGON
## Dimension:     XY
## Bounding box:  xmin: -82.85243 ymin: 27.82206 xmax: -80.44697 ymax: 28.79132
## Geodetic CRS:  NAD83
##    STATEFP COUNTYFP COUNTYNS       AFFGEOID GEOID     NAME
## 1       12      009 00295749 0500000US12009 12009 Brevard
## 2       12      101 00295739 0500000US12101 12101   Pasco
##          NAMELSAD STUSPS STATE_NAME LSAD      ALAND
## 1 Brevard County     FL    Florida   06 2628762626
## 2   Pasco County     FL    Florida   06 1933733392
##        AWATER                      geometry
## 1 1403940953 MULTIPOLYGON (((-80.98725 2...
## 2  694477432 MULTIPOLYGON (((-82.80493 2...
```

This time, the data frame gives one row per Florida county. We can use the same `ggplot` code to visualize the county-level shape data.

```
fl_county_sf |>
  ggplot() +
  geom_sf() +
  theme_void()
```

3.7.4 Choropleth maps

In a *choropleth* map, regions on a map are colored according to some quantity associated with that region. For instance, we can map color to the political candidate who won the most votes in the 2020 US presidential elections in each county.

The tibble `pres_election` contains county presidential election returns from 2000-2020. Let us filter the data to the 2020 election returns in Florida and, to see how the election map has changed over time, also collect the returns from the 2008 election.

```
library(edsdata)
fl_election_returns <- election |>
  filter(year %in% c(2008, 2020), state_po == "FL")
fl_election_returns
```

```
## # A tibble: 536 x 12
##     year state   state_po county_~1 count~2 office candi~3
##    <dbl> <chr>   <chr>    <chr>     <chr>   <chr>  <chr>
## 1   2008 FLORIDA FL       ALACHUA   12001   US PR~ BARACK~
## 2   2008 FLORIDA FL       ALACHUA   12001   US PR~ JOHN M~
## 3   2008 FLORIDA FL       ALACHUA   12001   US PR~ OTHER
## 4   2008 FLORIDA FL       BAKER     12003   US PR~ BARACK~
## 5   2008 FLORIDA FL       BAKER     12003   US PR~ JOHN M~
## 6   2008 FLORIDA FL       BAKER     12003   US PR~ OTHER
## 7   2008 FLORIDA FL       BAY       12005   US PR~ BARACK~
## 8   2008 FLORIDA FL       BAY       12005   US PR~ JOHN M~
## 9   2008 FLORIDA FL       BAY       12005   US PR~ OTHER
## 10  2008 FLORIDA FL       BRADFORD  12007   US PR~ BARACK~
## # ... with 526 more rows, 5 more variables: party <chr>,
## #   candidatevotes <dbl>, totalvotes <dbl>,
## #   version <dbl>, mode <chr>, and abbreviated variable
## #   names 1: county_name, 2: county_fips, 3: candidate
```

We apply some `dplyr` work to determine the candidate winner in each county. We convert the `candidate` variable to a *factor* so that the colors appear consistently with respect to political party in the following visualization.

```
candidates_levels <- c("DONALD J TRUMP", "JOHN MCCAIN",
                       "BARACK OBAMA", "JOSEPH R BIDEN JR")

fl_county_winner <- fl_election_returns |>
  group_by(year, county_name) |>
  slice_max(candidatevotes) |>
  ungroup() |>
  mutate(candidate = factor(candidate,
                    levels = candidates_levels))
fl_county_winner
```

```
## # A tibble: 134 x 12
##     year state   state_po county_~1 count~2 office candi~3
##    <dbl> <chr>   <chr>    <chr>     <chr>   <chr>  <fct>
## 1   2008 FLORIDA FL       ALACHUA   12001   US PR~ BARACK~
## 2   2008 FLORIDA FL       BAKER     12003   US PR~ JOHN M~
## 3   2008 FLORIDA FL       BAY       12005   US PR~ JOHN M~
## 4   2008 FLORIDA FL       BRADFORD  12007   US PR~ JOHN M~
## 5   2008 FLORIDA FL       BREVARD   12009   US PR~ JOHN M~
## 6   2008 FLORIDA FL       BROWARD   12011   US PR~ BARACK~
```

```
## 7   2008 FLORIDA FL      CALHOUN   12013   US PR~ JOHN M~
## 8   2008 FLORIDA FL      CHARLOTTE 12015   US PR~ JOHN M~
## 9   2008 FLORIDA FL      CITRUS    12017   US PR~ JOHN M~
## 10  2008 FLORIDA FL      CLAY      12019   US PR~ JOHN M~
## # ... with 124 more rows, 5 more variables: party <chr>,
## #    candidatevotes <dbl>, totalvotes <dbl>,
## #    version <dbl>, mode <chr>, and abbreviated variable
## #    names 1: county_name, 2: county_fips, 3: candidate
```

We can join the county shape data with the county-level election returns.

```
with_election <- fl_county_sf |>
  mutate(NAME = str_to_upper(NAME)) |>
  left_join(fl_county_winner, by = c("NAME" = "county_name"))
```

We can use the sf geom to plot the shape data and map the fill aesthetic to the candidate.

```
with_election |>
  filter(year == 2020) |>
  ggplot() +
  theme_void() +
  geom_sf(aes(fill = candidate))
```

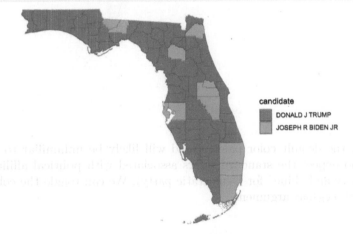

How do these results compare with the 2008 presidential election? The patchwork library allows us to panel two ggplot figures side-by-side.

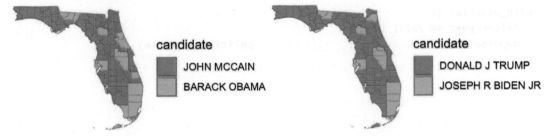

3.7.5   Interactive maps with `mapview`

`mapview` is a powerful package that can be used to generate an interactive map. We can use it to visualize the county-level data annotated with the 2020 election returns.

```
with_election |>
  filter(year == 2020) |>
  mapview(zcol = "candidate")
```

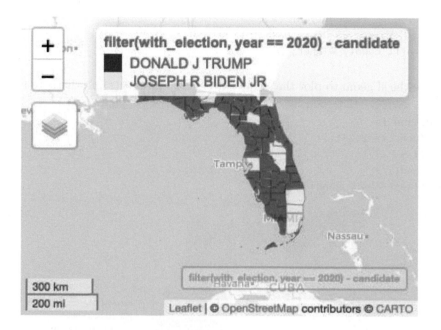

Unfortunately, the default color palette used will likely be unfamiliar to users of our visualization who expect the standard colors associated with political affiliation ("red" for Republican party and "blue" for Democratic party). We can toggle the colors used by setting the the `col.regions` argument.

```
political_palette <- colorRampPalette(c('red', 'blue'))

with_election |>
  filter(year == 2020) |>
  mapview(zcol = "candidate", col.regions = political_palette)
```

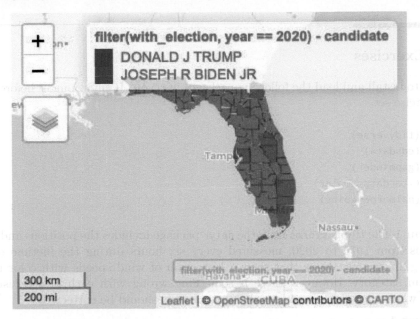

We can also zoom in on a particular county and show the results for just those areas.

```
with_election |>
  filter(year == 2020, NAME %in% c("HILLSBOROUGH", "MANATEE")) |>
  mapview(zcol = "candidate", col.regions = political_palette)
```

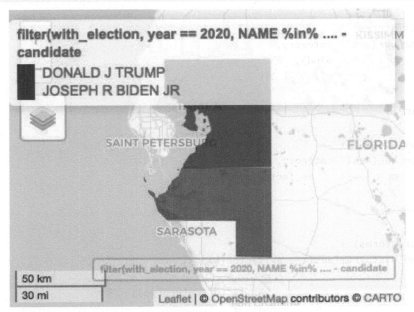

For instance, this map reveals election results for the Tampa Bay and Brandenton areas.

3.8   Exercises

Be sure to install and load the following packages into your R environment before beginning this exercise set.

```
library(tidyverse)
library(edsdata)
library(gapminder)
library(lterdatasampler)
library(palmerpenguins)
```

Question 1 The tibble `storms` from the `dplyr` package includes the positions and attributes of storms from 1975 to 2020, measured every six hours during the lifetime of a storm. Following are visualizations that show a histogram of wind speeds with color mapped to the storm category. However, there is something wrong with each of the visualizations. Explain what went wrong and how the `ggplot2` code should be corrected.

- **Figure 1**

```
ggplot(storms) +
  geom_bar(aes(x=wind,fill=category))
```

- **Figure 2**

```
ggplot(storms) +
  geom_bar(aes(x=category,
               y=wind),
          stat="identity")
```

- **Figure 3**

```
ggplot(storms) +
  geom_bar(aes(x=category,
               fill=as_factor(wind)))
```

- **Figure 4**

```
ggplot(storms) +
  geom_histogram(aes(x = as.factor(wind),
                     y = "Count",
                     fill = as.factor(category)),
                 stat='identity')
```

Question 2 The tibble `penguins` from the package `palmerpenguins` includes measurements for penguin species, island in Palmer Archipelago, size, and sex.

- **Question 2.1** Are any of these categorical variables and, if so, what kind (e.g., nominal, ordinal, or binary)? Are there any numerical variables and, if so, what kind (e.g., continuous, or discrete)?

- **Question 2.2** Generate a `ggplot` visualization that shows the number of penguins on each island. Fill your bars according to penguin species. Do the islands contain roughly the same proportion of each species?

- **Question 2.3** By default, `ggplot` uses the "stack" positional adjustment. Modify your code from **Question 2.2** to use different positional adjustments. Try "dodge" and then "identity" with an amount of alpha. Which adjustment allows you to address **Question 2.2** most effectively?

Question 3 The dataset `longjump` from the package `edsdata` contains results from the qualifier and finals in the men's long jump event in the 2012 Summer Olympic Games.

```
longjump
```

```
## # A tibble: 94 x 7
##     year event        rank name        country dista~1 status
##    <dbl> <chr>       <dbl> <chr>       <chr>     <dbl> <chr>
##  1  2012 qualifier       1 Mauro Vin~  Brazil     8.11 Q
##  2  2012 qualifier       2 Marquise ~  United~     8.11 Q
##  3  2012 qualifier       3 Aleksandr~  Russia     8.09 q
##  4  2012 qualifier       4 Greg RUTH~  Great ~    8.08 q
##  5  2012 qualifier       5 Christoph~  Great ~    8.06 q
##  6  2012 qualifier       6 Michel TO~  Sweden     8.03 q
##  7  2012 qualifier       7 Godfrey K~  South ~    8.02 q
##  8  2012 qualifier       8 Will CLAYE  United~     7.99 q
##  9  2012 qualifier       9 Mitchell ~  Austra~     7.99 q
## 10  2012 qualifier      10 Tyrone SM~  Bermuda    7.97 q
## # ... with 84 more rows, and abbreviated variable name
## #   1: distance
```

- **Question 3.1** Form a tibble called that contains only the results for the London 2012 Olympic Games.

- **Question 3.2** Create a histogram of the distances in the qualifier event for the London 2012 Olympic Games. Fill your bars using the `status` variable so you can see the bands of color corresponding to qualification status. Missing values correspond to participants who did not qualify.

- **Question 3.3** Repeat *Question 3.2* but make a histogram of the distances in the final event.

- **Question 3.4** Adjust your code in **Question 3.2** and **Question 3.3** to include the `identity` positional adjustment. You may wish to set an alpha[4] as well to better distinguish the differences. What do you observe when including/not including this adjustment?

- **Question 3.5** Following are some statements about the above two distributions. Select those that are *FALSE* by including its corresponding number in the following *vector* `jump_answers`.

    1. We used histograms because both of these variables are categorical.
    2. Both of these distributions are skewed.

---

[4]https://ggplot2.tidyverse.org/reference/aes_colour_fill_alpha.html#alpha

3. We observe the histogram for the qualifier event follows a left-tailed distribution.

4. We can color the different category storms using the variable `status` because it is a numerical variable.

- **Question 3.6** The following code visualizes a map of the world. Annotate this map with the countries that participated in the men's long jump event in the London 2012 Games using a point or polygon geom (use `long` for x and `lat` for y). You will first need to join the map data in `world` with `longjump2012`. Then extend the plot in the name `world_map` by adding a new geom layer; set the data for this new layer to use the joined data.

```
world <- map_data("world") |>
  mutate(name = region) |>
  select(long, lat, group, name)

world_map <- ggplot(world) +
  geom_polygon(aes(x = long, y = lat, group = group),
               fill = "white", color = "grey50") +
  coord_quickmap()
world_map
```

- **Question 3.7** Which country in South America participated in the men's long jump in the 2012 London Games? Indicate your answer by setting the name `north_america_participant` to the appropriate thing.

Question 4 In **Question 7** from Chapter 2 we computed the average annual compensation of New York local authorities. However, the average does not tell us everything about the amounts employees are paid. It is possible that only a few employees make the bulk of the money, even among this select group. We can use a *histogram* to visualize information about a set of numbers. We have prepared a tibble `nysalary_cleaned` that already contains the cleaned compensation data; recall the the `Total Compensation ($)` variable is in *tens of thousands* of dollars.

```
nysalary_cleaned
```

```
## # A tibble: 1,676 x 20
##    Authority~1 Fisca~2 Last ~3 Middl~4 First~5 Title Group
##    <chr>       <chr>   <chr>   <chr>   <chr>   <chr> <chr>
##  1 Albany Cou~ 12/31/~ Adding~ L       Ellen   Seni~ Admi~
##  2 Albany Cou~ 12/31/~ Boyea   <NA>    Kelly   Conf~ Admi~
##  3 Albany Cou~ 12/31/~ Calder~ <NA>    Philip  Chie~ Exec~
##  4 Albany Cou~ 12/31/~ Cannon  <NA>    Matthew Gove~ Admi~
##  5 Albany Cou~ 12/31/~ Cerrone A       Rima    Budg~ Mana~
##  6 Albany Cou~ 12/31/~ Chadde~ M       Helen   Mark~ Mana~
##  7 Albany Cou~ 12/31/~ Charla~ M       Elizab~ Dire~ Mana~
##  8 Albany Cou~ 12/31/~ Dickson C       Sara    Acco~ Admi~
##  9 Albany Cou~ 12/31/~ Finnig~ <NA>    James   Oper~ Admi~
## 10 Albany Cou~ 12/31/~ Greenw~ <NA>    Kathryn Dire~ Mana~
## # ... with 1,666 more rows, 13 more variables:
## #   Department <chr>, `Pay Type` <chr>,
## #   `Exempt Indicator` <chr>,
## #   `Base Annualized Salary` <chr>,
```

```
## #  `Actual Salary Paid` <chr>, `Overtime Paid` <chr>,
## #  `Performance Bonus` <chr>, `Extra Pay` <chr>,
## #  `Other Compensation` <chr>, ...
```

- **Question 4.1** Make a histogram of the compensation of employees in nysalary_cleaned in *density scale.* Use the sequence c(seq(0, 10,2), 15, 20, 40) for your bin breaks. Assign the resulting ggplot object to a name g.

  The later bins have very few individuals so the bar heights become too short to make out visually. The following code chunk overlays your histogram with a geom text layer that annotates each bin with the corresponding density in that bin.

  ```
  g +
    geom_text(stat = "bin", aes(y = stat(density),
                          label = round(stat(density),5)),
              vjust = -0.2, size=2, breaks=c(seq(0, 10,2), 15, 20, 40))
  ```

- **Question 4.2** Using the histogram, how many employees had a total compensation of more than 100K in the 2020 fiscal year? Answer the question manually using the density formulas presented in the textbook. You will need to do some arithmetic to find the answer; R can be used as a calculator.

- **Question 4.3** Answer the same question using dplyr code. Give your answer as a tibble containing a single row and a single column named n. Store this in a tibble called employees_more_than_100k.

- **Question 4.4** Do most New York employees make around the same amount, or are there some who make a lot more than the rest?

Question 5 The tibble gapminder from the gapminder library gives data on life expectancy, GDP per capita, and population by country. In this exercise we will visualize the relationship between life expectancy and GDP per capita.

- **Question 5.1** Form a tibble called gapminder_relevant that contains data only for Asia and Europe in the year 1987.

- **Question 5.2** On one graph, create a scatter plot of life expectancy versus GDP per capita. Color the points according to continent and vary the size of the points according to population.

- **Question 5.3** The textbook discussed using numeric position scales to transform an axis in a plot. Modify your ggplot2 code from **Question 5.2** to include a log transformation on the x-axis using scale_x_log10().

- **Question 5.4** Using the above two visualizations, select which of the following statements can be correctly inferred by including them in the following *vector* gap_answers. For those that you did not select, if any, explain why the statement cannot be made.

  - We observe more populous countries in the "Europe" cluster than in the "Asia" cluster.
  - Countries with higher GDP per capita are associated with higher life expectancies, and we can also observe that countries in Asia are more correlated with lower GDP per capita than countries in Europe in 1987.
  - Moving a unit of distance along the x-axis has the same effect in both visualizations.
  - The GDP per capita for the majority of countries in Asia are practically 0 according to the first visualization.

- **Question 5.5** The tibble `gapminder` contains five variables: `country`, `continent`, `year`, `lifeExp`, and `pop`. Are any of these categorical variables and, if so, what kind (e.g., nominal, ordinal, or binary)? Are there any numerical variables and, if so, what kind (e.g., continuous, or discrete)?

Question 6 This question is a continuation of the 2017 Australian Marriage Law Postal Survey[5] examined in **Question 10** from Chapter 2.

- **Question 6.1** After forming the tibble `with_response`, give `ggplot2` code that reproduces the following plot.

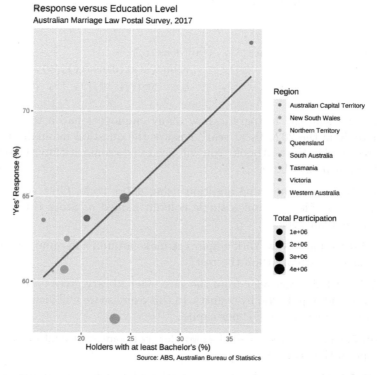

- **Question 6.2** Based on this figure, how would you respond to the statement:

"Because support for same-sex couples increases in areas with higher percentage of holders with at least Bachelor's degrees, we can say that if a state/territory has a higher percentage of holders with at least a Bachelor's degree, then the more that state/territory will support same-sex marriage."

Is this a fair statement? Why or why not?

[5]https://www.abs.gov.au/ausstats/abs@.nsf/mf/1800.0

Question 7 The dataset `ntl_icecover` from the package `lterdatasampler` gives data on ice freeze and thaw dates for lakes in the Madison, Wisconsin area from 1853 to 2019. The data includes lake names, dates of freeze-up and thaw, and duration of ice cover for Lakes Mendota and Monona. Pull up the help (`?ntl_icecover`) for more information on this dataset.

- **Question 7.1** The variable `ice_on` gives the freeze date for a given lake. Add a new variable to the tibble `ntl_icecover` called `days_to_freeze` that gives the number of days to the freeze date from the start of the calendar year. Assign the resulting tibble to the name `with_freeze`.

  HINT: The functions `year` and `yday` from the package `lubridate` can be helpful for determining the year and the day of the year from a date, respectively. Also note that for some years the freeze date may not occur until the next calendar year, e.g., the freeze date in Lake Mendota in 1875 was January 10, 1876. Calling `yday()` on this date would yield 10 when the correct figure is actually 375. Adjust the expression used in your `mutate()` call accordingly.

- **Question 7.2** Generate line plots showing days to the freeze date versus year, one for Lake Mendota and another for Lake Monona. The line plots should be given in a single overlaid figure and colored accordingly. Does there appear to be an association between the two variables?

- **Question 7.3** Let us place an arbitrary marker at the year 1936, approximately the half-way point between the year when data collection started and ended. Add a Boolean variable to `icecover_with_temp` called `before1936` that flags whether the year is before 1936. Then generate an overlaid histogram showing the distribution of annual mean air temperature for the periods 1853–1936 and 1937–2019.

- **Question 7.4** By comparing the "bulk" of the data in the distributions following the periods 1853–1936 and 1937–2019, does it appear that there are more days until the freezing date in recent history? Or is it more or less the same across both periods?

Question 8 Economists John R. Lott and Carlisle E. Moody published a widely circulated article in March 2020[6] on mass gun violence in the United States and the rest of the world. Its appendix references the following figure:

---

[6]https://econjwatch.org/articles/brought-into-the-open-how-the-us-compares-to-other-countries-in-the-rate-of-public-mass-shooters

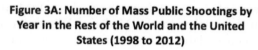

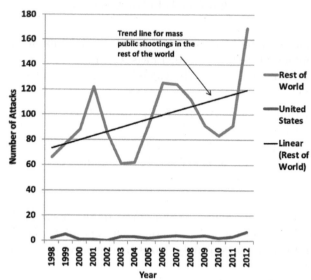

Read Lott and Moody's paper[7] and then the response paper[8] written by economist Lankford on potential flaws in the data, specifically regarding whether public mass shootings and other types of attacks, e.g., acts of terrorism, can be viewed as a single form of violence. We will try to recreate this plot as well as we can, and then address some questions about the reasonableness of this visualization.

- **Question 8.1** Use Lankford's spreadsheet data made available here[9] in his response paper[10]. This is the Lott and Moody data annotated with additional variables by Lankford.

  Load the sheets "Foreign Cases 1448" and ""US Cases_43" into a tibble using the `read_xlsx()` function from the `readxl` package. Assign these tibbles to the names `foreign` and `usa` which correspond to the foreign and United States figures, respectively.

- **Question 8.2** What is the observational unit in each of these datasets? Note briefly some of the measurements corresponding to this observational unit.

- **Question 8.3** Form a tibble named `usa_relevant` from the United States figures that gives the year of the incident, the country, and Lankford's two Boolean variables indicating (1) if the perpetrator killed 4 or more victims and (2) if the attack was committed by member(s) of a terrorist organization or genocidal group.

- **Question 8.4** Repeat **Question 8.3**, but this time for the figures corresponding to the rest of the world. Assign the resulting tibble to the name `foreign_relevant`.

  Run the following code chunk. It applies some basic preprocessing necessary to complete the following steps:

---

[7]https://econjwatch.org/articles/brought-into-the-open-how-the-us-compares-to-other-countries-in-the-rate-of-public-mass-shooters

[8]https://econjwatch.org/articles/the-importance-of-analyzing-public-mass-shooters-separately-from-other-attackers-when-estimating-the-prevalence-of-their-behavior-worldwide

[9]https://econjwatch.org/file_download/1131/LankfordMar2020AppendixB.xlsx

[10]https://econjwatch.org/articles/the-importance-of-analyzing-public-mass-shooters-separately-from-other-attackers-when-estimating-the-prevalence-of-their-behavior-worldwide

```
foreign_relevant <- foreign_relevant |>
  drop_na(Year) |>
  filter(if_any(everything(), function(x) x != "x")) |>
  mutate(
    Country = as.factor(Country),
    across(where(is.character), as.numeric))
usa_relevant <- usa_relevant |>
  drop_na(Year)
```

- **Question 8.5** Form a tibble named `full_relevant` that contains both the figures corresponding to the United States (in `usa_relevant`) and the rest of the world (in `foreign_relevant`).

- **Question 8.6** Lott and Moody's visualization draws a comparison between the United States and the rest of the world. Add a new column to `full_relevant` called `is_usa` where each value in the column is `TRUE` if the country is "United States" and `FALSE` otherwise. Assign the resulting tibble back to the name `full_relevant`.

- **Question 8.7** Form a tibble named `summarized_attacks` that gives the number of incidents that occurred in each year, with respect to the location of the incident (in the United States or in the rest of the world). Three variables should be present in the resulting tibble: `Year`, `is_usa`, and `n` (the count).

- **Question 8.8** Finally, time to visualize! We have everything we need to recreate Lott and Moody's visualization. Using what you know about `ggplot` and the layered grammar of graphics, recreate their plot.

- **Question 8.9** Let us now bring in Lankford's analysis. Filter the data to include (1) only those incidents that are known to have at least one perpetrator who killed 4 or more victims, and then (2) incidents known to have at least one perpetrator who killed 4 or more victims, *and* the attack was not committed by member(s) of a terrorist organization or genocidal group. For each case, repeat **Question 8.7** and **Question 8.8**. What do you observe?

- **Question 8.10** From your observations, do you find Lott and Moody's original visualization to be reasonable? Moreover, can it be used to say, in general terms, that the United States is a safer country when compared to the rest of the world?

Question 9 This question is a continuation of *Question 10* from Chapter 2.

- **Question 9.1** Using the tibble `top_10`, generate a bar chart that visualizes the top 10 states with the highest average unemployment rate; the bars should be presented in order of increasing unemployment rate. Fill your bars according to the candidate winner for that state. Consider using a position scale to avoid any overlap in text labels.

- **Question 9.2** Does the bar chart reveal any possible associations between average unemployment rate and the candidate that received the most votes in the top 10 states with the highest average unemployment rates?

Note: The following exercises correspond to material that appears only in the accompanying website, at: https://ds4world.cs.miami.edu/.

Question 10: Visualizing commutes using `tidycensus`. In this exercise, we will examine the way people commute for work using the census data and visualize the result. Let us begin by loading the libraries we need for the exercise.

```
library(tidyverse)
library(viridisLite)
library(viridis)
library(tidycensus)
```

As before, let us load the variables available for the 2019 1-year dataset. The destination is 'acs1vars1, as shown next.

```
acs1vars <- load_variables(year = 2019, dataset = "acs1")
acs1vars
```

- **Question 10.1** Since our purpose is to explore the way people commute to work, let us guess what is necessary to get to the data. Let us collect all the rows where the concept contains the word "TRANSPORTATION" and store it in the nametrans0.

- **Question 10.2** By using View() to open the data frame trans0, you can visually examine the collection of all the variables related to TRANSPORTATION. By scrolling in the View window, you see that the first group with the prefix B08006is the general population broken down with the means of transportation and with the gender. So, let us collect all the variables whose name starts with B08006 and store the variables in trans1. Do you have 51 rows in trans1?

- **Question 10.3** Let us further screen this by collecting those whose label *ends* with ":". Let us store it in the name trans2.

- **Question 10.4** Now we have a more manageable set of variables with just four in the three (male+female, male, and female) categories. Let us use the first four of the 12 collected. Store the four names in trans_names. You can pull the variable and take the prefix of size 4 or directly type the elements with the list construct c.

- **Question 10.5** Query the census data for California at the county level with the names in trans_names with 2019 as the year and store it in ca_trans. Specify "B08006_001" as the summary variable.

- **Question 10.6** Let us take the ratio of the four by dividing them by the summary variable. Save the result in a name ca_trans0.

- **Question 10.7** The suffix "002" is for transportation by cars. Let us generate a geographical presentation of how much of each county has workers "commuting by car". Use the color option "inferno" with scaled filling in a scale_fill_viridis_c layer. Add a title "People Commuting to/from Work by Car" to the plot.

- **Question 10.8** The suffix "004" is for car-pooling. Let us generate a geographical presentation of how much of each county has workers commuting by car. Use the color option "inferno" with scaled filling in scale_fill_viridis_c. Add a title "People Commuting to/from Work by Car-pooling" to the plot.

- **Question 10.9** Finally, the suffix "008" is for using public transportation. Let us generate a geographical presentation of how much of each county has workers "commuting by car. Use the color option"inferno" with scaled filling in scale_fill_viridis_c. Add a title "People Commuting to/from Work by Public Transportation" to the plot.

- **Question 10.10** Do your visualizations reveal any relationships between geographical area and the census variable examined (e.g., commuting by car, public transportation, car-pooling)?

Doing Statistics

4

Building Simulations

The previous chapters have shown the basics of tidying, transforming, and visualizing data using R and the tidyverse. Armed with these tools, this unit turns to how we can extract insights from our data. Key to this investigation is statistics and inherent to that is the notion of *randomness.*

We use the word *random* all the time on a regular basis. For instance, you may be familiar with the concept of *gachapon* or loot boxes[1] in PC and mobile games which gives players a chance to obtain prized items using real-world currency. While they have recently stirred up much controversy[2], the basic idea boils down to how to leverage *randomness.*

We often speak of randomly picking a number between 1 and 100. If someone chooses the number by adding the day of the month plus the minute of the hour showing on the clock of her smart watch, her choice is not random; once you have learned what she has chosen previously, you know what she will choose again. Therefore, anything generated by a systematic, predetermined, and deterministic procedure is not random.

Randomness is so essential to conducting experiments in statistics because it is what allows us to *simulate* physical processes in the real world, often thousands and hundreds-of-thousands of times. All this at zero cost of coordinating an actual physical experiment, which may not be feasible depending on the circumstances.

This chapter begins by exploring randomness using R. We will then leverage randomness to build out simulations of real-life phenomena – like *birthdays!* – and how we may extract insights from them.

4.1 The `sample` Function

We begin our study by learning how to generate random numbers using R. There are many functions that R has which involve random selection; one of these is called `sample()`. It picks one item at random from a list (i.e., vector), where the choice will likely occur at all positions. A prime example of randomness is tossing a coin with chance of heads 50% and chance of tails 50%.

```
fair_coin <- c("heads", "tails")
sample(fair_coin, size = 1)
```

```
## [1] "tails"
```

---

[1] https://en.wikipedia.org/wiki/Loot_box
[2] https://doi.org/10.1371/journal.pone.0206767

DOI: 10.1201/9781003320845-4

Run the cell a few times and observe how the output changes. The unpredictable nature makes the code, though short, stand out from all the R code we have written so far.

Note that the function has the form `sample(vector_name, size)`, where `vector_name` is the name of the vector from which we will select an item and `size` is how many items we want to select from the vector.

Here is another example: a football game ("American football" for non-US readers) begins each half by kicking a football from the 20-yard line of a team toward the goal of the opponent. The decision of which team gets to kick the ball is through a ritual that takes place three minutes prior to the "kick off". In the ritual, a referee tosses a coin and the visiting team calls "heads" or "tails".

If the visiting team calls correctly, the team gets to decide whether the team will kick or receive; otherwise, the home team makes the decision. Note that the ritual is somewhat redundant.

Suppose that the referee throwing the coin is not clairvoyant and has no powers to foretell "heads" or "tails" and that the coin is the same fair coin from our previous example. Then, the referee can simply throw the coin and make the visiting team kick the ball if the coin turns "heads" (or "tails").

Thus, the action of choosing the team boils down to selecting from a two-element vector consisting of "kick" and "receive" with chance of 50% for each.

```
two_groups <- c("kick", "receive")
sample(two_groups, size = 1)
```

```
## [1] "receive"
```

A nice feature of `sample` is that we can instruct it to repeat its element-choice action multiple times in sequence without influence from the outcome from the previous runs. For instance, we can select the kicking teams for 8 games.

```
sample(two_groups, size = 8, replace = TRUE)
```

```
## [1] "receive" "kick"    "kick"    "kick"    "kick"
## [6] "kick"    "kick"    "kick"
```

Note that a third argument `replace` is specified here. By setting it to `TRUE`, we allow the same selection (say, `receive`) from the `two_groups` vector to be made more than once. We call this method sampling *with replacement*. In contrast, toggling this argument to be `FALSE` would make choices from the previous executions unavailable. We call this way sampling *without replacement*. R does this by default.

What happens if the size of the vector is smaller than the number of repetitions?

```
sample(two_groups, size = 8)
```

```
Error in sample.int(length(x), size, replace, prob) :            ⚡ Show Traceback
  cannot take a sample larger than the population when 'replace = FALSE'
```

There are not enough elements to choose from!

Note that we have made an implicit interpretation of the code: we wrote the code assuming that it will tell the role that the visiting team will play at the kick off; that is, if the value the code generates is `kick` then the visiting team will kick and if it is `receive` they will receive. We must not mistake this interpretation. That is, the output the code produces is NOT the role the home team will play; it is for selecting the action of the *visiting* team. Implicit interpretations we make on the code often play an important role.

If the visiting team is very keen to start by kicking, we may choose to instead translate the outcome as `Yes` when it is `kick` and `No` otherwise. If we choose such a Yes/No interpretation, we view the random scheme producing `kick` as an "event" and interpret the output being `kick` as the event "occurring".

4.2 `if` Conditionals

Often times in programming work we need to take a certain action based on the outcome of an event. For instance, if a coin turns up heads, a friend wins a \$5 bet or a visiting team in football is designated as the "kicker". Such *conditional* statements and how to use them in programming are the subject for this section.

4.2.1 Remember logical data types?

We saw in an earlier chapter that one of the core data types is logical data. These were the easiest to remember of the bunch because logical data types can only have one of two values: `TRUE` and `FALSE`. We also often call such values *booleans*.

A major use of logical data is to direct computation based on the value of a boolean. In other words, we can write code that has two choices for some part of its action where which of the two actions actually occurs depends on the value of the boolean.

If we draw the choice-inducing boolean value from random generation, say using `sample`, we can randomly select among possible actions. We call such an "action selection" based on the value of a boolean *conditional execution*.

4.2.2 The `if` statement

A program unit in a conditional execution is a *conditional statement*. A *conditional statement* is one that allows selection of an action from multiple possibilities. In R and in many other languages, we usually write a conditional statement as a multi-line statement. It is entirely possible to put everything in one, but such style is confusing and prone to errors.

Conditional statements usually reside inside a function. This allows us to express alternative behavior depending on argument values.

In R, and in many programming languages, a conditional statement begins with an `if` header. What appears after the keyword is a pair of parentheses, in which a condition to examine appears. After the condition part appears an action to perform, which is a series of statements flanked by a pair of curly braces.

The syntax specifies that if the condition inside the pair of parentheses is true, the program executes the statements appearing in the ensuing pair of braces. We call a statement in this form an *if statement*.

Let us see our first example of an if statement, which is a function that returns the sign of a number.

```
sign <- function(x) {
  if (x > 0) {
    return("positive")
  }
}
sign(3)
```

```
## [1] "positive"
```

The function `sign` receives a number and returns a string representing the sign of the number. Actually, the return value of the function exists only if the number is strictly positive; otherwise, the function does not return anything.

```
print(sign(-3))
```

```
## NULL
```

What is the boundary separating `positive` and nothing? We know that the condition is `x > 0` so we can say that the boundary is the point 0, but 0 falls on the side that produces no return.

Can we make the return `positive` when x is equal 0? Sure can!

```
sign <- function(x) {
  if (x >= 0) {
    return("positive")
  }
}
sign(0)
```

```
## [1] "positive"
```

```
sign(0.1)
```

```
## [1] "positive"
```

```
print(sign(-0.1))  # force a print
```

```
## NULL
```

We can put a series of conditional statements in a function. If a conditional statement contains a return statement and R executes that statement, R skips the remainder of the code in the function.

```
sign <- function(x) {
  if (x > 0) {
    return("positive")
  }
```

```
  if (x < 0) {
    return("negative")
  }
}
sign(2)
```

```
## [1] "positive"
```

```
sign(-2)
```

```
## [1] "negative"
```

```
print(sign(0))  # force a print
```

```
## NULL
```

Instead of saying "if x is less than 0, it is negative", it would be more natural to say "*otherwise*, if x is less than 0, it is negative". We can express "otherwise" using the keyword else. Let's revise the above code.

```
sign <- function(x) {
  if (x > 0) {
    return("positive")
  } else if (x < 0) {
    return("negative")
  }
}
sign(3)
```

```
## [1] "positive"
```

```
sign(-3)
```

```
## [1] "negative"
```

```
print(sign(0))  # force a print
```

```
## NULL
```

What do we do about 0? By adding another else if block to the code, we can make the function return something in the case when the number is 0.

```
sign <- function(x) {
  if (x > 0) {
    return("positive")
  } else if (x < 0) {
    return("negative")
  } else if (x == 0) {
    return("neither positive nor negative")
```

```
  }
}
```

```
sign(0)
```

```
## [1] "neither positive nor negative"
```

Since the condition `x == 0` is exactly the condition `x` satisfies when the execution reaches the second `else if` block, we can jettison the `if` and the condition. The resulting block is what we call an `else` block.

```
sign <- function(x) {
  if (x > 0) {
    return("positive")
  } else if (x < 0) {
    return("negative")
  } else {
    return("neither positive nor negative")
  }
}
```

```
sign(0)
```

```
## [1] "neither positive nor negative"
```

4.2.3   The `if` statement: a general description

We are now ready to present a more general form of the `if` statement.

```
if (<expression>) {
    <body>
} else if (<expression>) {
    <body>
} else if (<expression>) {
    <body>
...
} else {
    <body>
}
```

The keyword `else` means "otherwise" and the keyword `else if` is a combination of `else` and `if`. We can stack up `else if` blocks after an initial `if` to define a series of alternative options.

Following are some notes to keep in mind:

- An `if`-block cannot begin with an `else` in the series.

- There must exist one `if` clause.
- When a series of `else if` blocks appear after an `if`, this represents a series of alternatives. We call this an `if` sequence.

- An if without a preceding else begins a new if sequence.
- R scans all the conditions appearing in an if sequence takes an action when it finds a condition that evaluates to TRUE. All other actions before and after the matching one are ignored.
- The else block is optional; it takes care of everything that does not have a match.

4.2.4   One more example: comparing strings

We end this section with another example, this time comparing strings.

```
get_capital <- function(x) {
  if (x == "Florida") {
    return("Talahassee")
  } else if (x == "Georgia") {
    return("Atlanta")
  } else if (x == "Alabama") {
    return("Montgomery")
  } else {
    return("Oops, don't know where that is")
  }
}
```

Here is a dataset containing some students and their state of residence.

```
some_students <- tibble(
  name = c("Xiao", "Renji", "Timmy", "Christina"),
  state = c("Florida", "Florida", "Alabama", "California")
)
some_students
```

```
## # A tibble: 4 x 2
##   name      state
##   <chr>     <chr>
## 1 Xiao      Florida
## 2 Renji     Florida
## 3 Timmy     Alabama
## 4 Christina California
```

We can annotate the tibble with a new column containing the capital information for each state using a purrr map.

```
some_students |>
  mutate(capitol = map_chr(state, get_capital))
```

```
## # A tibble: 4 x 3
##   name      state     capitol
##   <chr>     <chr>     <chr>
## 1 Xiao      Florida   Talahassee
```

```
## 2 Renji      Florida    Talahassee
## 3 Timmy      Alabama    Montgomery
## 4 Christina California Oops, don't know where that is
```

Note how Christina's capital information defaults to the result of the `else` condition.

4.3   **for** Loops

Let us now use conditional execution to simulate a simple betting game on a coin.

4.3.1   Prerequisites

We will need some functions from the tidyverse in this section, so let us load it in.

```
library(tidyverse)
```

4.3.2   Feeling lucky

We will use the same fair coin from the kick-off we saw in the football example. However, feeling lucky, you wager that you can make some money off this coin by betting a few dollars on heads – can we tell if your intuition is right? Let's find out!

We imagine a function that will receive a string argument representing the side the coin is showing, and returns the result of the bet. If the coin shows up `heads` you get 2 dollars. But, if it shows `tails` you lose 1 dollar.

```
one_flip <- function(x) {
  if (x == "heads") {
    return(2)
  } else if (x == "tails") {
    return(-1)
  }
}
```

Let us see how the function works.

```
c(one_flip("heads"), one_flip("tails"))
```

```
## [1]  2 -1
```

To play the game based on one flip of a coin, we can use `sample` again.

```
sides <- c("heads","tails")
one_flip(sample(sides, size = 1))
```

```
## [1] -1
```

We can avoid having to create a `sides` variable by including the vector directly as an argument.

```
one_flip(sample(c("heads","tails"), size = 1))
```

[1] 2

Now let us expand this and develop a multi-round betting game.

4.3.3 A multi-round betting game

Previously, we split the game into two actions: a function `one_flip` computing the gain/loss and a `sample` call simulating one round of the game using the gain/loss function. We now combine these two into one.

```
betting_one_round <- function() {
  # Net gain on one bet
  x <- sample(c("heads","tails"), size = 1)
  if (x == "heads") {
    return(2)
  } else if (x == "tails") {
    return(-1)
  }
}
```

Betting on one round is easy – just call the function!

```
betting_one_round()
```

[1] -1

Run the cell several times and observe how the value is sometimes 2 and sometimes -1. How often do we get to see 2 and how often do we get to see -1? Will your wager come out on top?

You could run this function multiple times and tally, of the runs you observed, how many times you won 2 dollars and how many times you lost a dollar. You could then compare the difference between the gains and losses. This is quite a tedious process and we still don't have a sense of how variable the results are. For that, we would need to run this process thousands or millions of times. Should we grab a good pencil and get tallying? No way! Let's use the power of R.

We can instruct R to take care of the repetitive work by repeating some action a number of times with specific instruction such as, "for round X, use this information". *Iteration* is the name we use in programming to refer to things that repeat. In R and in many other languages, a keyword that leads a code in *iteration* is `for`. We also use the word *loop* to refer to a process that repeats.

So a code that repeats a process with `for` as the header is a for-loop.

In R, the way to specify a for-loop is to say: "for each item appearing in the following sequence, starting from the first and towards its end, do this."

```
for (hand in c("rock", "paper", "scissors")) {
  print(hand)
}
```

```
## [1] "rock"
## [1] "paper"
## [1] "scissors"
```

```
for (season in c("spring", "summer", "fall", "winter")) {
  print(season)
}
```

```
## [1] "spring"
## [1] "summer"
## [1] "fall"
## [1] "winter"
```

In the two for loops, hand and season are the names we use to refer to the elements that the iteration picks from the lists. In other words, the first for-loop picks an item from the three-element sequence and we use the name hand to access the item. The same is true for the second for-loop.

Our present interest is in writing a program that repeats the betting game many times. To repeat an action a number of times, we use a pair of numbers with a colon in between. The expression is X:Y, where X represents the start and Y the end. The expression represents the series of integers starting from X and ending with Y.

```
1:5
```

```
## [1] 1 2 3 4 5
```

```
30:20
```

```
##  [1] 30 29 28 27 26 25 24 23 22 21 20
```

```
a <- 18
b <- 7
a:b
```

```
##  [1] 18 17 16 15 14 13 12 11 10  9  8  7
```

```
b:a
```

```
##  [1]  7  8  9 10 11 12 13 14 15 16 17 18
```

Wow! R is so smart that when the second number is smaller than the first number, it decreases the number by 1, instead of increasing it by 1.

Now we can use the sequence generation to write a for-loop. This one prints 1 through 5.

```
for (i in 1:5) {
  print(i)
}
```

```
## [1] 1
## [1] 2
## [1] 3
## [1] 4
## [1] 5
```

We can apply this to the betting game.

```
for (i in 1:10) {
  print(betting_one_round())
}
```

```
## [1] 2
## [1] -1
## [1] -1
## [1] 2
## [1] 2
## [1] -1
## [1] 2
## [1] -1
## [1] -1
## [1] 2
```

Note that the function `better_one_round` is self-contained, meaning not requiring an argument, and so the code that R runs is identical among the ten iterations. However, `betting_one_round` has a call to `sample` and that introduces randomness in the execution and, therefore, the results we see in the ten lines are not uniform and can be different each time we run the `for` loop.

4.3.4   Recording outcomes

You may have realized "the results of the ten runs disappear each time I run it; is there a way to record them?" The answer: yes!

We create an integer vector to store the results, where the vector has the same length as the number of times we issue a bet; each element in the vector stores the result of one bet.

```
rounds <- 10
outcomes <- vector("integer", rounds)
for (i in 1:rounds) {
  outcomes[i] <- betting_one_round()
}
```

```
outcomes
```

```
##  [1] -1 -1 -1 -1 -1 -1 -1  2  2  2
```

This will do the job. The body of this `for` loop contains two actions: (1) run the betting function `betting_one_round()`, and (2) store the result into ith slot of the `outcomes` vector. Both actions are executed for each item in the sequence `1:rounds`.

You may have noticed how stepping through the `outcomes` vector this way, individually assigning each element the result of one bet, can be cumbersome to write. If so, you would be in good company. The philosophy of R, and especially the tidyverse, prefers to eliminate the need to write many common `for` loops. We saw one example of this already when we used *map* from the `purrr` package to apply a function to a column of data. Here, we will use the function `replicate` to repeat a simulation many times.

Here is how we can rewrite our simulation using just two lines of code.

```
rounds <- 10
outcomes <- replicate(n = 10, betting_one_round())
```

```
outcomes
```

```
## [1]  2  2  2  2 -1  2 -1  2  2  2
```

We can use `sum` to count the number of times money changed hands.

```
sum(outcomes)
```

```
## [1] 14
```

Looks like we made some money!

Note that while `replicate` eliminates the need to write `for` loops in common situations, R internally must still perform a `for` loop, i.e., the code for `replicate` contains a `for` loop and it is not directly visible to us as the programmer who wrote the code. Therefore, the chief benefit of using constructs like `replicate` and map is not for its speed, but clarity: it is much easier to read (and write!).

If you are still not convinced, we defer to a prominent data scientist and an authority on tidyverse for an explanation.

Of course, someone has to write loops. It doesn't have to be you. — Jenny Bryan

---

4.3.5 Example: 1,000 tosses

Iteration using `replicate` is a handy technique. For example, we can see the variation in the results of 1,000 bets by running exactly the same code for 1,000 bets instead of ten.

```
rounds <- 1000
outcomes <- replicate(n = rounds, betting_one_round())
```

The vector `outcomes` contains the results of all 1000 bets.

```
length(outcomes)
```

```
## [1] 1000
```

How much money did we make?

```
sum(outcomes)
```

```
## [1] 545
```

To see how often the two different possible results appeared, we can create a tibble from `outcomes` and then use `ggplot2`.

```
outcome_df <- tibble(outcomes)
ggplot(outcome_df, aes(x = outcomes)) +
  geom_bar() +
  coord_flip()
```

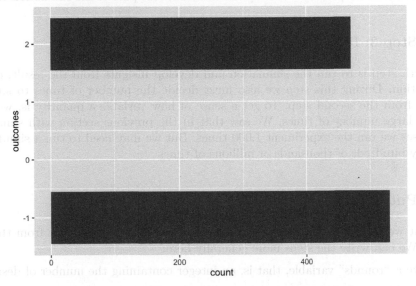

As we would expect, each of the two outcomes 2 and -1 appeared about 500 of the 1000 times. And, because we bet an extra dollar for every heads we get, we come out on top. Not bad!

4.4 A Recipe for Simulation

Simulation, in data science, refers to the use of programming to imitate a physical process. Sometimes we call it "computer simulation" to articulate that the computer is the author of such simulation. A simulation consists of three main steps.

4.4.1 Prerequisites

As before, let us load in the tidyverse.

```
library(tidyverse)
```

4.4.2 Step 1: Determine what to simulate

The first step is to figure out which part of the physical process we want to imitate using computation and decide how we will represent that part numerically.

4.4.3 Step 2: Figure out how to simulate one value

The second step is to figure out how to generate values for the numerical representation from the first step using a computer program. If some numbers require updating during the simulation procedure, we will figure out how to update them. Randomness is often a key ingredient in this step. This is usually the most difficult part of the simulation to complete.

4.4.4 Step 3: Run the simulation and visualize!

The fourth step is to run the simulation and develop insights from the result, often using visualization. During this step we also must decide the number of times to simulate the quantity from the second step. To get a sense of how variable a quantity is, we must run step 2 a large number of times. We saw that in the previous section with simulating bet wins/losses we ran the experiment 1,000 times. But we may need to run a simulation even more, say hundreds of thousands or millions of times.

4.4.5 Putting the steps together in R

Note that we have followed these same steps in the betting experiment from the previous section. We can write the steps more generally here:

- Create a "rounds" variable, that is, an integer containing the number of desired repetitions.
- Create an "experiment" function to simulate one value based on the code we developed.
- Call the function `replicate` to replicate the experiment function a great number of times.
  - Store the results to a variable. We call this the "outcomes" vector.
  - A general format takes the form: `outcomes <- replicate(n = rounds, experiment_func())`

The outcomes vector will contain all the simulated values. A good next step would be to visualize the distribution of the simulated values by counting the number of simulated values that fall into some category or, perhaps more directly, by using `ggplot`!

We now turn to some examples.

4.4.6   Difference in the number of heads and tails in 100 coin tosses

Let us return to the coin tosses. As we see powers of 10 as the easiest kind of numbers to deal with, let us back down from 300 to 100. If the coin is fair, we anticipate a half of the tosses we make is "Head". The simulation we have written receives as the number of repetitions, and returns a vector representing the results of the simulated coin tosses with an added interpretation of "heads" as 1 and "tails" as -1. We can go back to the process of flipping a coin and develop an insight as to when we flip a coin 300 times, how the number of "heads" is likely to look.

In this example we will simulate the number of heads in 300 tosses of a coin. The histogram of our results will give us some insight into how many heads are likely.

Let's get started on the simulation, following the steps above.

4.4.7   Step 1: Determine what to simulate

We want to simulate the process of tossing 300 fair coins, where each coin toss generates "heads" or "tails" as the outcome. There is only one number we are interested in the physical process - the number of "heads".

4.4.8   Step 2: Figure out how to simulate one value

Now we know that we want to know the number of "heads" in 100 coin tosses, we have to figure out how to make one set of 100 tosses and count the number of heads. Let's start by creating a coin. We eliminate the gain/loss calculation from the previous program in the ensuing simulation program. We start by stating the two possible outcomes of toss. What we define is a two-element vector, as before, and we call it `sides`.

```
sides <- c("heads", "tails")
```

We use `sample()` to sample from the two-element vector. Recall that we can specify the number of samples and if we want to replenish the vector with the item that the sample has chosen. The code below shows how we sample from `sides` 8 times with replacement.

```
tosses <- sample(sides, size = 8, replace = TRUE)
tosses
```

```
## [1] "tails" "heads" "tails" "tails" "heads" "tails"
## [7] "heads" "heads"
```

We can count the number of heads by using `sum()` as before:

```
sum(tosses == "heads")
```

```
## [1] 4
```

Our goal is to simulate the number of heads in 100 tosses. We have only to replace the 8 with 100.

```
outcomes <- sample(sides, size = 100, replace = TRUE)
num_heads <- sum(outcomes == "heads")
num_heads
```

```
## [1] 40
```

Play with the code a few times to see how close the number gets to the expected one half, 150.

```
one_trial <- function() {
  outcomes <- sample(sides, size = 100, replace = TRUE)
  num_heads <- sum(outcomes == "heads")
  return(num_heads)
}
```

You can simply call this function to generate an outcome of one experiment.

```
one_trial()
```

```
## [1] 54
```

4.4.9   Step 3: Run and visualize!

Here we face a critical question, "for our goal of developing an insight about coin tosses, how many times do we want to repeat it?" We can easily run the code 10,000 times, so let's choose 10,000 times.

We have programmed one_trial so that it executes 100 coin tosses and returns the number of "Heads". We now need a loop to repeat one_trial as many times we want.

To do that, we use the replicate construct.

```
# Number of repetitions
num_repetitions <- 10000

# simulate the experiment!
heads <- replicate(n = num_repetitions, one_trial())
```

By executing heads after this produces all elements of heads. That will be a lot of lines on the screen, so you may not want to do it! Instead, we can ask heads how many elements it has, using the length function we have seen before.

```
length(heads)
```

```
## [1] 10000
```

Aha! It has the desired number of elements. We can peek at some of the elements in the vector.

```
heads[1]
```

```
## [1] 50
```

```
heads[2]
```

```
## [1] 48
```

Using `tibble` we can collect the results as a table. Recall that `tibble` needs the index values and the data.

```
results <- tibble(
  repetition = 1:num_repetitions,
  num_heads = heads
)
```

```
results
```

```
## # A tibble: 10,000 x 2
##    repetition num_heads
##         <int>     <int>
## 1           1        50
## 2           2        48
## 3           3        53
## 4           4        53
## 5           5        46
## 6           6        49
## 7           7        51
## 8           8        54
## 9           9        52
## 10         10        45
## # ... with 9,990 more rows
```

```
ggplot(results) +
  geom_histogram(aes(x = num_heads, y = after_stat(density)),
                 color = "gray", fill = "darkcyan",
                 breaks = seq(30.5, 69.6, 1))
```

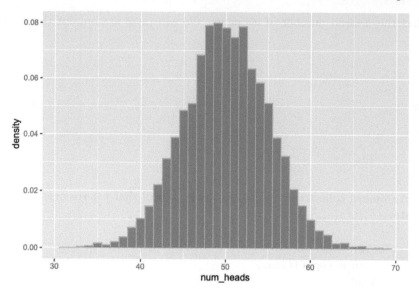

In this histogram, each bin has width 1 and we place it centered on the value. For example, the vertical bar on 50 is the number of times the simulation generated 50 as the result. We see that the histogram looks symmetric with 50 as the center, which is a good sign. Why? Our expectation is that the number 50 is the ideal number of "Heads" in 100 tosses of a fair coin. If the coin is fair, for all number d, the event that we see $50 - d$ "Heads" is as likely to happen as the event that we see $50 + d$ Heads. The symmetry that we observe confirms the hypothesis. We also see that about 8% (i.e., 0.08 on the y-axis) of the simulation results produced 50. Furthermore, we see that very few times we see occurrences of numbers less than 35 or greater than 65. We thus conclude from the experiment that the range of the number is likely to reside in the range $[35, 65]$.

4.4.10 Not-so-doubling rewards

There is a famous story of a king awarding a minister with doubling amount of grains. There are many versions of the story, but the gist is like this: one day a king has decided to award a minister for his great work.

- The kind asks, "Great work, what do you want from me as a reward? You name it, I will make your wish come true."
- The minister says, "Your Majesty, what an honor! If you are so kind as to indulge me, may I ask to receive grains of rice on a board of chess. We will start with one grain on a space on the board, given one day. The next day, I would like two grains on the next space. The following day, I would like four grains on the third space. Each day, I would like twice as many grains as you have given me on the previous day. In this manner, for the next 64 days, you would be so generous to give me grains of rice. Would that be too imposing to ask?"
- The king says, "You ask so little. That would be so easy to do. Of course, this great King will grant you your request."

The question is how many grains of rice will the minister receive at the end of the 64 day period?

We know the answer to the question. The daily amount doubles each day starting from 1. He would thus receive:

$$1 + 2 + 4 + 8 + 16 + \cdots = 2^0 + 2^1 + 2^2 + 2^3 + 2^4 + \cdots + 2^{63}$$

We can express this quantity more compactly as $2^{64} - 1$.

Why? Suppose he has one more grain in his pocket to add to the piles at the end of the 64th day. We have

$$1 + 1 + 2 + 4 + 8 + \cdots + 2^{63} - 1$$

as the same total amount. The first two occurrences of 1 are equal to 2. So, we can simplify the sum as

$$2 + 2 + 4 + 8 + \cdots + 2^{63} - 1$$

We have got rid of the twos. By joining the first two 2's, we get

$$4 + 4 + 8 + \cdots + 2^{63} - 1$$

At the end of calculation, we get

$$2^{63} + 2^{63} - 1$$

and this is equal to $2^{64} - 1$, which is, by the way, an obscenely large number.

We know the story as a fable that tells us we must think before promising something. This story took place hundreds of years ago, where there were no computers. For the king to provide the grains he had promised to give to the minister, he would have ordered a clerk to do the calculation.

If the clerk is super-human, her calculation would be perfect, and so the total amount she would provide would be exactly what we had anticipated. But, since she is human, she is prone to error. In the process of writing down numbers, there may be various errors, such as skipping a digit or writing a wrong digit.

If she does not notice an error, the minister would get a different number of grains at the end of the 64 days. Suppose we want to mimic the process of her calculation using a computer, with the chance factor in mind. This is how "randomness" comes in to play.

Let us consider two different scenarios for the source of error:

- The errors are **independent**. That is, the error the clerk makes on a day does **not** affect the bookkeeping the next day.

- The errors are **dependent**. That is, the error the clerk makes affects the next day's counting and has a lasting effect on the bookkeeping for the remaining days.

The first scenario is easier, so we will develop a simulation scheme for this first. Also, to simplify the number crunching, we consider the number of grains the minister receives at the end of 10th day. We know that this number should equal $2^{10} - 1$.

4.4.10.1 Step 1: Determine what to simulate

We are interested in simulating the number of grains the minister receives at the end of the 10th day, assuming independent errors in bookkeeping. We hypothesize that the number of grains should cluster around $2^{10} - 1 = 1023$.

4.4.10.2 Step 2: Figure out how to simulate one value

We imagine the variability in the clerk's calculation for the number of grains a minister receives. Based on what we know from the story, three actions are possible:

- The clerk counts one less grain.
- The clerk counts the right number of grains.
- The clerk counts one extra grain.

We will assume that "getting it right" has a slightly higher chance of occurring (2/4) with the other two actions having an equal chance of occurring (1/4). We can simulate the clerk's action using `sample` and setting the `prob` argument.

```
sample(c(-1, 0, 1), 1, prob=c(1/4, 2/4, 1/4))
```

```
## [1] 0
```

The following function receives a number of grains as an argument and returns the number of grains after the clerk's calculation. This is the amount the minister would receive after some day.

```
after_clerk_calculation <- function(grains) {
  grains + sample(c(-1, 0, 1), 1, prob=c(1/4, 2/4, 1/4))
}
```

We can try the function with some arbitrary number of grains, say, 10.

```
after_clerk_calculation(10)
```

```
## [1] 10
```

Try running this a few times to observe the different outcomes. Sometimes the clerk gets it right (10), counts one less (9), or counts one more (11).

The expected amount the minister should receive each day follows the form $2^{\text{day number}-1}$. We can write the following sequence for the amounts starting at day 1 and ending after day 10.

```
2 ** (0:9)
```

```
## [1]   1   2   4   8  16  32  64 128 256 512
```

Using a `purrr` map, we can simulate the amounts after the clerk's calculation by applying the function `after_clerk_calculation` to each of the elements in the above sequence.

```
map_dbl(2 ** (0:9), after_clerk_calculation)
```

```
## [1]   1   3   5   8  15  31  64 127 256 513
```

Therefore, the total number of grains the minister receives is the sum of these amounts.

```
map_dbl(2 ** (0:9), after_clerk_calculation) |>
  sum()
```

```
## [1] 1023
```

We now have enough machinery to write a function for simulating one value. This functions receives the number of days as an argument and returns the total grains received by the minister after the given number of days is over.

```
total_grains_received <- function(num_days) {
  map_dbl(2 ** (0:(num_days - 1)), after_clerk_calculation) |>
    sum()
}
```

For a 10-day scheme, we call the function as follows.

```
total_grains_received(10)
```

```
## [1] 1021
```

Run the cell a few times and observe the variability in the number of grains.

4.4.10.3 Step 3: Run and visualize!

We will use 10,000 repetitions of the simulation this time to get a sense of the variability. Each element of the grains vector stores the resulting number of grains at the end of the 10th day in each simulation of the story.

```
# Number of repetitions
num_repetitions <- 10000

# simulate the experiment!
grains <- replicate(n = num_repetitions, total_grains_received(10))
```

As before, we collect our results into a tibble.

```
results <- tibble(
  repetition = 1:num_repetitions,
  num_grains = grains
)
results
```

```
## # A tibble: 10,000 x 2
##    repetition num_grains
##         <int>      <dbl>
## 1           1       1022
## 2           2       1023
```

```
##   3           3        1019
##   4           4        1022
##   5           5        1021
##   6           6        1022
##   7           7        1023
##   8           8        1023
##   9           9        1021
## 10          10        1022
## # ... with 9,990 more rows
```

Finally, we visualize our results.

```
ggplot(results) +
    geom_histogram(aes(x = num_grains, y = after_stat(density)),
                   color = "gray", fill = "darkcyan", bins = 18) +
    geom_point(aes(x = 1023, y = 0), color = "salmon", size = 3)
```

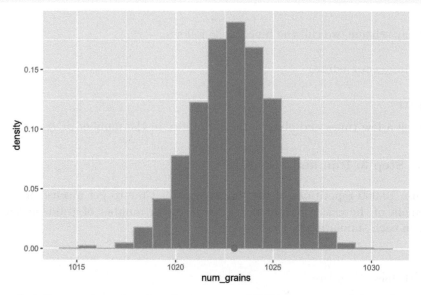

We observe that the number of grains cluster around 1023, as expected (see the orange dot). In general, each round after the first can create a difference of at most 1, so with N rounds, the difference is at most $N - 1$. Our simulation confirms this fact.

4.4.11   Accumulation

We now turn to the second scenario in the doubling grains story. Because the calculation for the next day *depends* on what happened the previous day, we will not be able to use the map or replicate operations as we did when assuming the errors were independent. We need a programming construct that allows us to update some value as we go along. We could use the for loop construct shown in Section 4.3[3] to achieve this work easily, but we would like to eliminate the need to write a loop as much as possible.

---

[3]https://ds4world.cs.miami.edu/building-simulations.html#for-loops

purrr offers another construct called `accumulate` that sequentially applies a 2-argument function to elements of a vector. A key aspect of its operation is that each application of the function uses the *result* of the previous application as the first argument.

Let's see an example. Consider the following character vector of fruits.

```
delicious_fruits <- c("apple", "banana", "pineapple", "mango")
```

We can use `accumulate` to implement string concatenation. Here, we provide an anonymous function that receives two arguments `acc` (an accumulator) and `nxt` (the next element in the input vector). The `str_c` function is called using the two arguments using the colon as a separator. The effect achieved is the joining, or "concatenation", of all strings in the vector `delicious_fruits` into a single string.

```
accumulate(delicious_fruits, \(acc, nxt) str_c(acc, nxt, sep = ":"))
```

```
## [1] "apple"
## [2] "apple:banana"
## [3] "apple:banana:pineapple"
## [4] "apple:banana:pineapple:mango"
```

There is a good deal of technical detail here so let us unpack what we just did. The accumulator (stored in the argument `acc`) stores the resulting string after concatenation with each element in the vector `delicious_fruits`. The accumulation begins with the first element, the string `"apple"`. After concatenation with the argument `nxt` (containing the string `"banana"`), the resulting string is `"apple:banana"` and the accumulator is updated with this value in the next step of the iteration, available in the argument `acc`.

The process continues until each element of the input vector has been exhausted. The result of the accumulation at each step is shown in the vector returned by the `accumulate` function. This vector has the same length as the input vector and the final result appears at the last index (index 4), a single string containing each fruit separated by a colon.

We could discard the intermediate results and extract just the final product by applying the function `last` from `dplyr`. Observe how this is equivalent to `str_c` when used with the collapse setting.

```
# str_c from stringr with collapse
str_c(delicious_fruits, collapse = ":")
```

```
## [1] "apple:banana:pineapple:mango"
```

```
# using accumulate!
accumulate(delicious_fruits, \(acc, nxt) str_c(acc, nxt, sep = ":")) |>
  last()
```

```
## [1] "apple:banana:pineapple:mango"
```

We can also set an initial value to use to begin the accumulation.

```
accumulate(delicious_fruits, \(acc, nxt) str_c(acc, nxt, sep = ":"),
           .init = "a")
```

```
## [1] "a"
## [2] "a:apple"
## [3] "a:apple:banana"
## [4] "a:apple:banana:pineapple"
## [5] "a:apple:banana:pineapple:mango"
```

This has the effect of extending the resulting vector length by 1.

In some cases, it is desirable to use the accumulator and ignore the elements in the input vector given (in the argument `nxt`). This can be useful when you care only about the accumulation and repeating this for some number of steps.

For instance, the following `accumulate` continuously adds 10 to an initial value 10. The length of the input vector controls the number of steps taken, but the vector contents are ignored.

```
accumulate(541:546, \(acc, nxt) acc + 10, .init = 10)
```

```
## [1] 10 20 30 40 50 60 70
```

It is also possible to terminate the accumulation early based on some condition being met using the done sentinel. This can also be useful depending on the simulation scheme. Question 4 from the exercise set explores this feature in greater depth.

Pop quiz: In the above `accumulate` example, would you expect the resulting vector to change if we had used the sequence `1:6` as the input vector? Why or why not?

4.4.12   Lasting effects of errors

We can use the `accumulate` construct in the doubling grains simulation. Recall that, under the second scenario, the error the clerk makes on some day *affects* the bookkeeping for the remaining days. That is, the amount the clerk gives on a day after the initial day is two times the amount she gave in the previous with a possible error of ± 1 grain.

We write a function `grains_after_day` that receives a single argument containing the current number of grains. It returns the sum of the current grains and the calculated grains after `after_clerk_calculation` is called. When things go right, this has the desired effect of *doubling* the current number of grains.

```
grains_after_day <- function(current_grains) {
  new_amount <- current_grains + after_clerk_calculation(current_grains)
  return(max(1, new_amount))
}
```

The second line is added as a sanity check to ensure the resulting grain amounts do not ever go negative.

We can use this function within an `accumulate` call to simulate one value in this experiment. Here we provide an initial value 1 to begin the accumulation with and use the input vector only to step the simulation 10 times.

```
accumulate(1:10, \(acc, nxt) grains_after_day(acc), .init = 1)
```

```
## [1]   1   2   3   6  11  23  47  94 188 375 749
```

We rewrite the function `total_grains_received` to use the `accumulate` call and retrieve the final value.

```
total_grains_received <- function(num_days) {
  accumulate(1:10, \(acc, nxt) grains_after_day(acc), .init = 1) |>
    last()
}
```

We can now call this function a large number of times, say, 10,000, to gauge the amount of variability. Each element of the `grains` vector stores the resulting number of grains after the 10th day, assuming *dependent* errors.

```
num_repetitions <- 10000
grains <- replicate(n = num_repetitions, total_grains_received(10))
```

Finally, we visualize the result.

```
grains |>
  tibble() |>
  ggplot() +
  geom_histogram(aes(x = grains, y = after_stat(density)),
                 bins = 18, color = "gray", fill = "darkcyan") +
  geom_point(aes(x = 1023, y = 0), color = "salmon", size = 3)
```

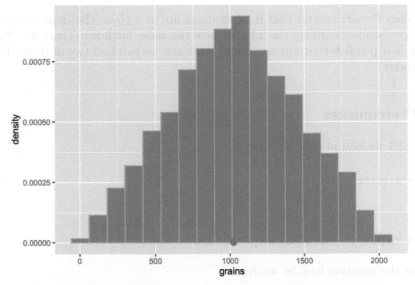

Sometimes the minister received few to no grains at the end of the 10th day and sometimes he received much more than he asked for! Like the first scenario, we see the simulated values cluster again around the expected amount.

4.4.13 Simulation round-up

This section discussed multiple iteration constructs for building simulations. The following list clarifies when to use each.

- `map_*`
  - **Arguments:** (1) a sequence (2) a function that receives as an argument an element from that sequence
  - **Returns:** List or vector the same length as the input sequence
  - **When to use?** Applying a function to a sequence or a tibble column
- `replicate`
  - **Arguments:** (1) Number of repetitions (`reps`) (2) a function to repeat
  - **Returns:** Vector of length `reps` that can be stored for later analysis
  - **When to use?** *Repeating* a function some fixed number of times
- `accumulate`
  - **Arguments:** (1) a sequence (2) a two-argument function; first argument is an *accumulator* and the second argument an element from the sequence
  - **Returns:** Vector the same length as the input sequence that contains the results of the accumulation at each step (or one longer if `.init` argument is set)
  - **When to use?** Applying a function where some value needs to be updated through the duration of the simulation

4.5 The Birthday Paradox

Happy birthday! Someone in your class has a birthday today! Or, should we say happy *birthdays*?

The *Birthday Paradox* states that if 23 students are in a class, the chances are 50/50 that there are two students among the 23 who have the same birthday. There are 365 days in a year. How is it possible that among just 23 students, we can find two of them that have the same birthday?

4.5.1 Prerequisites

As before, let us load in the tidyverse.

```
library(tidyverse)
```

4.5.2 A quick theoretical exploration

We answer the question first by analysis.

Assume that we will consider only a non-leap year (no February 29 – sorry leap year babies!) and each of the 365 birthdays are likely to occur. Since each birthday is likely to occur as any other birthday, we can look at the question at hand by counting the number of possible birthday combinations.

We have 23 students in the class. Each student gets to choose her birthday freely without considering the birthdays of the other 22 students. The 23 students make their choices and then check whether the choices fall into one of the no-duplicate selections.

With no restriction, the number of possible choices of birthdays for the 23 students is

$$365 * 365 * 365 * \cdots * 365 = 365^{23}$$

These possibilities contain the cases where birthdays may be shared among the people.

In comparison, the number of possibilities for selecting 23 birthdays so that no two are equal to each other requires a bit complicated analysis. We can view the counting problem using the following hypothetical process.

The 23 students in the class will pick their birthdays in order so that there will be no duplicates.

- The first student has complete freedom in her choice. She chooses one from the 365 possibilities.
- The next student has almost complete freedom. She can pick any birthday but the one the first student has chosen. There are now 364 possibilities.
- The third student has again almost complete freedom. She can pick any birthday but the ones the first two students have chosen. Since the first two students picked different birthdays, there are 363 possibilities for the third student.

We can generalize the action. The k-th person has $365 - k + 1$ choices.

By combining all of these for the 23 students, we have that the number of possibilities for choosing all-distinct birthdays is

$$365 \cdot 364 \cdot 363 \cdot \cdots \cdot 353.$$

Thus, the chances for the 23 students to make the selections so there are no duplicates are thus

$$(365 \cdot 364 \cdot 363 \cdot \cdots \cdot 343)/365^{23}.$$

Moving terms, we get that the chances are

$$\frac{365}{365} \cdot \frac{364}{365} \cdots \frac{343}{365}.$$

The quantity is approximately 0.4927. The chances we find a duplicate are 1 minus this quantity, which is approximately 0.5073. Pretty interesting, isn't it?

4.5.3   Simulating the paradox

The second method, an alternative to the formal mathematical analysis, is to use simulation to mimic the process of selecting 23 birthdays independently from each other. This simulation is slightly more difficult than the ones we have seen in the previous section.

4.5.4   Step 1: Determine what to simulate

In this simulation, we are interested in obtaining the *chance* or probability that two students in the class have the same birthday among a group of 23 individuals.

4.5.5   Step 2: Figure out how to simulate one value

We start by using `sample` to draw 23 random birthdays. The vector represented by the sequence `1:365` contains the days in the year one can pick from.

```
chosen_birthdays <- sample(1:365, size=23, replace=TRUE)
chosen_birthdays
```

```
##  [1] 282  51 118  33  86 197 166 289  57 193 179 271 288
## [14]  59 255  38 352  45  52 290 333 128 242
```

We now check how many duplicates there are in the class.

```
sum(duplicated(chosen_birthdays))
```

```
## [1] 0
```

Let's pull these pieces together into a function we can use. The function returns `TRUE` if there are any duplicates in the class; `FALSE` otherwise.

```
any_duplicates_in_class <- function() {
  chosen_bdays <- sample(1:365, size=23, replace=TRUE)
  num_duplicates <- sum(duplicated(chosen_bdays))
  if (num_duplicates > 0) {
    return(1)
  } else {
    return(0)
  }
}
```

```
any_duplicates_in_class()
```

```
## [1] 0
```

We now imagine multiple classrooms that each have 23 students. We will survey each of the classes for any birthday duplicates in the class. Luckily, we can use `any_duplicates_in_class` to help us with the surveying work. Once the surveying is done, we will calculate the *proportion* of duplicates among all the classrooms we surveyed. Let's assume there are 100 classrooms in the school.

```
one_birthday_trial <- function() {
  classrooms <- 100
  num_duplicates <- replicate(n = classrooms, any_duplicates_in_class())
  return(sum(num_duplicates) / classrooms)
}
```

We can check how we did after one survey.

```
one_birthday_trial()
```

```
## [1] 0.5
```

Run this cell a few times. Observe how this value is somewhat close to the theoretical 0.51. We now have one trial of our simulation.

4.5.6 Step 3: Run and visualize!

We generate 10,000 simulations and store the results in a vector. Since the value `one_birthday_trial` returns is a nonnegative integer, we create its simplified version, where we reduce all positive values to 1 and retain 0 as 0.

```
# Number of repetitions
num_repetitions <- 10000

# simulate the experiment!
bday_proportions <- replicate(n = num_repetitions, one_birthday_trial())

# and done!
```

As before, we collect our results into a tibble.

```
results <- tibble(
  repetition = 1:num_repetitions,
  proportions = bday_proportions
)
results
```

```
## # A tibble: 10,000 x 2
##    repetition proportions
##         <int>       <dbl>
## 1           1        0.46
## 2           2        0.47
## 3           3        0.47
## 4           4        0.46
## 5           5        0.53
## 6           6        0.49
## 7           7        0.44
## 8           8        0.37
## 9           9        0.49
## 10         10        0.52
## # ... with 9,990 more rows
```

Finally, we visualize our results.

```
ggplot(results) +
  geom_histogram(aes(x = proportions, y = after_stat(density)),
                 color = "gray", fill = "darkcyan",
                 breaks = seq(0.35, 0.65, 0.01)) +
  geom_point(aes(x = 0.51, y = 0), color = "salmon", size = 3)
```

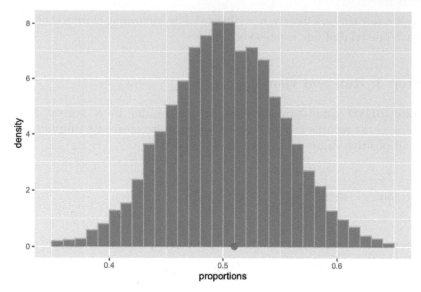

We see that the histogram looks symmetric and centered around 0.51 (see the orange dot), as expected. Neat!

4.6 Exercises

Be sure to install and load the following packages into your R environment before beginning this exercise set.

```
library(tidyverse)
library(edsdata)
library(gapminder)
```

Question 1. Everyday Alana, an amateur transcendentalist[4] and photographer, walks by a pond for one hour, rain or shine. During the walk she sometimes sees some animals. Notable ones among them: jumping fish, great blue heron, and squirrels. Over the past 80 days she has witnessed these 40 times, 40 times, and 40 times, respectively.

- **Question 1.1** What are the probabilities that she witnesses jumping fish, that she witnesses a blue heron, and that she witnesses a squirrel on any particular day individually? Write down three expressions that provide these probabilities and assign them to names `wit_fish`, `wit_heron`, and `wit_squirrel` respectively.

  Alana is suspect that when she observes a jumping fish on a given day, she is more likely to also encounter the other two animals on the same day. Likewise if she first sees a great blue heron or a squirrel.

  The tibble `alana` from the `edsdata` package is the actual record of Alana's witnessing events.

---

[4]https://en.wikipedia.org/wiki/Transcendentalism

```
alana

## # A tibble: 80 x 4
##       day heron  fish squirrel
##     <int> <dbl> <dbl>    <dbl>
## 1     1     0     0        0
## 2     2     0     0        0
## 3     3     1     1        1
## 4     4     1     1        1
## 5     5     1     1        1
## 6     6     0     0        0
## 7     7     0     0        0
## 8     8     0     1        1
## 9     9     1     1        1
## 10   10     1     1        1
## # ... with 70 more rows
```

The rows are the 80 days and the columns are witness (1) or non-witness (0) of the three animals.

- **Question 1.2** If the observational unit is an individual observation made by Alana during her walk, then the tidy data guidelines tell us that each row contains multiple observations. Let us tidy the `alana` tibble using a `tidyr` function so that the three variables become materialized: `day`, `animal_observed`, and `witness`. Call the resulting tibble `alana_tidy`.

- **Question 1.3** Form a tibble called `summarized_observations` that gives the total number of times that Alana saw each of the three animals during her trip (or, more technically, the total number of 1 witnessing events for each animal). The tibble should have two variables named `animal_observed` and `n`. Do the values match the stated counts of 40?

- **Question 1.4** From your tibble `alana_tidy`, extract the number of days in which Alana witnessed *all* three animals. Store your answer (a single double value) in the name `observed_witness3`.

Question 2 We can now apply *simulation* to help address Alana's question. That is, we will assume that the true probability of observing any of the three animals follows what you found in **Question 1** and that the observation of one animal has no influence on the observation of any of the other two animals (i.e., each observation is *independent* of the other).

We can then simulate Alana's 80-day trip a great number of times. More specifically, we will simulate the number of days Alana saw all three animals during the said "artificial" trip. If the actual record of observing all three animals is different from what our simulation shows, we have evidence that supports Alana's claim that the probability of observing some animal may be *dependent* on the probability of first observing any of the other two. That would be welcome news – no more waiting around for the great blue heron!

Let us approach this simulation in parts, using the same recipe for simulation we learned. Be sure you have read and understood the examples from the textbook before proceeding.

- **Question 2.1** Write a function `findings_from_one_walk` that takes a double day as an argument and returns a tibble giving the results after one simulated walk in Alana's trip. Here is what this tibble looks like after one possible run for day 5.

day	animal_observed	witness
5	heron	1
5	fish	0
5	squirrel	1

NOTE: The values in the `witness` column should be different every time this function is called! How to generate these values? See the `sample` function we saw in class.

```
findings_from_one_walk(5) # an example call for a simulated "day 5"
```

Using your function `findings_from_one_walk`, we can simulate a full 80-day trip as follows:

```
map(1:80, findings_from_one_walk) |>
  bind_rows()
```

- **Question 2.2** Write a function `one_alana_trial` that takes no arguments, simulates a full 80-day trip, and returns the number of days that all three animals were witnessed. The result should vary each time it is run.

  HINT: To complete this, you can re-use code you have already written, i.e., from Question 4 in Part I.

```
one_alana_trial() # an example call
```

The returned value from your function `one_alana_trial()` composes one simulated value in our simulation.

- **Question 2.3** We are now ready to put everything together and run the simulation. Set a name `num_repetitions` to 1,000. Run your function `one_alana_trial()` `num_repetitions` number of times and store the simulated values in a vector named `witness3_trials`.

  NOTE: This may take a few seconds to run.

- **Question 2.4** Construct a tibble named `results` by aligning `num_repetitions` (create a sequence from 1 to `num_repetitions`) and `witness3_trials` with the column names `repetition` and `witness3_trial`, respectively.

- **Question 2.5** Using `ggplot2`, generate a histogram of `witness3_trial` from the tibble `results`. Remember to plot in density scale; you may also wish to lower the number of bins to a smaller value, say, 10. Your plots should label all axes appropriately and include a title informing what is being visualized.

- **Question 2.6** Compare your histogram with the value `observed_witness3` computed earlier. Where does it fall in this histogram? Is it close to the center where the "bulk" of the simulated values are?

```
observed_witness3
```

- **Question 2.7** Based on what you see in the above histogram and how it compares with `observed_witness3`, what would you say to the following statement: "since each animal is observed 40 out of 80, the chance of seeing all three animals after first seeing one of them is still 50-50, about the same as the chance of heads or tails after a fair coin toss"?

Question 3. A friend recently gave Cathy a replica dollar-coin. Cathy noticed that the coin has a slight bias towards "Heads". She tossed it a few times to find that the Heads/Tails ratio is like 3 to 2; that is, if she tosses the coin five times, she would observe three Heads and two Tails among the five outcomes. Noticing the curious behavior, Cathy asked a friend Jodie to play the following game with her. For preparation, Cathy and Jodie each prepare a pile of ten lollipops. Then they would repeat the following **ten times**.

- Jodie predicts Heads or Tails and Cathy tosses the coin. If the prediction matches the outcome, Jodie takes one lollipop from Cathy's pile; if the prediction does not match the outcome, Cathy takes one lollipop from Jodie's pile.

Let us do some analysis of the game.

- **Question 3.1** We have collected the results from one round of the game in the vectors `jodie_predictions` and `outcomes`. How many lollipops does Jodie have now?

```
jodie_predictions <- c("Tails", "Tails", "Heads",
                       "Heads", "Tails", "Tails",
                       "Heads", "Tails", "Tails",
                       "Heads")
outcomes          <- c("Tails", "Heads", "Heads",
                       "Heads", "Heads", "Tails",
                       "Heads", "Heads", "Heads",
                       "Tails")
```

- **Question 3.2** Suppose Jodie selects either side of the coin with equal chance, i.e., she does not favor "Heads" anymore than she does "Tails" and vice versa. Based on the observation that Cathy's coin follows a "Heads"/"Tails" ratio of 3:2, after one round of the game how many lollipops is Jodie expected to lose to Cathy?

Let us use simulation to check our analysis of the problem. We will define two functions, `simulate_cathy_coin()` and `simulate_jodie_prediction()`, that simulates a single flip of Cathy's coin and Jodie's prediction of the side a coin lands on, respectively.

```
simulate_jodie_prediction <- function() {
  # Jodie chooses either side of a coin with equal chance.
  sample(c("Heads", "Tails"), prob = c(1/2, 1/2), size = 1)
}
simulate_cathy_coin <- function() {
  # Cathy's coin is known to be biased towards "Heads" in a 3:2 ratio.
  sample(c("Heads", "Tails"), prob = c(3/5, 2/5), size = 1)
}
```

- **Question 3.3** Write a function called `one_flip_lollipop_wins` that simulates the number of lollipops Jodie wins after one flip of Cathy's coin. The function receives two arguments, a prediction function `coin_flip_func` and a coin flip function `prediction_func`; these should be called in the duration of the function to simulate a flip of Cathy's coin and Jodie's prediction of that coin. The function returns either `1` or `-1`. We can interpret `1` to be Jodie winning one lollipop and `-1` to be Jodie losing one lollipop.

```
# An example call using the functions corresponding
```

```
# to the game Cathy and Jodie are playing.
one_flip_lollipop_wins(simulate_cathy_coin, simulate_jodie_prediction)
```

- **Question 3.4** One round of the game Cathy and Jodie are playing consists of 10 coin flips. Write a function named `gains_after_one_round()` that simulates the number of lollipops Jodie wins after one round of the game. The function receives three arguments, `flips`, the number of flips in a round, `coin_flip_func`, and `prediction_func`. The function returns the number of lollipops Jodie wins after the round is over. Use the `replicate` construct to call `one_flip_lollipop_wins()` a number of times.

```
# One round consists of 10 coin flips.
gains_after_one_round(10, simulate_cathy_coin,
                          simulate_jodie_prediction)
```

- **Question 3.5** Let us now simulate the game a large number of times. Using the `replicate` construct with the function `gains_after_one_round()`, simulate 10,000 games. Collect the lollipops won from each simulated game in a vector named `simulated_gains`.

  The following function `hist_from_simulation()` produces a histogram of the simulated gains you generated.

```
hist_from_simulation <- function(simulated_results) {
  wins_tibble <- tibble(
    repetition = 1:length(simulated_results),
    gain = simulated_results
  )

  g <- ggplot(wins_tibble) +
    geom_histogram(aes(x = gain, y = after_stat(density)),
                   bins=12,
                   color = "gray", fill = "darkcyan")
  return(g)
}


hist_from_simulation(simulated_gains)
```

- **Question 3.6** What value does the bulk of the data center around? Does this agree with or contradict our earlier analysis in **Question 3.2**?

- **Question 3.7** Cathy proposes to switch roles. In other words, Jodie will toss the coin, and Cathy will predict. Knowing about the bias of her coin, Cathy bets on predicting "Tails" every time. Write a function `simulate_cathy_prediction` that receives no arguments and simulates Cathy's prediction under this scheme.

- **Question 3.8** Repeat **Question 3.5** but now using `simulate_cathy_prediction()`. Assign your simulated values to the name `simulated_gains_switched_roles`. Then generate a histogram of the results using `hist_from_simulation()`.

- **Question 3.9** Where is the bulk of the data centered around now? Is the prior knowledge of the coin's bias helpful for winning more lollipops in the long run? Or does Cathy come out the same regardless of who is making the predictions? Explain your reasoning.

- **Question 3.10** It will be helpful to visualize the two histograms in the same plot. Using `simulated_gains` and `simulated_gains_switched_roles`, generate an overlaid histogram showing the simulated results when Jodie predicts the coin flip together with the results when Cathy predicts the coin flip. Your plot should include a legend showing which histogram corresponds to which player making the predictions.

  HINT: Before writing any `ggplot2` code, you will need to develop a tibble that contains the results from both. See the code in `hist_from_simulation()` for some hints on how to accomplish this.

Question 4 This question is a continuation of **Question 3**.

Suppose we make a further change to the rule of the game so that the game will continue until either player loses all their lollipops. If one round is defined as a single coin toss, how many rounds will it take for them to complete the game? Let us compute the number of rounds by simulation.

Suppose that Jodie and Cathy each begin with 10 lollipops in their pile. There are two different ways the game ends.

- One is by Jodie losing all her lollipops,
- The other is by Cathy losing all her lollipops.

We have seen that the present strategy (Cathy always predicts "Tails") slightly favors Cathy winning. The histogram that follows this strategy clusters around a small negative value, approximately -2. We thus use a positive round number if Jodie loses the lollipops and the round number with the sign flipped if Cathy loses the lollipops.

We can accomplish the simulation scheme using the incantation:

```
gains_after_one_round(1, simulate_cathy_coin, simulate_cathy_prediction)
```

We can then use the `accumulate` construct shown in Sections 4.3 and 4.4. Recall that we can use this to call some function repeatedly, each time using the result of the previous application as the argument.

- **Question 4.1** First, write a function `total_lollipops_after_play` that takes a single argument `jodie_lollipops` that gives the current number of lollipops **in Jodie's pile**. This function:
  - Checks if Jodie lost all her lollipops (Jodie's pile contains 0 lollipops) or Cathy lost all her lollipops (Jodie's pile contains 20 lollipops). If this condition succeeds, the game should be terminated by calling `done()` in a return call, i.e., `return(done())`
  - Otherwise, simulate one round of the game as shown above. Save the result to a name `gains`. The function returns the addition of `gains` and the argument `lollipops`.

```
total_lollipops_after_play <- function(jodie_lollipops) {

}
```

- **Question 4.2** Write another function `play_until_empty_pile` that takes no arguments. This function will:
  - Simulate `total_lollipops_after_play` using the `accumulate` function. Call `accumulate` for a maximum of 1,000 rounds (e.g., using the input sequence `1:1000`) using the number of lollipops Jodie starts with as the initial value.

- Determine how many lollipops remain in Jodie's pile after the game is over. This can be done by inspecting the *last* value in the vector returned by `accumulate`. Call this `num_remaining`.
- If Jodie lost (how can you tell?), the function returns the number of rounds played.
- Otherwise, if Jodie won (how can you tell?), then the functions the *negative* of the round number.

```
play_until_empty_pile <- function() {


}
```

The following is an example call of your function. The total number of rounds played will be different each time the function is called. Most of the time you will observe a positive value.

```
play_until_empty_pile()
```

- **Question 4.3** Simulate the game 10,000 times and collect the results into a vector named `simulated_durations`.

- **Question 4.4** What are the maximum and the minimum of `simulated_durations`? How about the mean? According to these results, who is the winner in the long run? Is it true that Cathy always wins when she predicts "Tails"?

- **Question 4.5** Plot a histogram of the simulated duration values. Set the number of bins equal to 25.

- **Question 4.6** What does the shape of the distribution you found tell you about who wins in the long run? Can this histogram be used to draw such a conclusion?

Question 5: The Paradox of the Chevalier De Méré French people used to love (maybe still do) gambling with dice. In the 17th century, Antoine Gombaud Chevalier De Méré stated that the following two events are equally probable.

- throwing a dice four times in a row and 1 turning up at least once, and
- throwing a pair of dice 24 times in a row and two 1's simultaneously turning up at least once.

His logic was as follows:

- There are six faces to a dice. The chance of observing the first event is $4 * (1/6) = 2/3$ because there is a $1/6$ probability of a 1 turning up on a single roll and the dice is rolled 4 times.

- There are six faces to each dice. The chance of observing the second event is $24 * (1/36) = 2/3$ because there is a $1/36$ probability of two 1's turning up after a single roll and the pair is rolled 24 times.
  Unfortunately, there is a flaw in his logic. Using a bit of probability analysis, we have:

- There are six faces to a dice, so the chances of observing the first event is the *opposite* of seeing anything but 1 four times in a row, which is $1 - (5^4/6^4) = 0.517746....$

- There are 35 outcomes that are not a double 1's when rolling a pair of dice (out of 36 possible). Thus, the chance of observing the second event is the *opposite* of seeing any of those 35 outcomes 24 times. So we have: $1 - (35^24/36^24) = 0.491403....$

Not only the two values are different from each other, but also they are different from 2/3.

We are lucky that the theoretical analysis here has been straightforward. That is because we are able to take advantage of the fact that the events "at least one 1 turns up" and "no 1s turns up" are *complementary*. This is also sometimes called the "rule of subtraction" or the "1 minus rule".

However, theoretical analysis can be cumbersome and, depending on the problem, is not always possible. Sometimes it is easier to find the answer using *simulation*! Let us try to disprove the Chevalier's estimation by means of simulation.

Following is a function `simulate_dice_roll` that simulates one dice roll:

```
simulate_dice_roll <- function() {
  sample(1:6, size = 1)
}
simulate_dice_roll()
```

- **Question 5.1** Write a function `has_one_appeared` that simulates whether a 1 appears after a six-sided dice roll. The function takes no arguments and returns `TRUE` if the simulated roll turns up 1 and `FALSE` otherwise.

```
has_one_appeared() # an example call
```

- **Question 5.2** Write another function called `has_double_ones_appeared` that simulates whether two 1s appear after rolling a pair of six-sided dice. The function takes no arguments and returns `TRUE` if double 1s turn up after the simulated rolls and `FALSE` otherwise.

```
has_double_ones_appeared() # an example call
```

- **Question 5.3** Write a function `chavalier_first_event` that simulates the Chavalier's first event. The function takes no arguments. It should execute `has_one_appeared` four times and return a *Boolean* indicating whether a 1 has appeared.

```
chavalier_first_event() # example call
```

- **Question 5.4** Write a function `chavalier_second_event` that simulates the Chavalier's second event. This function takes no arguments. It should execute `has_double_ones_appeared` 24 times and return a *Boolean* indicating whether a pair of dice both turned up 1.

- **Question 5.5** Simulate each of the Chavalier's two events 10,000 times. Store the resulting values from each of the simulations into two vectors named `simulated_values_first_event` and `simulated_values_second_event`, respectively.

The following code chunk puts together your results into a tibble named `sim_results`.

```
sim_results <- tibble(
  first_event = simulated_values_first_event,
  second_event = simulated_values_second_event)
sim_results
```

- **Question 5.6** Tidy the tibble `sim_results` using a `tidyr` function so that each observation in the tibble refers to the outcome of a single simulation and materializes two variables:

event (referring to either the first or second event) and `outcome` (the outcome of that event). The resulting tibble should contain 20,000 observations. Assign this tibble to the name `sim_results_tidy`.

- **Question 5.7** Generate a bar chart using `sim_results_tidy`. Fill your bars according to the event in the variable `event`. We recommend using the "dodge" positional adjustment so that comparisons are easier to make.

  Repeat **Question 5.5** through **Question 5.7** to observe differences in the resulting histogram, if any. Then proceed to the following question:

- **Question 5.8** A fellow classmate claims that the above simulation is unable to disprove the Chavalier's flawed reasoning. He cites two reasons for his claim:

  - The "first event" and "second event" bars are too close to each other to say anything with confidence.
  - Because we repeated the simulation a large number of times, there is too much variability in the results.

  Which of his reasons, if any, are valid? Explain your reasoning.

5

Sampling

In the previous chapter we learned that we can make selections by chance using randomness and that we can encapsulate randomness by means of simulation. In this chapter we will use sampling more aggressively and learn to exploit random samples for drawing meaningful conclusions from data.

Specifically, we will learn that:

- By drawing random samples, we are able to observe what an unknown distribution might look like
- By conducting experiments using random sampling we can assess how unlikely or likely a phenomenon is at hand.

In both cases, the number of samples we draw and the manner of sampling play an important role.

5.1 To Sample or Not to Sample?

This chapter introduces some sampling preliminaries that we need for building our experiments.

5.1.1 Prerequisites

Before starting, let's load the tidyverse as usual.

```
library(tidyverse)
```

We will familiarize ourselves with the use of sampling from tibbles, or data frames.

In this chapter specifically, we will use the mpg data set that comes packaged with the tidyverse. Here is a preview of it:

```
mpg
```

```
## # A tibble: 234 x 11
##    manuf~1 model displ  year   cyl trans drv     cty   hwy
##    <chr>   <chr> <dbl> <int> <int> <chr> <chr> <int> <int>
## 1 audi    a4      1.8  1999     4 auto~ f        18    29
## 2 audi    a4      1.8  1999     4 manu~ f        21    29
## 3 audi    a4      2    2008     4 manu~ f        20    31
```

DOI: 10.1201/9781003320845-5

```
##  4 audi     a4        2     2008    4 auto~ f        21      30
##  5 audi     a4        2.8   1999    6 auto~ f        16      26
##  6 audi     a4        2.8   1999    6 manu~ f        18      26
##  7 audi     a4        3.1   2008    6 auto~ f        18      27
##  8 audi     a4 q~     1.8   1999    4 manu~ 4        18      26
##  9 audi     a4 q~     1.8   1999    4 auto~ 4        16      25
## 10 audi     a4 q~     2     2008    4 manu~ 4        20      28
## # ... with 224 more rows, 2 more variables: fl <chr>,
## #   class <chr>, and abbreviated variable name
## #   1: manufacturer
```

Each row of the tibble represents an individual; in the mpg tibble, each individual is a car model. Sampling individuals can thus be achieved by sampling the rows of a table.

The contents of a row are the values of different variables measured on the same individual. So the contents of the sampled rows form samples of values for each of the variables.

5.1.2 The existential questions: Shakespeare ponders sampling

When we get down to the business of sampling, some critical decisions must be decided before beginning to sample. Not unlike the famous soliloquy given by Prince Hamlet[1] in Shakespeare's *Hamlet*, the questions of how to sample are not always straightforward to answer and merit consideration. Here are the main ones to consider:

- Is the sampling to be done deterministically or probabilistically?
- Is the sampling to be done systematically or not?
- Can the same record appear more than once in the sampled data frame?
- Do the records have a equal chance of becoming a sample?

5.1.3 Deterministic samples

Recall that when working with data frames, we often anticipate the columns of the table to represent the properties we can obtain from an individual object. A collection of the properties representing an individual is a *record* or a *data object*. The rows of a data frame are the data records. The row X and column Y of the data frame is the property Y of the data object X.

By "sampling" we mean to select rows from the data frame. If we want to select distinct rows of the data frame, there is a convenient function slice for the action, which you may recall. The first argument of the function call specifies the source for sampling, and the other specifies the rows we draw using vector representation. The following generates a new tibble with rows 3, 25, and 100 of mpg.

```
mpg_sub <- mpg |>
  slice(c(3, 25, 100))
mpg_sub
```

```
## # A tibble: 3 x 11
##   manufa~1 model displ  year   cyl trans drv    cty   hwy
##   <chr>    <chr> <dbl> <int> <int> <chr> <chr> <int> <int>
```

---

[1] https://www.poetryfoundation.org/poems/56965/speech-to-be-or-not-to-be-that-is-the-question

```
## 1 audi      a4     2    2008    4 manu~ f        20    31
## 2 chevrol~ corv~  5.7  1999    8 auto~ r        15    23
## 3 honda    civic  1.6  1999    4 manu~ f        28    33
## # ... with 2 more variables: fl <chr>, class <chr>, and
## #   abbreviated variable name 1: manufacturer
```

Note that slice does not care if the row numbers appearing in the second argument are all different or if the row numbers are given in non-decreasing order. For instance, we can create a tibble containing four instances of Audi A4's and Chevrolet Corvette's.

```
mpg_sub <- mpg |>
  slice(c(3, 25, 3, 25, 3, 25, 3, 25))
mpg_sub
```

```
## # A tibble: 8 x 11
##    manufa~1 model displ year   cyl trans drv      cty   hwy
##    <chr>    <chr> <dbl> <int> <int> <chr> <chr> <int> <int>
## 1 audi     a4     2    2008    4 manu~ f        20    31
## 2 chevrol~ corv~  5.7  1999    8 auto~ r        15    23
## 3 audi     a4     2    2008    4 manu~ f        20    31
## 4 chevrol~ corv~  5.7  1999    8 auto~ r        15    23
## 5 audi     a4     2    2008    4 manu~ f        20    31
## 6 chevrol~ corv~  5.7  1999    8 auto~ r        15    23
## 7 audi     a4     2    2008    4 manu~ f        20    31
## 8 chevrol~ corv~  5.7  1999    8 auto~ r        15    23
## # ... with 2 more variables: fl <chr>, class <chr>, and
## #   abbreviated variable name 1: manufacturer
```

In the above examples, we knew beforehand which rows would appear in the sampled data frame because we specified explicitly the corresponding index of rows to include (e.g., four repeats of row 3 and row 25). We call such a sampling process with a selection vector a *deterministic sample*. The determinism refers to the non-existence of chance during the selection process.

An alternative to directly specifying the row numbers is specifying a condition on a variable for a record to be in the sample. We have also seen this before. This is the dplyr verb filter.

```
mpg_sub <- mpg |>
  filter(manufacturer == "land rover")
mpg_sub
```

```
## # A tibble: 4 x 11
##    manufa~1 model displ year   cyl trans drv      cty   hwy
##    <chr>    <chr> <dbl> <int> <int> <chr> <chr> <int> <int>
## 1 land ro~ rang~  4    1999    8 auto~ 4        11    15
## 2 land ro~ rang~  4.2  2008    8 auto~ 4        12    18
## 3 land ro~ rang~  4.4  2008    8 auto~ 4        12    18
## 4 land ro~ rang~  4.6  1999    8 auto~ 4        11    15
## # ... with 2 more variables: fl <chr>, class <chr>, and
## #   abbreviated variable name 1: manufacturer
```

The condition on the variable of a record either puts it in the subset or it does not. Exactly one of the two must occur. Like `slice`, `filter` also selects the records deterministically. If you are not convinced, you could run through each of the 234 rows of `mpg` by hand and manually check whether the manufacturer is indeed a `land rover`. Those that pass the check will end up in the sample data frame and, by the end of it all, you would end up with the same result as `filter` – no manual effort needed!

5.1.4 Random sampling

The antonym of "deterministic" is "non-deterministic", which is a concept that plays an important role in computer science. The meaning of it is quite obvious: it refers to a process that is not deterministic. However, it is quite vague in that it does not state *how* to draw the non-determinism. Data science prefers a more concrete form of non-determinism called *randomness*, which we have seen before. We call the process of sampling that leverages randomness to make its draws *random sampling*.

The pool of subjects from which we draw samples is called the *population*. There are multiple types of populations and the determining factors of the types is its quantity and the way the samples are generated.

When we draw a *random sample*, there are two questions that must be answered before the sampling is done: (1) what is the *population* being measured, and (2) what is the the *chance of selection* for each group in that population? If either of these points are not known beforehand, the sample obtained is **NOT** a random sample! This is an important point to emphasize because sometimes a sample can appear "random" even though it is not.

Following are some (non-comprehensive) examples of populations and how sampling might be carried out.

- *Accessible Data with Succinct Definition* We want to study the choice of major at a college. The registrar's office can generate a complete list of all the students currently attending the college. We could form a random sample by selecting at random a student from the enrollment database where each student has an equal chance of appearing in the sample. Alternatively, a deterministic sample may apply some filtering. This can be done to narrow down the population to a specific group (e.g., the full-time sophomore students). We can then form a random sample from just this group or we can opt to include all the students in the group in our study.
- *Continuous Population Requiring Discretization* We want to study the quality of air based on how "blue" the skies are. There is no clear definition of the population. The geographical location of measurement equipment, the area to cover in the sky, and the time of measurement can be factors in determining the samples. The determination of these factors essentially *discretize* the continuous data. After determining all these factors succinctly, a technician can make a measurement.
- *Data with Succinct Definition Beyond Reach* We want to study the relationship between the height and weight of the people living in the United States. The population appears to have a clear definition, but it is difficult to determine who gets to be included in the population because the population is transient (due to babies coming to life, people moving out of the country, etc.). By specifying at which point of time the person must be living in the United States, the definition can be succinct. The problem is that it is impossible to include everyone in the population. There are more than 300 million people, and we have no way of knowing who those people are.

- *Non-existing Population Requiring Active Generation* We want to examine the fairness of a coin. Each time we throw the coin, we generate a record. Throwing the coin N times you get a population of N records, but you may wish to use the entire population for the analysis.

5.1.5 To sample systematically or not?

In systematic samples, we choose samples in a systematic manner. For example, we can select all students whose university-issued ID number has an odd number in the second-to-last position from the population of all students in a university. Also, we could select, from the entire population of the United States, all voters whose mailing address has a postal code ending in 1, whose street address has a digit 2, owns a car, and whose license plate has either the letter T or the numeral 6. Systematic samples are usually *convenient* to carry out and are often called *convenience samples*.

Here is an example of making systematic sampling on the mpg data set using dplyr verbs. Let's start by preparing a version of the data frame that includes the row index for better visual inspection.

```
mpg_with_index <- mpg |>
  mutate(row_index = row_number()) %>%
  relocate(row_index, .before = manufacturer)
mpg_with_index
```

```
## # A tibble: 234 x 12
##    row_index manufac~1 model displ  year   cyl trans drv
##        <int> <chr>     <chr> <dbl> <int> <int> <chr> <chr>
## 1          1 audi      a4      1.8  1999     4 auto~ f
## 2          2 audi      a4      1.8  1999     4 manu~ f
## 3          3 audi      a4      2    2008     4 manu~ f
## 4          4 audi      a4      2    2008     4 auto~ f
## 5          5 audi      a4      2.8  1999     6 auto~ f
## 6          6 audi      a4      2.8  1999     6 manu~ f
## 7          7 audi      a4      3.1  2008     6 auto~ f
## 8          8 audi      a4 q~   1.8  1999     4 manu~ 4
## 9          9 audi      a4 q~   1.8  1999     4 auto~ 4
## 10        10 audi      a4 q~   2    2008     4 manu~ 4
## # ... with 224 more rows, 4 more variables: cty <int>,
## #   hwy <int>, fl <chr>, class <chr>, and abbreviated
## #   variable name 1: manufacturer
```

We will now pick one of the first 10 rows at random, and then we will select every 10th row after that.

```
# Pick random start among rows 0 through 9; then every 10th row.
start <- sample(1:10, size = 1)
mpg_with_index |>
  slice(seq(start, n(), by = 10))
```

```
## # A tibble: 24 x 12
##    row_index manufac~1 model displ  year   cyl trans drv
##        <int> <chr>     <chr> <dbl> <int> <int> <chr> <chr>
## 1          2 audi      a4      1.8  1999     4 manu~ f
## 2         12 audi      a4 q~   2.8  1999     6 auto~ 4
## 3         22 chevrolet c150~   5.7  1999     8 auto~ r
## 4         32 chevrolet k150~   6.5  1999     8 auto~ 4
## 5         42 dodge     cara~   3.3  2008     6 auto~ f
## 6         52 dodge     dako~   3.9  1999     6 manu~ 4
## 7         62 dodge     dura~   5.2  1999     8 auto~ 4
## 8         72 dodge     ram ~   5.2  1999     8 manu~ 4
## 9         82 ford      expl~   4.6  2008     8 auto~ 4
## 10        92 ford      must~   3.8  1999     6 auto~ r
## # ... with 14 more rows, 4 more variables: cty <int>,
## #   hwy <int>, fl <chr>, class <chr>, and abbreviated
## #   variable name 1: manufacturer
```

Run the code a few times to see how the output varies.

This attempts to be a combination of systematic sampling and random sampling. The starting point is random and there are 10 possibilities, and we pick one from these 10 possibilities with probability 10% (i.e., one in ten chances). The selection after determining the starting point is deterministic; we select every tenth element from the starting point. Therefore, there are just ten different sample data sets out of the original data set.

There are many other ways of selecting one out of ten records from the dataset. We could expand the systematic selection to assembling every 10 rows into groups and selecting exactly one from each group. This selection would open up more possibilities.

Despite the use of the random initial point, we would not consider the resulting sample drawn to be a *random sample*. The systematic selection of every 10 rows after the random initial choice may disproportionately favor one or more groups from the population. Without further consideration, we would label this a *convenience* sample. Convenience samples are examined in greater depth in Section 5.6[2].

Note, also, that there exist combinations that can never be generated. For instance, in both of the schemes we just described, it is impossible to generate a sample that contains the very first two records. Can you see why?

5.1.6 To sample with replacement or not?

When we conduct random sampling, there are two main strategies. One is to prohibit any record from appearing more than once in the sampled data frame. The other is, of course, not to impose such a restriction.

We call the first strategy *sampling without replacement* and the second *sampling with replacement*. Most of the sampling we conduct is sampling *with* replacement. The reasons for this will follow in the next section.

---

[2]https://ds4world.cs.miami.edu/sampling.html#convenience-sampling

5.1.7   To select samples uniformly or not?

There are situations in which the population we want to observe is a mixture of several groups, but the representation is not equal among them. For example, a university may have a 50:50 ratio among male and female students but a dataset covering some of the student population has a 20:80 ratio instead. We may want to sample four male students for every female student sampled in order to remain true to the gender distribution in the student population.

5.2   Distribution of a Sample

One of the most important applications of sampling is to obtain an approximation for a "true" distribution, which we often do not know. Think of a question like: "what percent of people in the United States are taller than six feet?" If we had access to a census that contained all qualifying individuals in the population, we could compute the answer directly. Without a census, we would need to not only define who are "the people in the United States," but engage in an enormous effort to collect and record all the heights – a task that is not only tedious, but likely impossible to accomplish.

Enter sampling. Instead of trying to account for every individual in the country, we can *sample* people from the population that we determine and approximate the height of people in the United States from the sample we collect. Our hope is that any histogram we construct from our sample will be close enough to the one we would have if we could record the heights from the entire population. All the heights from the U.S. population follows what we call the *true distribution* and those from our sampled population the *sampling distribution*.

Let's examine what an sampling distribution looks like with an example of throwing die.

5.2.1   Prerequisites

As before, let's load in the tidyverse as usual.

```
library(tidyverse)
```

5.2.2   Throwing a 6-sided die

There are six faces to a (fair) die. The outcome of throwing a die is the face that turns up.

```
die <- tibble(face = 1:6)
die

## # A tibble: 6 x 1
##    face
##   <int>
## 1     1
## 2     2
```

```
## 3        3
## 4        4
## 5        5
## 6        6
```

Since the die is fair, the probability of the faces are all equal to each other and each is exactly 1/6. The *true distribution* then is the probability where each of the faces is exactly 1/6. Here is what that distribution looks like. Observe that the height is equal for all bars, at the level of 1/6.

```
ggplot(die) +
    geom_histogram(aes(x = face, y = after_stat(density)),
                   bins = 6, color = "gray")
```

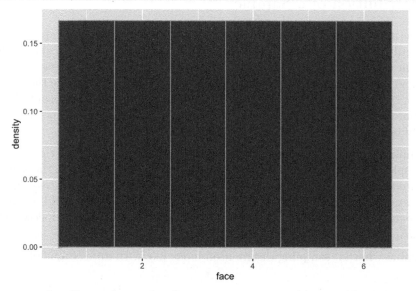

Does that mean that if you throw the die six times, you would see each of the faces exactly once? Not at all!

Here is a quick counterargument: Assume that your intuition is correct. After throwing the die five times you observed five different faces. You could then predict the face of the sixth one to be the one that appears next. In fact, you could apply the same logic to each consecutive five throws to predict the next one. What have we done here? Our observation is leading to a realization that, after the first five throws, the remaining throws actually become *deterministic*. This contradicts the *randomness* we are expecting from the die.

Therefore, the proportion of faces you see after throwing a die multiple times can be substantially different from the expected "1/6 for each face".

Note that the sampling we are about to conduct is different from the example of sampling heights from the U.S. population on three counts.

- The fair die may not exist in the real world, so we use a tibble that represents a fair die.
- The population exists only in our throwing of the die; that is, each time we throw the die, the throw and its outcome becomes a new member of the population.
- Because we can generate a sample any number of times, the population is actually infinite in size.

Nevertheless, we know what the true distribution looks like. It is our histogram shown above.

5.2.2.1   A sampling distribution

We now generate a sampling distribution using simulation. We used previously `sample` for generating samples from a vector. Here we will use the `dplyr` function `slice_sample` for sampling from a data frame. It draws at random from the rows of a data frame, with an argument to sample with replacement. It also receives an argument `n` for the sample size, and it returns a data frame consisting of the rows that were selected.

Here are the results of 10 rolls of a die.

```
die |>
  slice_sample(n = 10, replace = TRUE)
```

```
## # A tibble: 10 x 1
##       face
##      <int>
## 1       2
## 2       1
## 3       2
## 4       6
## 5       2
## 6       4
## 7       6
## 8       3
## 9       2
## 10      4
```

Run the cell above a few times and observe how the faces selected changes.

We can adjust the sample size by changing the number given to the `n` argument. Let's generalize the call by writing a function that receives a sample size `n` and generates the sampling histogram for this sample size.

```
sample_hist <- function(n) {
  die_sample <- die |>
    slice_sample(n = n, replace = TRUE)
  ggplot(die_sample, aes(x = face, y = after_stat(density))) +
    geom_histogram(bins = 6, color = "gray")
}
```

Here is a histogram of 10 rolls. Note how it does not look anything like the true distribution from above. Run the cell a few times to see how it varies.

```
sample_hist(10)
```

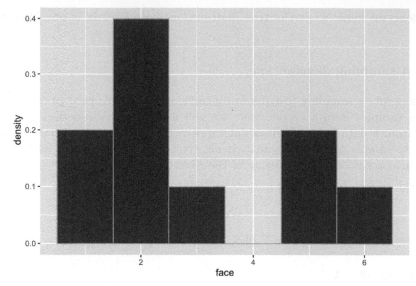

When we increase the sample size, we see that the distribution gets closer to the true distribution.

```
sample_hist(100)
```

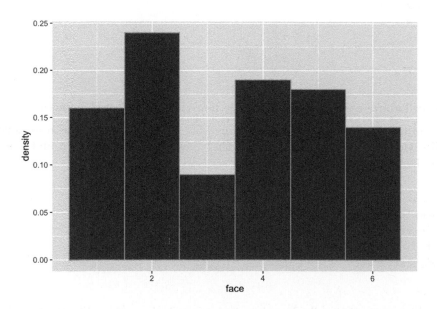

```
sample_hist(1000)
```

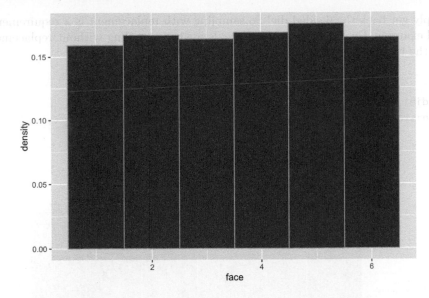

The phenomenon we have observed - the sampling distribution growing closer and closer to the true distribution as the sample size is increased - is an important concept in data science. In fact, it is so important that it has a special name: we call it the *Law of Averages*.

A critical requirement for the Law of Averages to be applicable is that the samples come from the same underlying distribution and do not depend on the drawing of other samples, e.g. the rolling of a 2 does not make the rolling of a 6 more likely on the next roll. This idea is sometimes called sampling "independently and under identical conditions", where the resulting distribution is "independently and identically distributed".

5.2.3 Pop quiz: why sample with replacement?

The careful reader may have noticed that in the calls to slice_sample in this section, the replace argument has been set to TRUE, i.e. the sampling is done with replacement. Why not sample without?

You may have already guessed at an answer: if we sample without replacement, we would not be able to make more than 6 draws since we would have run out of faces on the die to choose from! Here is what happens when trying to sample 6 times without replacement.

```
die |>
  slice_sample(n = 6, replace = FALSE)
```

```
## # A tibble: 6 x 1
##      face
##     <int>
## 1      6
## 2      5
## 3      2
## 4      3
## 5      4
## 6      1
```

We simply get back the 6-sided die! So sampling with replacement is a requirement for the dice roll example. But, there is another problem when sampling without replacement. Let's look at the true distribution again.

```
ggplot(die) +
  geom_histogram(aes(x = face, y =  after_stat(density)), bins = 6,
            color = "gray")
```

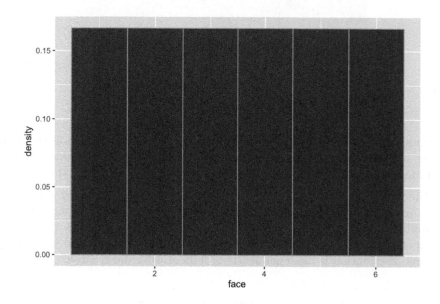

When we sample *without* replacement, we are effectively removing the possibility of that event happening from future draws. Put another way, this would be the same as somehow erasing or "deleting" a face from the die before drawing again. For instance, let's check the true distribution after rolling a 2.

```
selected_face <- 2   # assume a 2 was rolled
slice(die, -selected_face) |>
  ggplot() +
  geom_histogram(aes(x = face, y =  after_stat(density)), bins = 6,
            color = "gray")
```

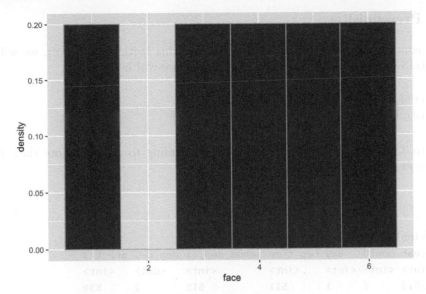

If we were to sample again from this die, the probability of rolling any of the faces is no longer the same as when we rolled the 2. For one thing, the probability of rolling the faces $1, 3, 4, 5, 6$ has increased to 20% (or $1/5$) and it is impossible to roll a 2. The underlying distribution has fundamentally changed, confirmed by the above ggplot.

Recall that a prerequisite for the Law of Averages to work is that the drawing of samples must be done "independently and under identical conditions". Sampling without replacement is a clear violation of this assumption. And yet, the story does not end there. In our example, there were only 6 individuals to choose from – the faces of a 6-sided die. If we were to increase the number of individuals that we could sample from, say the entire U.S. population, then the effect observed here actually becomes negligible. Sampling without replacement remains an important method for sampling, especially in drug studies with treatment/control groups where it is physically not possible to sample with replacement. Cloning people remains the imagination of science fiction, at least for now. Therefore, our hope is that both sampling plans will generate similar results in practice.

5.3 Populations

The Law of Averages can be useful when the population from which to draw samples is very large.

As an example, we will study a population of flight delay times. The tibble `flights` contains all 336,776 flights that departed from New York City in 2013. It is drawn from the Bureau of Transportation Statistics[3] in the United States. The help page `?flights` contains more documentation about the dataset.

---

[3]https://www.transtats.bts.gov/Homepage.asp

5.3.1 Prerequisites

We will continue to make use of the tidyverse in this section. Moreover, we will load the nycflights13 package, which has the flights table we will be using.

```
library(tidyverse)
library(nycflights13)
```

There are 336,776 rows in flights, each corresponding to a flight. Note that some delay times are negative; those flights left early.

```
flights
```

```
## # A tibble: 336,776 x 19
##     year month   day dep_time sched_dep_~1 dep_d~2 arr_t~3
##    <int> <int> <int>    <int>        <int>   <dbl>   <int>
## 1   2013     1     1      517          515       2     830
## 2   2013     1     1      533          529       4     850
## 3   2013     1     1      542          540       2     923
## 4   2013     1     1      544          545      -1    1004
## 5   2013     1     1      554          600      -6     812
## 6   2013     1     1      554          558      -4     740
## 7   2013     1     1      555          600      -5     913
## 8   2013     1     1      557          600      -3     709
## 9   2013     1     1      557          600      -3     838
## 10  2013     1     1      558          600      -2     753
## # ... with 336,766 more rows, 12 more variables:
## #   sched_arr_time <int>, arr_delay <dbl>, carrier <chr>,
## #   flight <int>, tailnum <chr>, origin <chr>,
## #   dest <chr>, air_time <dbl>, distance <dbl>,
## #   hour <dbl>, minute <dbl>, time_hour <dttm>, and
## #   abbreviated variable names 1: sched_dep_time,
## #   2: dep_delay, 3: arr_time
```

One flight departed 43 minutes early, and one was 1301 minutes late.

```
slice_min(flights, dep_delay)
```

```
## # A tibble: 1 x 19
##    year month   day dep_time sched_dep_t~1 dep_d~2 arr_t~3
##   <int> <int> <int>    <int>         <int>   <dbl>   <int>
## 1  2013    12     7     2040          2123     -43      40
## # ... with 12 more variables: sched_arr_time <int>,
## #   arr_delay <dbl>, carrier <chr>, flight <int>,
## #   tailnum <chr>, origin <chr>, dest <chr>,
## #   air_time <dbl>, distance <dbl>, hour <dbl>,
## #   minute <dbl>, time_hour <dttm>, and abbreviated
## #   variable names 1: sched_dep_time, 2: dep_delay,
## #   3: arr_time
```

```
slice_max(flights, dep_delay)
```

```
## # A tibble: 1 x 19
##     year month   day dep_time sched_dep_t~1 dep_d~2 arr_t~3
##    <int> <int> <int>    <int>         <int>   <dbl>   <int>
## 1   2013     1     9      641           900    1301    1242
## # ... with 12 more variables: sched_arr_time <int>,
## #   arr_delay <dbl>, carrier <chr>, flight <int>,
## #   tailnum <chr>, origin <chr>, dest <chr>,
## #   air_time <dbl>, distance <dbl>, hour <dbl>,
## #   minute <dbl>, time_hour <dttm>, and abbreviated
## #   variable names 1: sched_dep_time, 2: dep_delay,
## #   3: arr_time
```

If we visualize the distribution of delay times using a histogram, we can see that they lie almost entirely between -10 minutes and 200 minutes.

```
ggplot(flights, aes(x = dep_delay, y = after_stat(density))) +
  geom_histogram(col="grey", breaks = seq(-50, 200, 1))
```

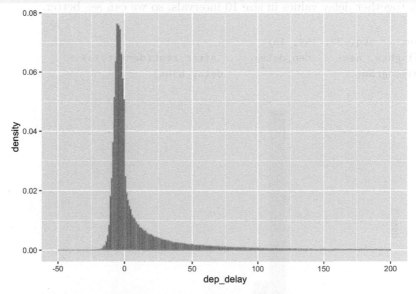

We are interested in the bulk of the data here, so we can ignore the 1.83% of flights with delays of more than 150 minutes. We will only use the core part of the data excluding the 1.83% using the filter method as we show below. Note that the table flight still has all the data.

```
nrow(filter(flights, dep_delay > 150)) / nrow(flights)
```

```
## [1] 0.0183386
```

```
delay_bins <- seq(-50, 150, 1)
ggplot(flights, aes(x = dep_delay, y = after_stat(density))) +
  geom_histogram(col="grey", breaks = delay_bins)
```

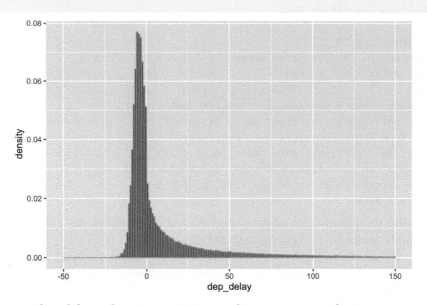

We group together delay values in size 10 intervals, so we can see better.

```
delay_bins <- seq(-50, 150, 10)
ggplot(flights, aes(x = dep_delay, y = after_stat(density))) +
  geom_histogram(col="grey", breaks = delay_bins)
```

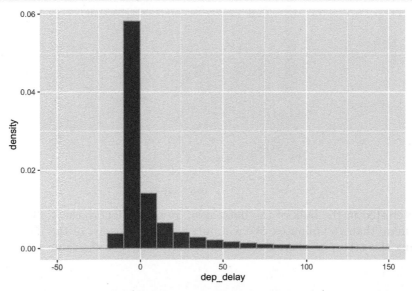

The tallest bar in the histogram is the $(-10, 0]$ bar is just below 0.06, which is equal to 6%. There are ten values of delay minutes in the bin, so we multiply this density value by 10 to assess that the delays in the interval occupy slightly below 60% of the data. We can confirm the visual assessment by counting rows:

```
nrow(filter(flights, dep_delay > -10 & dep_delay <= 0)) / nrow(flights)
```

```
## [1] 0.5571062
```

5.3.2 Sampling distribution of departure delays

Let us now think of the 336,776 flights as a *population*, and draw random samples from it with replacement. As we may try using various values for the number of samples, let us define a function for sampling and plotting.

The function `sample_hist` takes the sample size as its argument, which we call n, and draws a histogram of the results.

```
sample_hist <- function(n) {
  flights_sample <- slice_sample(flights, n = n, replace = TRUE)
  ggplot(flights_sample, aes(x = dep_delay, y = after_stat(density))) +
    geom_histogram(breaks = delay_bins, color = "gray")
}
```

As we saw in the simulation for throwing a die, the more samples we have, the closer the histogram of the sample approximates the histogram of the population.

Let us compare two sample sizes 10 and 100.

```
sample_hist(10)
```

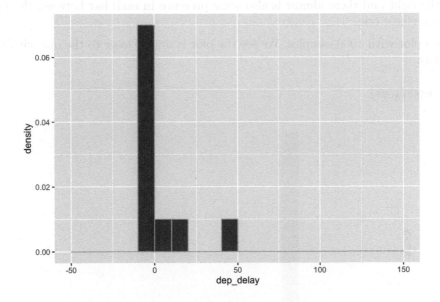

`sample_hist(100)`

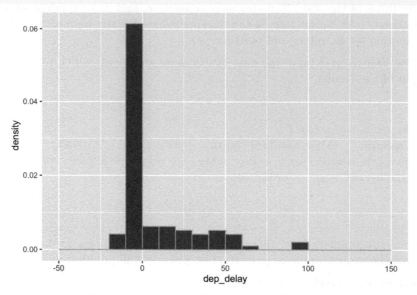

We may notice various differences between the two plots. The plots may vary each time we execute the code. Of all the differences that emerge when we execute the two codes and compare the plots, the most notable are the number of bars and how far the bars reach to the right without a gap. We notice that in the case of 100 samples, there are always bars far to the right and there almost is also some presence in each bar between the tallest bar and the farthest one.

Here is a plot with 1000 samples. We see the plot is much closer to the one with the entire flight data.

`sample_hist(1000)`

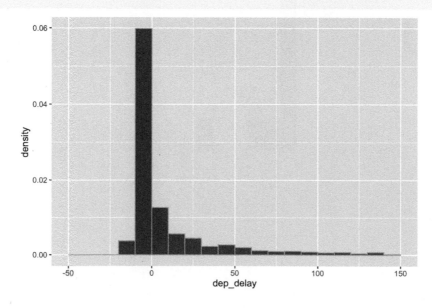

5.3.3   Summary: Histogram of the sample

From the experiments we have conducted with 10, 100, and 1000 samples, we have observed that:

- The larger the sample gets, the closer the sampling distribution becomes to the true distribution, which we obtain from the true population.
- At some large size, with high probability, the approximation of the true distribution we produce through sampling becomes almost indistinguishable from the true one.

Because of the two properties, we can use random sampling in *statistical inference*. The indistinguishable nature of sampling distributions is quite powerful. When the population is large, we can turn to sampling to generate a sample data set of a reasonably large size, which saves computation time.

5.4   The Mean and Median

There are quantitative evaluations that we commonly use in statistical analysis. For instance, in a population of flights that departed New York City (NYC) in 2013, the *median* departure delay can tell us something about the central tendency of the data – are most departure flights delayed, on time, or are they ahead of schedule? Similarly, in baseball, the ability of a player to produce a hit is measured by the player's *average* (or *mean*) hit rate; the hitting rate is the number of hits divided by the number of times the player stood in the batter box.

The criteria *median* and *mean* are useful for understanding properties of a population. When we calculate a quantity from a population, we call such quantities *parameters* of the population. Thus, the median and mean are two useful parameters we wish to estimate somehow. We will see how to in the next section, but for now we will study some properties of the mean and median to develop an appreciation for its usefulness.

First some vocabulary:

- The *mean* or *average* (we will use both interchangeably) of a vector of numbers is the sum of the elements divded by the number of elements in the vector.
- The *median* is the value at the middle position in the data after reordering the values. It is also called an *order statistic*, which we will learn about later.

5.4.1   Prerequisites

We will make use of the tidyverse in this chapter, so let's load it in as usual.

```
library(tidyverse)
```

5.4.2   Properties of the mean

The function `mean` returns the mean of a vector.

```
some_numbers <- c(8, 1, 8, 7, 9)
mean(some_numbers)
```

[1] 6.6

We can note some properties of the mean based on this example.

- The mean is not necessarily an element in the vector.
- The mean must be between the smallest and largest values in the vector and it need not be exactly in the middle between these two extremes.
- If the vector is measured in some unit (e.g. ft^2), the mean carries the same unit.
- The mean is a "smoother". For instance, imagine that the numbers in the vector above are dollar amounts owned by five friends. They pool the money together and deal out the money in even amounts among the friends. The amount each person will have is given in the above output: $6.6.
- Proportions are means. If a vector consists of only 1s and 0s, the sum of the vector is the number of 1s in it, and the mean is the *proportion* of 1s in the vector. Following is an example:

```
ones_and_zeros <- c(1, 1, 0, 0)
sum(ones_and_zeros)
```

[1] 2

```
mean(ones_and_zeros)
```

[1] 0.5

Thus, the mean tells us that 50% of the values in ones_and_zeros are 1s.

5.4.3   The mean: a measure of central tendency

Let us visualize the distribution of some_numbers using a histogram.

```
tibble(some_numbers) |>
  ggplot() +
  geom_histogram(aes(x = some_numbers, y = after_stat(density)),
                 color = "gray", fill = "darkcyan", binwidth = 1)
```

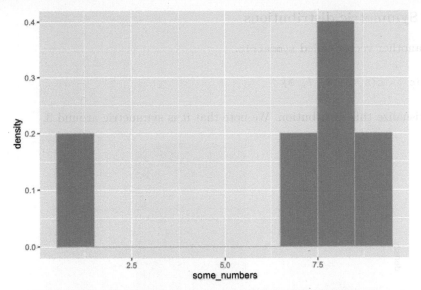

If we imagine the base of a histogram as a seesaw, then the mean is the *pivot* by which the seesaw is supported by. We can denote this point visually using a triangle.

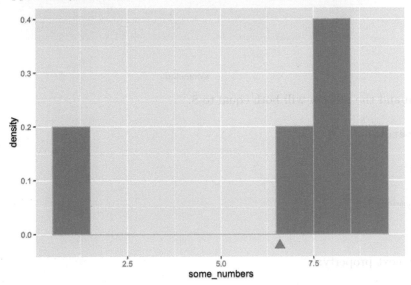

The pivot is the point at which this "seesaw" is balanced. If we nudge the point in any direction, say toward 7.5, the seesaw tips over to the right; if it is closer to 5, the seesaw tips to the left. That pivot is the mean. Thus,

---

The mean is the "pivot" or "balancing point" of the histogram.

5.4.4 Symmetric distributions

Here is another vector called `symmetric`.

```
symmetric <- c(8, 8, 8, 7, 9)
```

Let us visualize this distribution. We note that it is symmetric around 3.

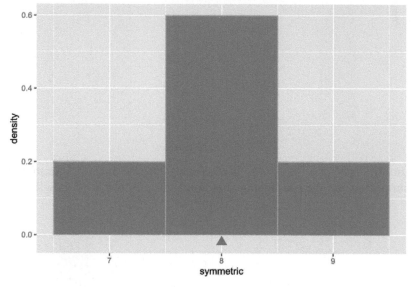

The mean and the median will both equal to 8.

```
mean(symmetric)
```

```
## [1] 8
```

```
median(symmetric)
```

```
## [1] 8
```

Thus, our next property:

For symmetric distributions, in general, the mean and the median will equal.

What if the distribution was not symmetric? We already have an example of one in the numbers contained by `not_symmetric`. Let us overlay both on the same histogram.

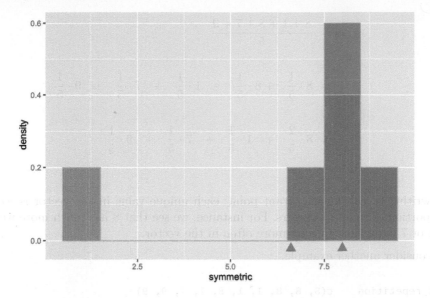

The cyan histogram shows the distribution represented by symmetric and the orange the distribution represented by some_numbers. If we again imagine the x-axis as a seesaw, we see that the cyan distribution balances around the pivot at 8. Both the mean and median of the cyan distribution is equal to 8.

The orange histogram of some_numbers starts out the same as the cyan at the right end but its left bar is at the value 1; the darker shading shows where the two histograms overlap. The median of the orange distribution is also 8. However, in order to keep this distribution "balanced" on the seesaw, we need to scoot the pivot to the left, to 6.6. That is the mean of the orange histogram.

```
median(some_numbers)
```

```
## [1] 8
```

```
mean(some_numbers)
```

```
## [1] 6.6
```

5.4.5   The mean of two identical distributions is identical

We end this section with one more important property about means. We said earlier that the mean of a vector is just the sum of its elements divided by the number of elements. However, observe that we could have calculated it in different ways:

$$\text{mean} = \frac{8 + 1 + 8 + 7 + 9}{5}$$

$$= 8 \cdot \frac{1}{5} + 8 \cdot \frac{1}{5} + 1 \cdot \frac{1}{5} + 7 \cdot \frac{1}{5} + 9 \cdot \frac{1}{5}$$

$$= 8 \cdot \frac{2}{5} + 1 \cdot \frac{1}{5} + 7 \cdot \frac{1}{5} + 9 \cdot \frac{1}{5}$$

This rewriting reveals an important point: each unique value in the vector is *weighted* by the proportion of times it appears. For instance, we see that 8 has much more weight than either 1 or 7 because it appears more often in the vector.

Let us consider another example.

```
lots_of_repetition <- c(8, 8, 8, 1, 1, 8, 7, 7, 9, 9)
mean(lots_of_repetition)
```

[1] 6.6

This vector has the same mean as some_numbers. What is the point here? The mean of a vector depends only on the distinct values and their proportions, *not* on the number of elements in the vector. Put differently, the mean depends only on the *distribution* of values.

Thus, our final property:

If two vectors have the same distribution, their means will equal.

5.5   Simulating a Statistic

In many situations we do not know the value of a parameter of a population. Yet, it turns out that random sampling is a reliable tool we can use to find a good estimate of a parameter. Since large-scale sampling produces a sampling distribution that approximates the true distribution, we hope that the value we compute from it will also be close enough to the parameter we could obtain from the population directly when the sample size is large enough.

Before jumping into the tidyverse, first some vocabulary:

- A *parameter* is some useful quantitative evaluation criteria about a population. For instance, a parameter is the mean height of all individuals in the United States. This is typically unknown to us.

- A *statistic* is a value obtained from an sampling distribution (which we can generate). The purpose of a statistic is to *estimate* a parameter. When we compute the mean or median from an sampling distribution, we call these statistics the *sample* mean and median, respectively.

5.5.1 Prerequisites

We continue to make use of the tidyverse in this section. We will first look at a toy example to demonstrate different properties of a statistic and then simulate a statistic in practice using flights data from `nycflights13`.

```
library(tidyverse)
library(nycflights13)
```

5.5.2 The variability of statistics

One thing we need to keep in mind is that the sampling distribution can look quite different between runs. Recall the histogram of departure delays in `flights`.

```
delay_bins <- seq(-50, 150, 10)
ggplot(flights, aes(x = dep_delay, y = after_stat(density))) +
  geom_histogram(col="grey", breaks = delay_bins)
```

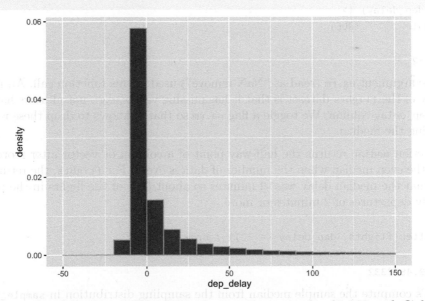

Here is a sampling histogram of the delays in a random sample of 1,000 such flights.

```
sample_1000 <- slice_sample(flights, n = 1000, replace = TRUE)
ggplot(sample_1000, aes(x = dep_delay, y = after_stat(density))) +
  geom_histogram(col="grey", breaks = delay_bins)
```

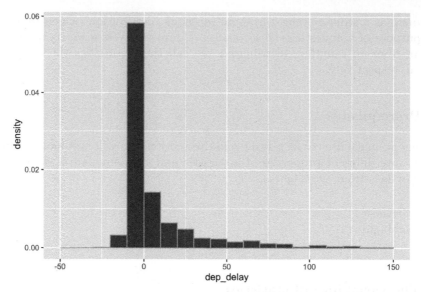

Since we are assuming that the population of all departed NYC flights in 2013 is available to us in flights, we can compute directly the value of the parameter "median flight delay." This is a rare luxury not usually possible using real data.

We can check the value we obtain through sampling against the value we obtain from the entire population.

```
flights |>
  pull(dep_delay) |>
  median(na.rm = TRUE)
```

[1] -2

Note the argument na.rm (read as "NaN remove") used in this function call. An interesting property of the flights dataset is that it has missing values, e.g., not all rows have a value in the dep_delay column. We toggle a flag na.rm so that R knows to drop these rows before computing the median.

The function median returns the half-way point of a column or vector after reordering the values (the even median when the number of data is even). For flights, the returned value means that the median delay was -2 minutes so about 50% of the flights in the population had early departures of 2 minutes or more.

```
nrow(filter(flights, dep_delay <= -2)) / nrow(flights)
```

[1] 0.4892332

Now let's compute the sample median from the sampling distribution in sample_1000.

```
sample_1000 |>
  pull(dep_delay) |>
  median(na.rm = TRUE)
```

[1] -1

Let us check what happens if we tried another random sample of 1,000 flights.

```
slice_sample(flights, n = 1000, replace = TRUE) |>
  pull(dep_delay) |>
  median(na.rm = TRUE)
```

```
## [1] -2
```

You can run the code a few times to see that the value that appears as the result is not consistent, but always either -1 or -2. While the two values are close, you cannot assert that the median is -1 or -2.

Is there any way out of this? Can we say anything about how likely the median is -1 and how likely the median is -2? We thus turn to simulation. We compute the sample median statistic many times using sampling and then see a histogram of the numbers we obtain.

5.5.3   Simulating a statistic

Before getting down to the business of estimating the median distribution, let us set up the steps for obtaining a distribution of a statistic.

Step 1: Select the statistic to estimate. Two questions are at hand. What statistic do we want to estimate? How many samples do we use in each estimate?

Step 2: Write the code for estimation. We need to write the code for sampling and then computing a statistic from the sample. Typically we encapsulate such steps in a function that we can use in a call to the function `replicate`.

Step 3: Generate estimations and visualize. A question at hand is how many times do we repeat the experiment? We may not have an answer to the question. We will write the code for repeating experiments, collecting the results in a vector whose length is equal to the number of repetitions, and then generating a visualization of the results we have collected in the vector. The code we write may be a function that takes the number of repetitions as an argument.

5.5.4   Guessing a "lucky" number

Before continuing on with the `flights` data, let us first see an example of simulating a statistic using a toy problem.

Your friend invites you to a game of rolling die. He rolls a special 30-sided die (sometimes called a "D30" die) 10 times and, after each roll, adds some "lucky" number of his choosing. He does not say what his lucky number is or show you the dice rolls before he adds the number to the roll.

Let us suppose his lucky number is 6. We can use `sample` to simulate the total rolls we get to see. This is what one set of total rolls might look like after the game is over:

```
lucky_number <- 6
d30_dice <- 1:30
sample(d30_dice, size = 10, replace=TRUE) + lucky_number
```

```
## [1] 11 32 18 13 10 32 14 17 14 26
```

Can you figure out your friend's lucky number knowing only these total rolls?

5.5.5   Step 1: Select the statistic to estimate

Let us play with two different statistics:

- From the total rolls shown to you, use the smallest one as your guess. Since the minimum value on a 30-sided die is 1, a good estimate is the smallest roll minus 1. Let us call this the *min-based estimator.*
- From the total rolls shown to you, leverage the *mean* of the rolls as your guess. Let us call this the *mean-based estimator.*

We simulate 10 rolls of a 30-sided die and add the lucky number to each roll. The following function one_game simulates the action.

```
one_game <- function(lucky_number) {
  sample(1:30, size = 10, replace=TRUE) + lucky_number
}
one_game(6)
```

```
## [1] 16 29 14 22 14 14 11  8 35 18
```

We can call one_game to generate a sample and then compute a statistic from it. This is a process we can repeat a large number of times to simulate a large number of sample statistics.

5.5.6   Step 2: Write code for estimation

The following function min_based implements the min-based estimator. As noted earlier, we use the smallest total roll as our estimate and subtract 1.

```
min_based <- function(sample) {
  min(sample) - 1
}
```

A mean-based estimate uses the *mean* total roll instead. This will look similar to our min-based estimator. However, to guess the lucky number from the mean-based estimate, we need to account for the average, or *expected*, value of a fair 30-sided dice. This can be computed using probability theory, but simulation is often easier!

Let us simulate a 30-sided die over a big run, say, 10,000 rolls.

```
sample_rolls <- sample(d30_dice, size = 10000, replace = TRUE)
```

We then take the *mean* of these rolls.

```
d30_expected_value <- mean(sample_rolls)
d30_expected_value
```

```
## [1] 15.5154
```

This value we just computed has a special name: the *expected* value. Statisticians give it this name because it is the number that we can *expect* to get in the long run, say, after doing an arbitrarily large number of trials. For a 30-sided die, probability theory dictates this should be 15.5. Our answer in d30_expected_value comes pretty close!

We are now ready to implement the mean-based estimator.

```
mean_based <- function(sample) {
  mean(sample) - d30_expected_value
}
```

The following function receives as an argument a functional `estimator`, which can be either the min-based or mean-based estimator. The function simulates one game, computes a statistic using the `estimator` function passed in, and returns the calculated value.

```
simulate_one_stat <- function(estimator) {
  one_game(lucky_number = 6) |>
    estimator()
}
```

Here is an example call using the min-based estimator. Run the cell a few times and observe the variability in the result. Does it guess correctly the lucky number?

```
simulate_one_stat(min_based) # an example call
```

```
## [1] 7
```

5.5.7   Step 3: Generate estimations and visualize

We issue 10,000 repetitions of the simulation. Here is the `replicate` call that makes use of the function `simulate_one_stat`. This is done twice, once for the mean-based estimator and again for the min-based estimator.

```
reps <- 10000

mean_estimates <- replicate(n = reps, simulate_one_stat(mean_based))
min_estimates <- replicate(n = reps, simulate_one_stat(min_based))
```

We collect the results together into a tibble `estimate_tibble`. Note that a pivot transformation is applied here so that one row corresponds to exactly one simulated statistic.

```
estimate_tibble <- tibble(mean_est = mean_estimates,
                          min_est = min_estimates) %>%
  pivot_longer(c(mean_est, min_est),
               names_to = "estimator", values_to = "estimate")
estimate_tibble
```

```
## # A tibble: 20,000 x 2
##     estimator estimate
##     <chr>        <dbl>
## 1 mean_est      6.28
## 2 min_est       6
## 3 mean_est      5.98
## 4 min_est       7
```

```
##  5 mean_est    4.08
##  6 min_est     8
##  7 mean_est    0.985
##  8 min_est     9
##  9 mean_est    3.38
## 10 min_est     7
## # ... with 19,990 more rows
```

We can visualize the sampling distribution of the two estimators using an overlaid histogram. We also plot the "balancing point" for each of these distributions (shown using a triangle) by computing the mean of the corresponding simulated statistics.

```
bins <- seq(0, 22, 1)

dist_mean_tib <- tibble(
  estimator = c("min_est", "mean_est"),
  mean = c(min_estimates |> mean(),
           mean_estimates |> mean()))

ggplot(estimate_tibble) +
  geom_histogram(aes(x = estimate, y = after_stat(density),
                     fill = estimator),
                 position = "identity", alpha = 0.5,
                 color = "gray", breaks = bins) +
  geom_point(data = dist_mean_tib,
             aes(x = mean, y = 0, color = estimator),
             shape = "triangle", size = 3) +
  scale_x_continuous(breaks = bins)
```

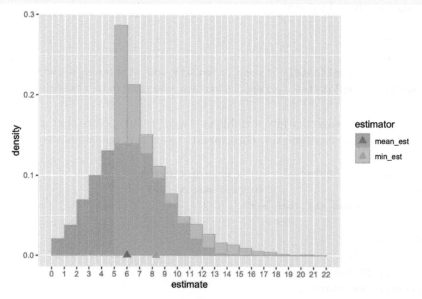

Let us examine the min-based estimator. First, observe that its balancing point (in cyan) is a number larger than the lucky number 6. Because this number is *larger* than the true value (the lucky number 6), we conclude that the min-based estimator *overestimates* and is, therefore, *biased*. This should not be surprising considering that, by design, this estimator can only predict a number equal to the lucky number or greater. This is the "bottoming out" effect that we observe in the $(5, 6]$ bin.

In contrast, the mean-based estimator appears to overestimate about as often as it underestimates. This gives rise to a familiar "bell-shaped" curve. We can see the effect of this is a balancing point that is about equal to the true value. Therefore, we can say that the estimator is *unbiased*.

The downside of the mean-based estimator is that it trades off bias for more *variability*. This turns out to be one of the advantages of the min-based estimator: high bias but low variability. This raises an important trade-off between *variance* and *bias* when working with statistics.

Which estimator would you choose for this problem? Are there other situations where you might prefer one more than the other, given the bias-variance trade-off?

5.5.8 Median flight delay in `flights`

Let us now return to estimating the median flight delay in the `flights` data frame. Recall that, following our assumptions, we know the value of the parameter the statistic is trying to estimate. It is the value -2.

5.5.9 Step 1: Select the statistic to estimate

We will draw random samples of size 1,000 from the population of flights and simulate the median.

5.5.10 Step 2: Write code for estimation

We know how to generate a random sample of 1,000 flights.

```
sampled <- flights |>
  slice_sample(n = 1000, replace = TRUE)
```

We also know how to compute the median of this sample.

```
sampled |>
  pull(dep_delay) |>
  median(na.rm = TRUE)
```

```
## [1] -2
```

Let's wrap this up into a function we can use in a `replicate` call.

```
one_sample_median <- function() {
  sample_median <- flights |>
    slice_sample(n = 1000, replace = TRUE) |>
    pull(dep_delay) |>
    median(na.rm = TRUE)
```

```
    return(sample_median)
}
```

5.5.11 Step 3: Generate estimations and visualize

We will issue 5,000 repetitions of our simulation. Here is the call to `replicate` that makes use of the function `one_sample_median`. Note that this simulation takes a bit more time to run: we are repeating an experiment where we draw 1,000 random samples a total of 5,000 times!

```
num_repetitions <- 5000
medians <- replicate(n = num_repetitions, one_sample_median())
```

Here are what some of the sample medians look like.

```
medians_df <- tibble(medians)
medians_df
```

```
## # A tibble: 5,000 x 1
##      medians
##        <dbl>
##  1        -2
##  2        -1
##  3        -2
##  4        -1
##  5        -2
##  6        -2
##  7        -2
##  8        -2
##  9        -2
## 10        -1
## # ... with 4,990 more rows
```

Of course, it would be much better to visualize these results using a histogram. This histogram displays the *sampling distribution* of the statistic. As before, let us also annotate this histogram with the balancing point of the distribution.

```
ggplot(medians_df) +
  geom_histogram(aes(x = medians, y = after_stat(density)),
                 color = "gray", fill = "darkcyan", bins = 3) +
  geom_point(aes(x = mean(medians), y = 0),
             shape = "triangle",
             color = "salmon", size = 3)
```

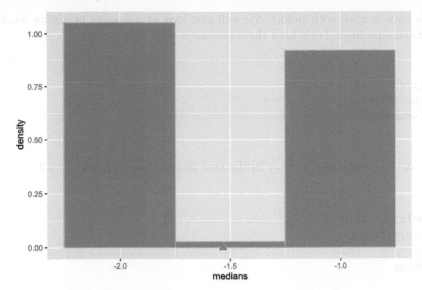

The sample median is very likely to be about -2, which is exactly the value of the population median. This is because the sampling distribution of 1,000 flight departure delays is close to what we know about the true distribution. We can guess that in running the experiment, each sampling distribution with 1,000 flights would have looked like the true one.

We can also observe that the balancing point of this distribution is pulled to the right of the true value. This is likely due to the original distribution having a long right-tail. Hence, we can say that the median statistic is *biased*. However, the bias in this case does not bear much practically as the simulated statistics are so close to the population parameter. This may be a case where we say that the result is "good enough."

The interested reader can extend the analysis to try a mean-based estimate for this example instead of the median-based used here. Would this estimator also be biased? How about its variability?

5.6 Convenience Sampling

The story goes that we can derive meaningful conclusions about a population using sampling distributions. Such distributions are formed by the application of random sampling. The one we have used so far has been drawing random samples with (or without) replacement. We refer to such a scheme as *simple random sampling*. However, simple random sampling is not the only way to generate a sampling distribution. We briefly discussed another method at the start of this chapter: *convenience sampling*. In this section, we examine convenience sampling in greater detail and discuss its suitability for statistical analysis.

5.6.1 Prerequisites

We will continue to make use of the tidyverse in this section. Our example in this section will be data about New York City flight delays in 2013 from the package nycflights13, which we have worked with before. We will also look at a sample of flights made available in the edsdata package. Let's load these packages.

```
library(tidyverse)
library(nycflights13)
library(edsdata)
```

Recall that the median flight delay in the true distribution of flights is −2.

```
median_delay <- flights |>
  pull(dep_delay) |>
  median(na.rm = TRUE)
median_delay
```

```
## [1] -2
```

5.6.2   Systematic selection

Recall that systematic selection is a kind of convenience sample. In systematic selection, we pick some random pivot (say, the 4th row), and then select every i-th row after that. This method of sampling is quite popular because of its sheer simplicity. For example, if we wanted to sample a student population at a university, we could select all students based on the digits in their university-issued ID number, e.g., selecting all students who have an odd number in the second-to-last position of their ID.

We can explore this sampling strategy using the flights tibble. We will select a random row in the tibble as the pivot, and then select every 100th row after that (recall that the dataset is quite large, with ~330K rows!).

```
start <- sample(1:nrow(flights), size = 1)
start
```

```
## [1] 260042
```

Does the generated distribution mirror what we know about the true distribution?

```
selected_rows <- seq(start, nrow(flights), 100)

slice(flights, selected_rows) |>
  ggplot(aes(x = dep_delay, y = after_stat(density))) +
  geom_histogram(fill = "darkcyan", color = "gray",
```

```
                      breaks = seq(-50, 150, 10)) +
        ggtitle(str_c("starting row = ", start))
```

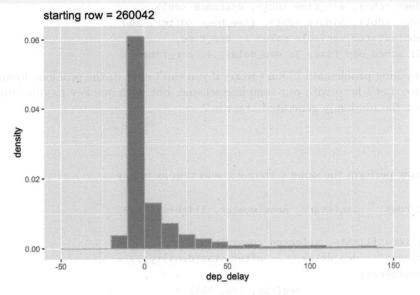

Looks good! The sampling distribution still looks a whole lot like the true distribution of flight delays. However, the number of samples that appear in the tail varies greatly across runs – why might that be? Run the cell a few times and observe the effect of the selected start variable on the resulting distribution.

From our initial exploration, it looks like systematic sampling is a good bet. Any statistics we compute using this sampling strategy is likely to provide a good estimate of the population parameter in question.

Now suppose that we reorganized the flights data a bit into a tibble called mystery_flights.

```
mystery_flights <- mystery_flights |>
  relocate(ID, .before = year)
mystery_flights
```

```
## # A tibble: 336,776 x 20
##       ID  year month   day dep_t~1 sched~2 dep_d~3 arr_t~4
##    <int> <int> <int> <int>   <int>   <int>   <dbl>   <int>
## 1      1  2013     1     1     517     515       2     830
## 2      2  2013     1     1     533     529      -6     850
## 3      3  2013     1     1     542     540       2     923
## 4      4  2013     1     1     544     545      -4    1004
## 5      5  2013     1     1     554     600      -6     812
## 6      6  2013     1     1     554     558      -3     740
## 7      7  2013     1     1     555     600      -5     913
## 8      8  2013     1     1     557     600      -6     709
## 9      9  2013     1     1     557     600      -3     838
## 10    10  2013     1     1     558     600      -2     753
```

```
## # ... with 336,766 more rows, 12 more variables:
## #   sched_arr_time <int>, arr_delay <dbl>, carrier <chr>,
## #   flight <int>, tailnum <chr>, origin <chr>,
## #   dest <chr>, air_time <dbl>, distance <dbl>,
## #   hour <dbl>, minute <dbl>, time_hour <dttm>, and
## #   abbreviated variable names 1: dep_time,
## #   2: sched_dep_time, 3: dep_delay, 4: arr_time
```

Notice anything problematic? Don't worry if you can't spot it; the problem doesn't jump at first glance. Let's keep with our sampling scheme, but with one key modification. We will assert that the randomly generated start is 2.

```
start <- 2
```

We can now perform the same systematic selection as before.

```
selected_rows <- seq(start, nrow(mystery_flights), 100)

slice(mystery_flights, selected_rows) |>
  ggplot(aes(x = dep_delay, y = after_stat(density))) +
  geom_histogram(fill = "darkcyan", color = "gray",
                 breaks = seq(-50, 150, 10)) +
  ggtitle(str_c("starting row = ", start))
```

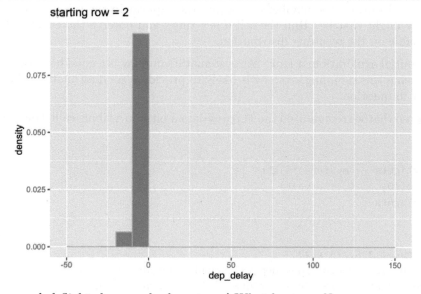

All of the sampled flights have early departures! What happened?

Let's break down the steps we took. The first row selected is at index 2, as told by start, and each row after increases by increments of 100. If we write out some of these indices, we would select rows:

```
tibble(row_index = seq(2, nrow(mystery_flights), 100))
```

```
## # A tibble: 3,368 x 1
##    row_index
```

```
##        <dbl>
## 1          2
## 2        102
## 3        202
## 4        302
## 5        402
## 6        502
## 7        602
## 8        702
## 9        802
## 10       902
## # ... with 3,358 more rows
```

There are several patterns that can be gleaned from this listing, but we will direct your attention to one in particular: these row numbers are all even! If we pick out some of these rows from mystery_flights, we find something revealing.

```
mystery_flights |>
  slice(c(2, 102, 202, 302, 402))
```

```
## # A tibble: 5 x 20
##      ID  year month   day dep_time sched~1 dep_d~2 arr_t~3
##   <int> <int> <int> <int>    <int>   <int>   <dbl>   <int>
## 1     2  2013     1     1      533     529      -6     850
## 2   102  2013     1     1      754     759      -4    1039
## 3   202  2013     1     1      933     937      -3    1057
## 4   302  2013     1     1     1157    1200     -10    1452
## 5   402  2013     1     1     1418    1419     -14    1726
## # ... with 12 more variables: sched_arr_time <int>,
## #   arr_delay <dbl>, carrier <chr>, flight <int>,
## #   tailnum <chr>, origin <chr>, dest <chr>,
## #   air_time <dbl>, distance <dbl>, hour <dbl>,
## #   minute <dbl>, time_hour <dttm>, and abbreviated
## #   variable names 1: sched_dep_time, 2: dep_delay,
## #   3: arr_time
```

Compare this with the rows just before these.

```
mystery_flights |>
  slice(c(1, 101, 201, 301, 401))
```

```
## # A tibble: 5 x 20
##      ID  year month   day dep_time sched~1 dep_d~2 arr_t~3
##   <int> <int> <int> <int>    <int>   <int>   <dbl>   <int>
## 1     1  2013     1     1      517     515       2     830
## 2   101  2013     1     1      753     755      -2    1056
## 3   201  2013     1     1      932     930       2    1219
## 4   301  2013     1     1     1157    1205      -8    1342
## 5   401  2013     1     1     1416    1411       5    1603
## # ... with 12 more variables: sched_arr_time <int>,
## #   arr_delay <dbl>, carrier <chr>, flight <int>,
## #   tailnum <chr>, origin <chr>, dest <chr>,
```

```
## #    air_time <dbl>, distance <dbl>, hour <dbl>,
## #    minute <dbl>, time_hour <dttm>, and abbreviated
## #    variable names 1: sched_dep_time, 2: dep_delay,
## #    3: arr_time
```

It turns out that the flights with even row numbers all have early departures. By fixing `start` to be an even value, our systematic sampling scheme was "fooled" into always choosing flights that are ahead of schedule. Under such circumstances, we conclude that this sample is *not* a random sample!

5.6.3   Beware: the presence of patterns

The `mystery_flights` tibble is a contrived example that required careful reorganization of the rows to create a setup where *every* flight with an even row index among the approximately 330K flights present in the dataset had an early departure. While it is quite unlikely that a real-world dataset would contain such an anomaly, the example points to valuable lessons that can occur in practice, especially when dealing with a convenience sample.

Real-world datasets are rife with patterns. Manufacturing errors due to a particular malfunctioning machine that assemble every n-th product; software engineering teams where every tenth member is designated the product manager; university-issued ID's where ID numbers ending in 0 are reserved for faculty members. Any systematic sampling scheme is much more prone to selecting samples that follow (unexpected) patterns than a simple random sample would be. Not unlike like the story with `mystery_flights`, a biased sample can result and lead to misleading (and likely erroneous) findings. Even more, some may be willing to exploit such patterns for the sole possibility of increasing the significance of their results – we would hardly call them data scientists!

Sampling strategies demand prudence on the part of the data scientist. It is for this reason that random sampling is the most principled approach.

5.7   Exercises

Be sure to install and load the following packages into your R environment before beginning this exercise set.

```
library(tidyverse)
library(edsdata)
library(gapminder)
```

Question 1 Following are some samples of students in a Data Science course named "CSC100":

1. All CSC100 students who attended office hours in the third week of classes
2. All undergraduate freshmen in CSC100
3. Every 11th person starting with the first person in the classroom on a random day of lecture
4. 11 students picked randomly from the course roster

Which of these samples are *random samples*, if any?

Question 2. The following function `twenty_sided_die_roll()` simulates one roll of a 20-sided die. Run it a few times to see how the rolls vary.

```
twenty_sided_die_roll <- function() {
  return(sample(seq(1:20), size = 1))
}
twenty_sided_die_roll()
```

Let us determine the *mean* value of a twenty-sided die by simulation.

- **Question 2.1** Create a vector `sample_rolls` that contains 10,000 simulated rolls of the 20-sided die. Use `replicate()`.

- **Question 2.2** Compute the mean of the vector `sample_rolls` you computed and assign it to the name `twenty_sided_die_expected_value`.

- **Question 2.3** Which of the following statements, if any, are true?

    1. The value of `twenty_sided_die_expected_value` is an element in `sample_rolls`.
    2. The distribution of `sample_rolls` is roughly uniform and symmetric around the mean.
    3. The value of `twenty_sided_die_expected_value` is at the midpoint between 1 and 20.
    4. The computed mean does not carry a unit.

Question 3: Convenience sampling. The tibble `sf_salary` from the package `edsdata` gives compensation information (names, job titles, salaries) of all employees of the City of San Francisco from 2011 to 2014 at annual intervals. The data is sourced from the Nevada Policy Research Institute's Transparent California[4] database and then tidied by Kaggle[5].

Let us preview the data:

```
library(edsdata)
sf_salary
```

```
## # A tibble: 148,654 x 13
##       Id Employe~1 JobTi~2 BasePay Overt~3 Other~4 Benef~5
##    <dbl> <chr>     <chr>     <dbl>   <dbl>   <dbl>   <dbl>
## 1      1 NATHANIE~ GENERA~ 167411.       0 400184.      NA
## 2      2 GARY JIM~ CAPTAI~ 155966. 245132. 137811.      NA
## 3      3 ALBERT P~ CAPTAI~ 212739. 106088.  16453.      NA
## 4      4 CHRISTOP~ WIRE R~   77916  56121. 198307.      NA
## 5      5 PATRICK ~ DEPUTY~ 134402.    9737 182235.      NA
## 6      6 DAVID SU~ ASSIST~ 118602    8601 189083.      NA
## 7      7 ALSON LEE BATTAL~  92492.  89063. 134426.      NA
## 8      8 DAVID KU~ DEPUTY~ 256577.       0  51322.      NA
## 9      9 MICHAEL ~ BATTAL~ 176933.  86363.  40132.      NA
## 10    10 JOANNE H~ CHIEF ~  285262       0  17116.      NA
## # ... with 148,644 more rows, 6 more variables:
## #   TotalPay <dbl>, TotalPayBenefits <dbl>, Year <dbl>,
## #   Notes <lgl>, Agency <chr>, Status <chr>, and
```

---

[4]https://transparentcalifornia.com/salaries/san-francisco/
[5]https://www.kaggle.com/datasets/kaggle/sf-salaries

```
## #   abbreviated variable names 1: EmployeeName,
## #    2: JobTitle, 3: OvertimePay, 4: OtherPay, 5: Benefits
```

Suppose we are interested in examining the mean total compensation of San Francisco employees in 2011, where total compensation data is available in the variable `TotalPay`.

- **Question 3.1** Apply the following three tidying steps:
    - Include only those observations from the year 2011.
    - Add a new variable `TotalPay (10K)` that contains the total compensation in amounts of tens of thousands.
    - Add a new variable `dataset` that contains the string `"population"` for all observations.

    Assign the resulting tibble to the name `sf_salary11`.

In most statistical analyses, it is often difficult (if not impossible) to obtain data on every individual from the underlying population. We instead prefer to draw some smaller sample from the population and estimate *parameters* of the larger population using the collected sample.

- **Question 3.2** Let us treat the 36,159 employees available in `sf_salary11` as the *population* of San Francisco city employees in 2011. What is the annual mean salary according to this tibble (with respect to `TotalPay`)? Assign your answer to the name `pop_mean_salary11`.

In Section 5.6[6] we learned that we need to be careful about the selection of observations when sampling data. One (generally bad) plan is to sample employees that are somehow *convenient* to sample. Suppose you randomly pick two letters from the English alphabet, say "G" and "X", and decide to form two samples: employees whose name starts with "G" and employees whose name starts with "X". Perhaps you are convinced that such a sample should be "random" enough...

- **Question 3.3** Explain why a sample drawn under this sampling plan would *not* be a *random sample*.

- **Question 3.4** Add a new variable to `sf_salary11` named `first_letter` that gives the first letter of an employee's name (provided in `EmployeeName`). Assign the resulting tibble to the name `with_first_letter`.

- **Question 3.5** Generate a bar plot using `with_first_letter` that shows the number of employees whose name begins with a given letter. For instance, the names of 2854 employees start with the letter "A". Fill your bars according to whether the bar corresponds to the letters "G" or "X".

The following function `plot_salary_histogram()` receives a tibble as an argument and compares the compensation distribution from the sample with the population using an overlaid histogram. No changes are needed in the following chunk; just run the code.

```
# an example call using the full data
plot_salary_histogram(with_first_letter)
```

- **Question 3.6** Write a function `plot_and_compute_mean_stat()` that receives a tibble as an argument and:
    - Plots an overlaid histogram of the compensation distribution with that of the population.
    - Returns the mean salary from the sample in the `TotalPay (10K)` variable as a *double*.

---

[6]https://ds4world.cs.miami.edu/sampling.html#convenience-sampling

```
plot_and_compute_mean_stat(with_first_letter) # an example call
```

The value reported by `plot_and_compute_mean_stat(with_first_letter)` is the *parameter* (because we computed it directly from the population) we hope to estimate by our sampling plan.

- **Question 3.7** Write a function `filter_letter()` that receives a character `letter` (e.g., "X") and returns the data consisting of all rows whose `EmployeeName` starts with the uppercase letter matching `letter`. Moreover, the value in the variable `dataset` should be mutated to the string `letter` for all observations in the resulting tibble.

```
filter_letter("Z") # an example call
```

- **Question 3.8** Let us now compare the convenience sample salaries with the full data salaries by calling your function `plot_compute_mean_stat()`. Call this function twice, once for the sample corresponding to "G" and another for the sample corresponding to "X".

- **Question 3.9** We have now examined two convenience samples. Do these give an accurate representation of the compensation distribution of the population of San Francisco employees? Why or why not?

- **Question 3.10** As we learned, a more principled approach is to use random sampling. Let us form a simple random sample by sampling at random *without replacement*. Target 1% of the observations in the population. The value in the variable `dataset` should be mutated to the string `"sample 1"` for all observations. Assign the resulting tibble to the name `random_sample1`.

- **Question 3.11** Repeat **Question 3.10**, but now target 5% of the rows. This time, rename the values in the variable `dataset` to the string `"sample 2"`. Assign the resulting tibble to the name `random_sample2`.

- **Question 3.12** Call your function `plot_compute_mean_stat()` on the two random samples you have formed.

  You should repeat **Question 3.10** through **Question 3.12** a few times to get a sense of how much the statistic changes with each random sample.

- **Question 3.13** Do the statistics vary more or less in random samples that target 1% of the observations than in samples that target 5%? Do the random samples offer a better estimate of the mean salary value than the convenience samples we tried? Are these results surprising or is this what you expected? Explain your answer.

Question 4 The tibble `penguins` from the package `palmerpenguins` includes measurements for penguin species, island in Palmer Archipelago, size, and sex. Suppose you are part of a conservation effort interested in surveying annually the population of penguins on Dream island. You have been tasked with estimating the number of penguins currently residing on the island.

To make the analysis easier, we will assume that each penguin has already been identified through some attached ID chip. This identifier starts at 1 and counts up to N, where N is the total number of observations. We will also examine the data only for one of the recorded years, 2007.

- **Question 4.1** Apply the following tidying steps to `penguins`:

  − Filter the data to include only those observations from Dream island in the year 2007.

  – Create a new variable named `penguin_id` that assigns an identifier to the resulting observations using the above scheme. (Hint: you can use `seq()`.)

Assign the resulting tibble to the name `dream_with_id`.

- **Question 4.2** In each survey you stop after observing 20 penguins and writing down their ID. It is possible for a penguin to be observed more than once. Generate a *vector* named `one_sample` that consists of the penguin IDs observed after one survey. Simulate this sample from the tibble `dream_with_id`.

- **Question 4.3** Generate a histogram in *density scale* of the observed IDs from the sample you found in `one_sample`. We suggest using the bins `seq(1,46, 1)`.

We will try to estimate the population of penguins on Dream island from this sample. More specifically, we would like to estimate N, where N is the largest ID number (recall this number is unknown to us!). We will try to estimate this value by trying two different statistics: a *max*-based estimator and a *mean*-based estimator.

- **Question 4.4** A max-based estimator simply returns the largest ID observed from the sample. Write a function called `max_based_estimate` that receives a vector x, computes the *max*-based estimate, and returns this value.

```
max_based_estimate(one_sample) # an example call
```

- **Question 4.5** The mean of the observed penguin IDs is likely halfway between 1 and N. We have that the midpoint between any two numbers is $\frac{1+N}{2}$. Solving this for N yields our mean-based estimate. Using this, write a function called `mean_based_estimate` that receives a vector x, computes the *mean*-based estimate, and returns the computed value.

```
mean_based_estimate(one_sample) # an example call
```

- **Question 4.6** Analyze several samples and histograms by repeating **Question 7.2** through **Question 7.5**. Which estimates, if any, capture the correct value for N?

- **Question 4.7** Write a function `simulate_one_stat` that receives two arguments, a tibble `tib` and a function `estimator` (that can be either be your `mean_based_estimate()` or `max_based_estimate()`). The function simulates a survey using `tib`, computes the statistic from this sample using the function `estimator`, and returns the computed statistic as a double.

```
simulate_one_stat(dream_with_id, mean_based_estimate) # example call
```

- **Question 4.8** Simulate 10,000 *max* estimates and 10,000 *mean* estimates. Store the results in the vectors `max_estimates` and `mean_estimates`, respectively.

- **Question 4.9** The following code creates a tibble named `stats_tibble` using the estimates you generated in **Question 4.8**.

```
stats_tibble <- tibble(max_est = max_estimates,
                       mean_est = mean_estimates) |>
  pivot_longer(c(max_est, mean_est),
               names_to = "estimator", values_to = "estimate")
stats_tibble
```

Generate a histogram of the *sampling distributions* of both statistics. This should be a single plot containing *two* histograms. You will need to use a positional adjustment to see both distributions together.

- **Question 4.10** How come the mean-based estimator has estimates larger than 46 while the max-based estimator doesn't?

- **Question 4.11** Consider the following statements about the two estimators. For each of these statements, state whether or not you think it is correct and explain your reasoning.

  - The max-based estimator is *biased*, that is, it almost always underestimates.
  - The max-based estimator has *higher* variability than the mean-based estimator.
  - The mean-based estimator is *unbiased*, that is, it overestimates about as often as it underestimates.

Question 5 This question is a continuation of the City of San Francisco compensation data from *Question 3* and assumes that the function `filter_letter()` exists and that the names `sf_salary11` and `pop_mean_salary11` have already been assigned.

- **Question 5.1** Let us write a function `sim_random_sample()` that samples `size` rows from tibble `tib` by sampling at random *with replacement*. The function returns a tibble containing the sample.

- **Question 5.2** Write a function `mean_stat_from_sample()` that receives a sample tibble `tib`. The function computes the mean of the variable `TotalPay` (`10K`) and returns this value as a double.

```
mean_stat_from_sample(sf_salary11) # using the full data
```

- **Question 5.3** Generate 10,000 simulated mean statistics using `sim_random_sample()` and `mean_stat_from_sample()`. Each simulated mean statistic should be generated from the tibble `sf_salary11` using a sample size of 100.

  The following code chunk puts the simulated values you found into a tibble named `stat_tibble`.

```
stat_tibble <- tibble(rep=1:10000,
                      mean=mean_stats)
```

- **Question 5.4** Generate a histogram showing the sampling distribution of these simulated mean statistics. Then, attach, to the histogram, a square at the population mean for 2011 at $y = 0$, with a size 2 square as the point.

  We see that the population mean lies in the "bulk" of the simulated mean statistics. Now that we have learned about different sampling plans, we can compare the statistics generated by these plans with this histogram.

- **Question 5.5** Compute a mean statistic called `head_stat` using the *first* 1,000 rows of `sf_salary`. Then compute a mean statistic called `tail_stat` using the *last* 1,000 rows.

- **Question 5.6** We saw that we can form *convenience* samples by partitioning the observations using the first letter of `EmployeeName`. Using a `purrr` map function, generate a *list* of tibbles (each corresponding to all employees whose name starts with a given letter) by mapping the 26 letters of the English alphabet to the function `filter_letter()`. Assign the resulting list to the name `by_letter`.

  Hint: `LETTERS` is a vector containing the letters of the English alphabet in uppercase.

- **Question 5.7** Use another `purrr` map function to map the list of tibbles `by_letter` to the function `mean_stat_from_sample` to obtain a vector of mean statistics for each sample. Assign the result to the name `letter_group_stats`.

  The following code chunk organizes your results into a tibble `letter_tibble`.

  ```
  letter_tibble <- tibble(
    letter = LETTERS[1:26],
    stat = letter_group_stats)
  ```

- **Question 5.8** Augment your histogram from **Question 5.2** with the computed statistics you found from the head sample, tail sample, and the convenience samples. Use a point geom for each statistic.

- **Question 5.9** Do you notice that some of the averages among the 26 letter samples are very close to the population mean while others are quite far away? How about for the statistics generated using the tail and head samples? Why does this happen?

6

Hypothesis Testing

In the previous chapters, we learned about randomness and sampling. Quite often, a data scientist receives some data and must make some assertion about it. There are typically two kinds of situations:

- She has one dataset and a model that the data should have followed. She needs to decide if it is likely that the dataset indeed follows the model?
- She has two datasets. She needs to decide if it is possible to explain the two datasets using a single model.

Here are examples of the two situations.

- A company making coins is testing the fairness of a coin. Using some machine, the company tosses the coin 10,000 times. They record the face that turns up. By examining the record of the 10,000 tosses, can you tell how likely it is that the coin is fair?
- Is the proportion of enrolled Asian American students at Harvard University disproportionately less than the pool of Harvard-admissible applicants? (*SFFA v. Harvard*)

For both situations, an important consideration to make is in terms of how to compare the differences and figuring out how a sample at hand was generated.

6.1 Testing a Model

Suppose 10,000 tosses of a coin generate the following counts for "Heads" and "Tails":

```
mystery_coin <- tibble(face = c("Heads", "Tails"),
                       face_counts = c(4953, 5047))
mystery_coin
```

```
## # A tibble: 2 x 2
##   face  face_counts
##   <chr>       <dbl>
## 1 Heads        4953
## 2 Tails        5047
```

By dividing each number by 10,000, we get the *proportion* of the occurrence of each face.

```
mystery_coin <- mystery_coin |>
  mutate(face_prop = face_counts / 10000)
mystery_coin
```

```
## # A tibble: 2 x 3
```

```
##    face  face_counts face_prop
##    <chr>       <dbl>     <dbl>
## 1 Heads        4953     0.495
## 2 Tails        5047     0.505
```

We notice that the values are not exactly 0.5 ($= 1/2$). How far away is that from what we know about the distribution of a fair coin?

We know that the probability of each face in a fair coin is $1/2$. By subtracting $1/2$ from each and obtaining the absolute difference values from them by removing any negative sign we have:

```
mystery_coin <- mystery_coin |>
  mutate(fair_prop = 1/2,
         abs_diff = abs(face_prop - fair_prop))
mystery_coin
```

```
## # A tibble: 2 x 5
##    face  face_counts face_prop fair_prop abs_diff
##    <chr>       <dbl>     <dbl>     <dbl>    <dbl>
## 1 Heads        4953     0.495       0.5  0.00470
## 2 Tails        5047     0.505       0.5  0.00470
```

We then compute the sum and take one half of this value.

```
mystery_coin |>
  summarize(tvd_value = sum(abs_diff) / 2)
```

```
## # A tibble: 1 x 1
##    tvd_value
##        <dbl>
## 1   0.00470
```

The number found by following the above steps is called the *test statistic*. The test statistic is a statistic used to evaluate a hypothesis, i.e., whether a coin at hand is fair or not. When we compute the test statistic from the data given to us, we call this the *observed value of the test statistic*.

There are many possible test statistics we could have tried for this problem. This one goes by a special name: the *total variation distance* (or, for short, TVD). The total variation distance serves as the measure for the difference between two distributions, namely, the difference between some given distribution (e.g., following the coin handed to us) and a sampling distribution (e.g., following a fair coin).

Another possible, and perhaps straightforward, test statistic is to simply count the number of heads that appear in the sample.

```
observed_heads <- mystery_coin |>
  filter(face == "Heads") |>
  pull(face_counts)
observed_heads
```

```
## [1] 4953
```

Is 4953 heads *too small* to be due to chance? It is hard to tell without knowing the number of heads we get by chance.

We can conduct some simulation to obtain a sampling distribution of the number of heads produced by a fair coin. As mentioned earlier, the proportion that each face turns up is not constant, even for a fair coin. So, by simulating tosses of a fair coin, we must expect to see a range of the number of heads seen.

6.1.1 Prerequisites

Before starting, let us load the `tidyverse` as usual.

```
library(tidyverse)
```

6.1.2 The `rmultinom` function

We saw before that we can generate a sampling distribution by putting in place some sampling strategy. Perhaps the most straightforward is simple random sampling with replacement. This will be the approach we continue to make use of here, as well as throughout the rest of the text.

We also learned about two different ways to sample with replacement. `sample` samples items from a *vector*, which we used when simulating the expected amount of grains a minister receives after some number of days. `slice_sample` functions identically, but instead samples from rows of a *data frame* or tibble. To generate a sampling distribution for this experiment, we could just use `sample` again. But there is a quicker way, using a function called `rmultinom`, which is tailored for sampling at random from *categorical distributions*. We introduce it here and will use it several times this chapter.

Here is how we can use it to generate a sampling distribution of 100 tosses of a fair coin.

```
fair_coin <- c(1/2, 1/2)
sample_vector <- rmultinom(n = 1, size = 100, prob = fair_coin)
sample_vector
```

```
##      [,1]
## [1,]   55
## [2,]   45
```

For generating a sampling distribution, we are more interested in the *proportion* of resulting heads and tails. Thus, we should divide by the number of tosses. Note how the probability of heads and tails is about equal.

```
sample_vector / 100
```

```
##      [,1]
## [1,] 0.55
## [2,] 0.45
```

So we can just as easily simulate proportions instead of counts.

The classic interpretation of `rmultinom` is that you have some marbles to put into boxes of size `size`, each with some probability `prob`; the length of `prob` determines the number of boxes. The result shows the number of marbles that end up in each box. Thus, the function takes the following arguments:

- `size`, the total number of marbles that are put into the boxes.
- `prob`, the distribution of the categories in the population, as a vector of proportions that add up to 1.
- `n`, the number of samples to draw from this distribution. We will typically leave this at 1 to make things easier to work with later on.

It returns a vector containing the number of marbles in each category in a random sample of the given size taken from the population. Because this distribution is so special, statisticians have given it a name: the *multinomial distribution*.

Let us see how we can use it to assess the model for 10,000 tosses of a coin.

6.1.3   A model for 10,000 coin tosses

We can extend our coin toss example code to incorporate the `rmultinom` function:

```
sample_tibble <- tibble(
  face = c("Heads", "Tails"),
  fair_probs = 1/2,
  sample_counts = rmultinom(n = 1,
                            size = 10000,
                            prob = fair_probs))
sample_tibble
```

```
## # A tibble: 2 x 3
##    face  fair_probs sample_counts[,1]
##    <chr>      <dbl>             <int>
## 1 Heads        0.5              5023
## 2 Tails        0.5              4977
```

How many heads are in this sample?

```
sample_heads <- sample_tibble |>
  filter(face == "Heads") |>
  pull(sample_counts)
sample_heads
```

```
##       [,1]
## [1,] 5023
```

We can use this to generate a sampling distribution for the number of heads produced by a fair coin. Let's wrap this up into a function we can call. This will produce one simulated test statistic under the assumption of a fair coin.

```
one_simulated_statistic <- function() {
  sample_tibble <- tibble(
    face = c("Heads", "Tails"),
```

```
        fair_probs = 1/2,
        sample_counts = rmultinom(n = 1,
                              size = 10000,
                              prob = fair_probs))

  sample_heads <- sample_tibble |>
    filter(face == "Heads") |>
    pull(sample_counts)

  return(sample_heads)
}
```

Next, we create a vector `sample_stats` containing 1,000 simulated test statistics. As before, we will use `replicate` to do the work.

```
num_repetitions <- 1000
sample_stats  <- replicate(n = num_repetitions,
                           one_simulated_statistic())
```

6.1.4   Chance of the observed value of the test statistic occurring

To interpret the results of our simulation, we start by visualizing the results using a histogram of the samples.

```
ggplot(tibble(sample_stats)) +
  geom_histogram(aes(x = sample_stats, y = after_stat(density)),
                 fill = "darkcyan", color = 'gray', bins=14) +
  labs(x = "Number of heads")
```

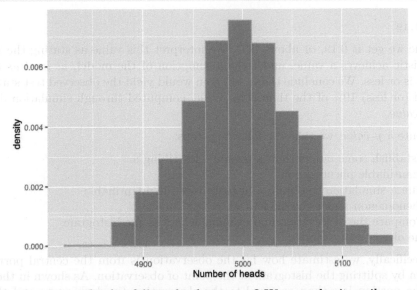

Where does the observed value fall in this histogram? We can plot it easily.

```
ggplot(tibble(sample_stats)) +
  geom_histogram(aes(x = sample_stats, y = after_stat(density)),
                 fill = "darkcyan", color = 'gray', bins = 12) +
  geom_point(aes(x = observed_heads, y = 0), size = 3,
             color = "salmon") +
  labs(x = "Number of heads")
```

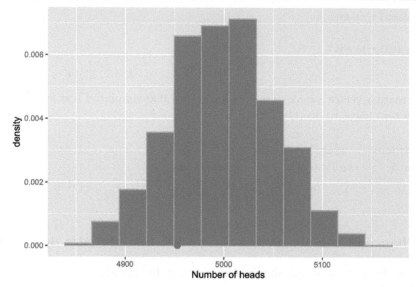

Let us look at the proportion of the elements in the vector `sample_stats` that are at least as small as the observed number of heads or more extreme, whose value we have stored in `observed_heads`. We simply count the elements matching the requirement and then divide the count by the length of the vector.

```
sum(sample_stats <= observed_heads) / length(sample_stats)
```

```
## [1] 0.19
```

The value we get is 0.19, or about 19%. We interpret this value as stating the *chance* the test statistic achieves a value, under the assumption of the model, *at least as extreme* as 4953 heads or less. We conclude that a fair coin would yield the observed test statistic value we found (or less) 19% of the time. This value, computed through simulation, is what we call a *p-value*.

To compute a *p-value*, we take the following steps:

1. Establish some model that is possibly describing a quantifiable phenomenon.
2. Run a simulation to obtain a histogram of the quantifiable phenomenon under the model.
3. Compare the observation and examine where in the histogram the observation stands.

More specifically, we estimate how far the observation is from the central portion of the histogram by splitting the histogram at the point of observation. As shown in the following figure, the portion less than or equal to the observation (in dark cyan) and the portion greater than or equal to the observation (in orange). Note that we can include the equality, when the sampled value equals the observed value, on either side.

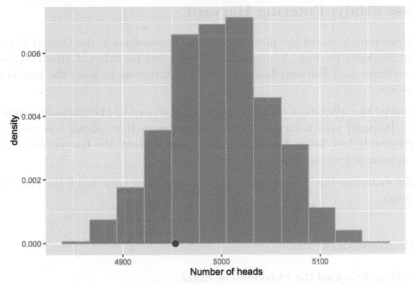

The more orange bars that are visible in the histogram, the higher the likelihood of the observation. Conversely, the less orange bars there are, the lower the likelihood of seeing such an observation. The area covered by the orange bars is formally called the "area in the tail." The area in the tail is designated a special name: **the p-value.**

When we compute a p-value, we have in mind two possible interpretations. We call them the *Null Hypothesis* and the *Alternative Hypothesis*.

- **Null Hypothesis (NH):** The hypothesis we use to create the model for simulation. For example, we assume that the coin we have is a fair coin, about 50% equal chance to see either face. Any variation from what we expect is because of chance variation.
- **Alternative Hypothesis (AH):** The opposite, or counterpart, of the Null Hypothesis. For example, the AH states that the coin is *biased* towards tails. The difference we observed is caused by something other than randomness.

It is important that your null hypothesis acknowledges differences in the data. For example, if the null hypothesis states that a die is fair, why did you not get any 3's when rolling the die 6 times?

We can provide one more definition of the p-value in the language of these hypotheses:

The chance, under the *null hypothesis*, of getting a test statistic equal to the observed test statistic or more extreme in the direction of the *alternative hypothesis*.

6.2 Case Study: Entering Harvard

Harvard University is one of the most prestigious universities in the United States. A recent lawsuit by Students for Fair Admissions (SFFA)[1] led by Edward Blum against Harvard University alleges that Harvard has used soft racial quotas to keep the numbers of Asian-Americans low.

Put differently, the allegation claims that, from the pool of Harvard-admissible American applicants, Harvard uses a racial discriminatory score that allows the college to choose disproportionately less Asian-Americans. As of this writing, the lawsuit has been appealed and is to appear before the Supreme Court.

This section examines the "Harvard model" to assess, at a basic level, the claim put forward by the lawsuit.

6.2.1 Prerequisites

Before starting, let's load the `tidyverse` as usual.

```
library(tidyverse)
```

6.2.2 Students for Fair Admissions

Harvard University publishes some statistics[2] on the class of 2024. According to the data, the proportions of the class for Whites, Blacks, Hispanics, Asians, and Others (International and Native Americans) are respectively 46.1%, 14.7%, 12.7%, 24.4%, and 2.1%.

We do not have the data of the students admissible to enter Harvard so, in lieu of this, we refer to student demographics enrolled in a four-year college. According to Chronicle of Higher Education[3] 2020-2021 Almanac, the racial demographics of full-time students in 4-year private non-profit institutions – Harvard is one of them – in Fall 2018 are: 63.6% White, 11.5% Black, 12.3% Hispanic, 8.1% Asian, and 4.5% Other.

Let's compile this information into a tibble.

```
class_props <- tribble(~Race, ~Harvard, ~Almanac,
                       "White", 46.1, 63.6,
                       "Black", 14.7, 11.5,
                       "Hispanic", 12.7, 12.3,
                       "Asian", 24.4, 8.1,
                       "Other", 2.1, 4.5)
class_props
```

```
## # A tibble: 5 x 3
##    Race      Harvard Almanac
```

---

[1] https://studentsforfairadmissions.org/
[2] https://college.harvard.edu/admissions/admissions-statistics
[3] http://www.chronicle.com

```
##     <chr>      <dbl>   <dbl>
## 1 White        46.1    63.6
## 2 Black        14.7    11.5
## 3 Hispanic     12.7    12.3
## 4 Asian        24.4     8.1
## 5 Other         2.1     4.5
```

The distributions may look quite different from each other. Of course, the demographics from the Almanac includes students who did not apply to Harvard, those who might not have got into Harvard, those applied and did not get in, and international students. Moreover, the Almanac data covers all full-time students in 2018 but not students who entered college in 2020.

Notwithstanding these differences, let us conduct an experiment to see how the demographics from Harvard look different from those given by the Almanac in terms of a sampling distribution.

As we will be handling proportions, let us scale the numbers down (by dividing each element by 100) so that they are expressed as percentages.

```
class_props <- class_props |>
  mutate(Harvard = Harvard / 100,
         Almanac = Almanac / 100)
class_props
```

```
## # A tibble: 5 x 3
##     Race      Harvard Almanac
##     <chr>       <dbl>   <dbl>
## 1 White       0.461   0.636
## 2 Black       0.147   0.115
## 3 Hispanic    0.127   0.123
## 4 Asian       0.244   0.081
## 5 Other       0.021   0.045
```

We will also write a function that computes the total variation distance (TVD) between two vectors.

```
compute_tvd <- function(x, y) {
  return(sum(abs(x - y)) / 2)
}
```

In this study, our observed value is the TVD between the distribution of students in the Harvard class and the Almanac.

```
harvard_diff <- class_props |>
  summarize(compute_tvd(Harvard, Almanac)) |>
  pull()
harvard_diff
```

```
## [1] 0.199
```

The Harvard class of 2024 has 2015 students. We can think of the process of sampling 2015 people to fill the "Harvard class" from those who "were attending" a four-year non-profit college in Fall 2018 and then examining their racial distribution. Of course, we cannot reach out to those individuals specifically, but we know the distribution of the entire population from which we want to sample. Therefore, we can simulate a large number of times what this "Harvard class" looks like and compare it with what we know about the actual Harvard class distribution, available in `harvard_diff`.

Our sampling plan can be framed as a "boxes and marbles" problem, as we saw in the previous section. There are five "boxes" to choose from, where each corresponds to a race: White, Black, Hispanic, Asian, and Other. The goal is to place marbles (which correspond to students) in each of the boxes, where the probability of ending up in any of the boxes is given by the Almanac.

This is an excellent fit for the `rmultinom` function. For example, here is one simulation of the proportion of races found in a "Harvard class."

```
sample_vector <- rmultinom(n=1, size=2015,
                           prob = pull(class_props, Almanac)) / 2015
sample_vector
```

```
##              [,1]
## [1,] 0.63374690
## [2,] 0.11761787
## [3,] 0.12406948
## [4,] 0.08784119
## [5,] 0.03672457
```

How far is our simulated class from the Almanac? We compute the TVD to find out.

```
class_props |>
  mutate(sample = sample_vector) |>
  summarize(compute_tvd(sample, Almanac)) |>
  pull()
```

```
## [1] 0.01052854
```

We wrap our work into a function.

```
one_simulated_class <- function(props) {
  props |>
    mutate(sample = rmultinom(n=1, size=2015, prob = Almanac) / 2015) |>
    summarize(compute_tvd(sample, Almanac)) |>
    pull()
}
```

Let us simulate what 10,000 classes could look like. This will be contained in a vector called `sample_class_tvds`. Also, as before, we will use `replicate` to do the work.

```
num_repetitions <- 10000
sample_class_tvds <- replicate(n = num_repetitions,
                               one_simulated_class(class_props))
```

We can now visualize the result.

```
ggplot(tibble(sample_class_tvds)) +
  geom_histogram(aes(x = sample_class_tvds, y = after_stat(density)),
           bins = 30, fill = "darkcyan", color = 'gray')
```

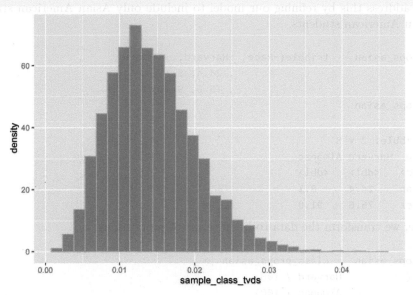

Where does the true Harvard class lie on this histogram? We can plot a point geom to find out.

```
ggplot(tibble(sample_class_tvds)) +
  geom_histogram(aes(x = sample_class_tvds, y = after_stat(density)),
           bins = 70, fill = "darkcyan", color = 'gray') +
  geom_point(aes(x = harvard_diff, y = 0), size = 3, color = "salmon")
```

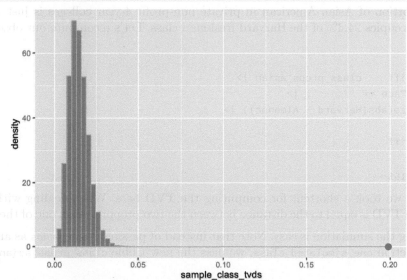

The orange dot shows the distance value of the Harvard value from the Almanac value. What we see is that the proportion of the races at Harvard is nothing like the national proportion.

6.2.3   Proportion of Asian American students

The prior experiment looked at the proportion of all races. However, the claim given by
the lawsuit is specifically about Harvard-admissible students who are Asian American. We
can now address this by refining our model to include only Asian American students and
non-Asian American students.

```
class_props_asian <- tribble(~Race, ~Harvard, ~Almanac,
                             "Asian", 24.4, 8.1,
                             "Other", 75.6, 91.9)
class_props_asian
```

```
## # A tibble: 2 x 3
##    Race  Harvard Almanac
##    <chr>   <dbl>   <dbl>
## 1 Asian    24.4     8.1
## 2 Other    75.6    91.9
```

As before, we transform the data to be in terms of proportions.

```
class_props_asian <- class_props_asian |>
  mutate(Harvard = Harvard / 100,
         Almanac = Almanac / 100)
class_props_asian
```

```
## # A tibble: 2 x 3
##    Race  Harvard Almanac
##    <chr>   <dbl>   <dbl>
## 1 Asian   0.244   0.081
## 2 Other   0.756   0.919
```

The proportion of Asian American in private non-profit 4-year colleges is just 8.1% while
the race occupies 24.4% of the Harvard freshman class. Let's recompute our observed TVD
value.

```
harvard_diff <- class_props_asian |>
  filter(Race == "Asian") |>
  summarize(abs(Harvard - Almanac)) |>
  pull()
harvard_diff
```

```
## [1] 0.163
```

Note that we took a shortcut for computing the TVD here. When dealing with two categories, the TVD is equal to the distance between the two proportions in one of the categories.

Re-running the simulation is easy. Note that instead of passing `class_props` as an argument
to the function `one_simulated_class`, we pass the new tibble `class_props_asian`.

```
num_repetitions <- 10000
sample_class_tvds <- replicate(n = num_repetitions,
                      one_simulated_class(class_props_asian))
```

We again visualize the result.

```
ggplot(tibble(sample_class_tvds)) +
  geom_histogram(aes(x = sample_class_tvds, y = after_stat(density)),
                 binwidth=0.002, fill = "darkcyan", color = 'gray')
```

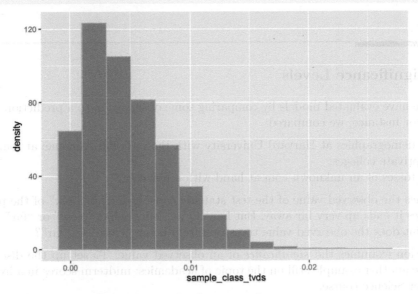

Where does the observed value fall in this histogram?

```
ggplot(tibble(sample_class_tvds)) +
  geom_histogram(aes(x = sample_class_tvds, y = after_stat(density)),
                 binwidth=0.002, fill = "darkcyan", color = 'gray') +
  geom_point(aes(x = harvard_diff, y = 0), size = 3, color = "salmon")
```

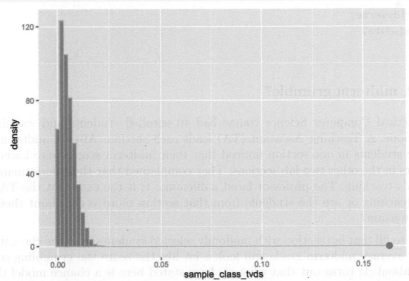

We find that the result is the same; the Harvard proportion of Asian American students is not at all like the national value. We can state, with great confidence, that Harvard enrolls much more Asian students than the national average.

Important note: The reader should be cautioned not to accept these results as direct evidence against the suit's case. As noted at the outset of this section, we do not know the proportion of Harvard-admissible students and must instead rely on a national Almanac for reference. The base population of students can be very much different which is, in fact, something we anticipated.

6.3   Significance Levels

So far we have evaluated models by comparing some observation to a prediction made by a model. For instance, we compared:

- Racial demographics at Harvard University with the national Almanac at four-year non-profit private colleges.
- 10,000 tosses of an unknown coin at hand with a fair coin.

Sometimes the observed value of the test statistic ends up in the "bulk" of the predictions; sometimes it ends up very far away. But how do we define what "close" or "far" is? And at what point does the observed value transition from being "close" to "far"?

This section examines the *significance* of an observed value. To set up the discussion, we introduce another example still on the topic of academics: midterm scores in a hypothetical Computer Science course.

6.3.1   Prerequisites

Before starting, let's load the `tidyverse` as usual. We will also use a dataset from the `edsdata` package, so let us load this in as well.

```
library(tidyverse)
library(edsdata)
```

6.3.2   A midterm grumble?

A hypothetical Computer Science course had 40 enrolled students and was divided into 3 lab sections. A Teaching Assistant (TA) leads each section. After a midterm exam was given, the students in one section noticed that their midterm scores were lower compared to students in the other two lab sections. They complained that their performance was due to the TA's teaching. The professor faced a dilemma: is it the case that the TA is at fault for poor teaching or are the students from that section more vocal about their grumbles following a exam?

If we were to fill that lab section with randomly selected students from the class, it is possible that their average midterm grade will look a lot like the score the grumbling students are unhappy about. It turns out that what we have stated here is a chance model that we can simulate.

Let's have a look at the data from each student in the course. The following tibble `csc_labs` contains midterm and final scores, and the lab section the student is enrolled in.

```
csc_labs
```

```
## # A tibble: 40 x 3
##    midterm final section
##      <dbl> <dbl> <chr>
##  1      73 79     F
##  2      30  0     D
##  3      91 77.2   D
##  4      89 76.5   D
##  5      71 76.5   H
##  6      28  0     H
##  7      32  0     F
##  8      54 88     D
##  9      88 76     D
## 10      59 68     F
## # ... with 30 more rows
```

We can use the dplyr verb `group_by` to examine the mean midterm grade as well as the number of students in each section.

```
lab_stats <- csc_labs |>
  group_by(section) |>
  summarize(midterm_avg = mean(midterm),
            count = n())
lab_stats
```

```
## # A tibble: 3 x 3
##   section midterm_avg count
##   <chr>         <dbl> <int>
## 1 D              70      14
## 2 F              74.3    17
## 3 H              68.6     9
```

Indeed, it seems that the section H students fared the worst, albeit by a small margin, among the three sections. Our statistic then is the mean grade of students in the lab section. Thus, our observed statistic is the mean grade from section H, which is about 68.56.

```
observed_statistic <- lab_stats |>
  filter(section == "H") |>
  pull(midterm_avg)
observed_statistic
```

```
## [1] 68.55556
```

We formally state our null and alternative hypothesis.

Null Hypothesis: The mean midterm grades of students in lab section H looks like the mean grades of a "section H" that is generated by randomly sampling the same number of students from the class.

Alternative Hypothesis: The section H midterm grades are too low.

To form a random sample, we will need to sample *without* replacement 9 students from the course to fill up the theoretical "lab section H".

```
random_sample <- csc_labs |>
  select(midterm) |>
  slice_sample(n = 9, replace = FALSE)
random_sample
```

```
## # A tibble: 9 x 1
##    midterm
##      <dbl>
## 1       89
## 2       92
## 3       73
## 4       75
## 5       82
## 6       68
## 7       67
## 8      100
## 9       88
```

We can look at the mean midterm score for this randomly sampled section.

```
random_sample |>
  summarize(mean(midterm)) |>
  pull()
```

```
## [1] 81.55556
```

Now that we know how to simulate one value, we can wrap this into a function.

```
one_simulated_mean <- function() {
  random_sample <- csc_labs |>
    select(midterm) |>
    slice_sample(n = 9, replace = FALSE)

  random_sample |>
    summarize(mean(midterm)) |>
    pull()
}
```

We will simulate the "section H lab" 10,000 times. Let us run the simulation!

```
num_repetitions <- 10000
sample_means <- replicate(n = num_repetitions, one_simulated_mean())
```

As before, we visualize the resulting distribution of grades.

```
ggplot(tibble(sample_means)) +
  geom_histogram(aes(x = sample_means, y = after_stat(density)),
                 bins = 15, fill = "darkcyan", color = 'gray')
```

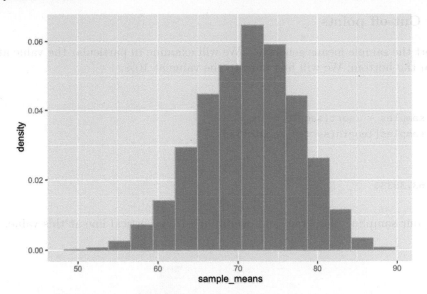

It seems that the grades cluster around 72. Where does the actual section H section lie? Recall that this value is available in the variable observed_statistic. We overlay a point geom to our above plot.

```
ggplot(tibble(sample_means)) +
  geom_histogram(aes(x = sample_means, y = after_stat(density)),
                 bins = 15, fill = "darkcyan", color = 'gray') +
  geom_point(aes(x = observed_statistic, y = 0),
             size = 3, color = "salmon")
```

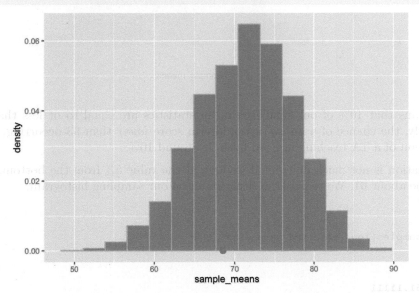

It seems that the observed statistic is "close" to the center of randomly sampled scores.

6.3.3 Cut-off points

Let's sort the sample means generated. We will examine in particular the value at 10% and 5% from the bottom. We will first turn to the value at 10%.

```
sorted_samples <- sort(sample_means)
sorted_samples[length(sorted_samples)*0.1]
```

```
## [1] 63.33333
```

We plot our sampling histogram and overlay it with a vertical line at this value.

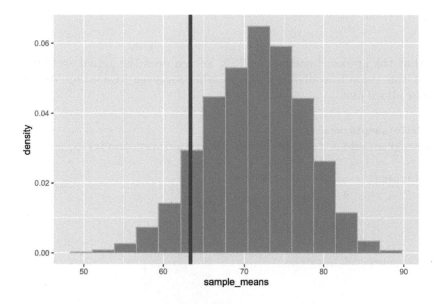

This means that 10% of our simulated mean statistics are equal to or less than 63. Put differently, the chance of a an average midterm score *lower* than 63 occurring, under the assumption of a TA teaching in good faith, is around 10%.

The situation is not much different if we look at the value 5% from the bottom, which we find to be about 61. We redraw the situation in on our sampling histogram.

```
sorted_samples[length(sorted_samples)*0.05]
```

```
## [1] 61.11111
```

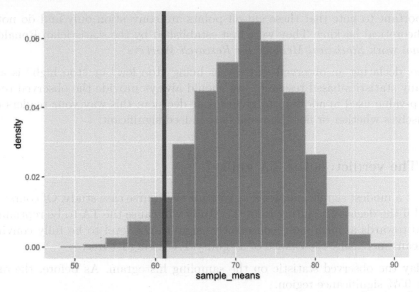

This time, the chance of obtaining a simulated mean midterm score *at least* as low as 61 is about 5%.

We say that the threshold point 63 is at the 90% *significance level* and the threshold point 61 is at the 95% significance level.

6.3.4   The significance level is an error probability

The last figure shows that, although rare, a lab section with a "TA teaching in good faith" can still produce mean midterm scores that are at least as low as 61. How often does that occur? The figure gives the answer for that as well: it does so with about 5% chance.

Therefore, if the TA is *teaching in good faith* and our test uses a 95% significance level to decide whether or not the TA is guilty, then there is about a 5% chance that the test will wrongly conclude that the mean midterm scores are too low and, consequently, the TA is at fault. This example points to a general fact about significance levels:

> If you use a $p\%$ significance level for the p-value, and the null hypothesis happens to be true, there is about a $1-p\%$ chance that the test will incorrectly conclude the alternative hypothesis.

Statistical inferences, unlike logical inferences, can be wrong! But the power of statistics is its ability to *quantify* how often this error will occur. In fact, we can control the chance of wrongly convicting the TA by choosing a higher significance level. We could look at the 99% significance level or even the 99.9% and 99.99% levels; these are commonly referred to in the area of physics which rely on enormous evidence to prove something axiomatic.

Here, too, are trade-offs. By minimizing the error of wrongly convicting a TA teaching in good faith, we increase the chance of another kind of error occurring: our test concluding nothing when in fact there is something unusual about the lab section's midterm grades.

It is important to note that these cut-off points are convention only and do not have any strong theoretical backing. They were first established by the statistician Ronald Fisher in his seminal work *Statistical Methods for Research Workers*[4].

Therefore, declaring an observed statistic as being "too low" or "too high" is a judgment call. In any statistics-based research, you should always provide the observed test statistic and the p-value used in addition to giving your decision; this way your readers can decide for themselves whether or not the results are indeed significant.

6.3.5   The verdict: is the TA guilty?

We can set a modest significance level at 95% for the course case study. Of course, judgment is needed if the decision resulting from this study will cause the TA to be reprimanded – we may tend towards a much more conservative significance level to be fully convinced, even if this means increasing the chance of a "guilty TA" being let free.

We overlay the observed statistic on the sampling histogram. As before, the orange bars show the 95% significance region.

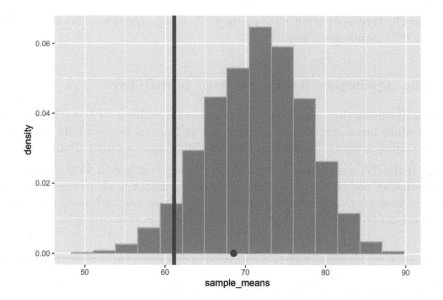

We see that the point does not cross the vertical purple line. We can check numerically how much area "is in the tail".

---

[4]https://en.wikipedia.org/wiki/Statistical_Methods_for_Research_Workers

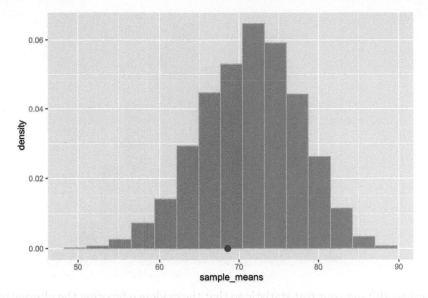

```
sum(sample_means <= observed_statistic) / num_repetitions
```

```
## [1] 0.3192
```

We conclude the TA's defense holds up pretty well: the average lab section H scores are not any different from those generated by chance.

A note, also, on drawing conclusions from a hypothesis test. Even if there was significant evidence to reject the null hypothesis at some conventional cut-off, caution must be exercised in interpreting the poor performance as being directly *caused* by the TA's instruction. There could be other variables at play that we did not account for that can affect the significance, e.g., the background of the students enrolled in this particular lab section (did they have less prior programming experience compared to the other sections?). We call these **confounding variables**, which we will examine in more depth in a later chapter.

In any case, it would be prudent to check in with the TA to get their take on the story.

6.3.6 Choosing a test statistic

By this point, we have been introduced to a few different test statistics. A common challenge when developing a hypothesis test is to first define what a "good" test statistic is for the problem.

Consider your alternative hypothesis and what evidence favors it over the null. If only "large values" or "small values" of the test statistic favor the alternative, then we recommend using the test statistic. For instance, in the midterm example, we considered only "small values" of the sample mean statistic to determine if the lab section H scores are "too low." In the Harvard admissions example, we considered "large values" of the TVD test statistic to determine if the TVD of the Harvard proportions is "too big" to have been generated by a model under the null hypothesis.

Avoid choosing test statistics where "both big values and small values" favor the alternative. In this case, the area that supports the alternative includes both the left and right "tails". Consider the following sampling histogram of the test statistic and note the tails as indicated by the orange bars.

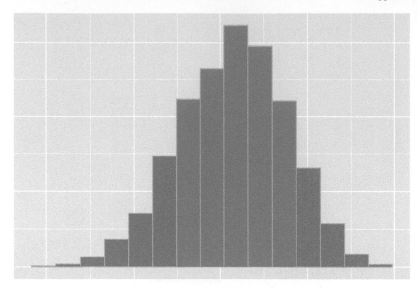

We suggest modifying your test statistic so that the evidence favoring the alternative involves only one tail.

Finally, we present a table with some common test statistics and when to use each.

Test statistic	When to use?
Total Variation Distance (TVD)	**Categorical data**; compare your sample with the distribution it was drawn from. `rmultinom`
Number of heads, difference in means	**Numerical data**; direction matters, e.g., "too few heads" or "too many heads"
Absolute difference, mean absolute difference	**Numerical data**; direction does not matter, only distance, e.g., "number of heads seen is different from a chance flip"

6.4 Permutation Testing

In the previous section, we study the use of hypothesis testing. In this section we learn a simple method to compare two distributions using a method we call *permutation testing*. This allows us to decide if the two distributions come from the same underlying distribution.

6.4.1 Prerequisites

Let us begin by loading the tidyverse. We will also use a dataset from the edsdata package.

```
library(tidyverse)
library(edsdata)
```

6.4.2 The effect of a tutoring program

The tibble `finals` from the `edsdata` package contains final exam grades in a hypothetical Computer Science course for 105 students. They are divided into two groups, based on two different offerings of the course labeled A and B. The more recent offering B featured a tutoring program for students to receive help on assignments and exams. The course instructor is interested in finding out if the tutoring program boosted overall performance in the class, measured by a final exam. This could help the instructor and department decide if the program should continue or even be expanded. Suppose that the dataset is collected over two semesters from the same Computer Science course.

Let's first load the dataset.

```
finals
```

```
## # A tibble: 102 x 2
##     grade class
##     <dbl> <chr>
## 1      89 A
## 2      17 A
## 3      94 A
## 4      51 A
## 5      49 A
## 6      93 A
## 7      52 A
## 8      54 A
## 9      57 A
## 10     65 A
## # ... with 92 more rows
```

We can examine the number of enrolled students in each of the two offerings.

```
finals |>
  group_by(class) |>
  count()
```

```
## # A tibble: 2 x 2
## # Groups:   class [2]
##   class     n
##   <chr> <int>
## 1 A        49
## 2 B        53
```

It appears they are about equal. Let's now turn to a distribution of the students in the offering that featured the tutoring program (class B) compared to those in the offering without the program (class A). To generate an overlaid histogram, we use the positional adjustment argument `identity` and set an `alpha` so that the bars are drawn with slight transparency.

```
ggplot(finals) +
  geom_histogram(aes(x = grade, y = after_stat(density), fill = class),
```

```
              bins = 10, color = "gray",
              alpha = 0.7, position = "identity")
```

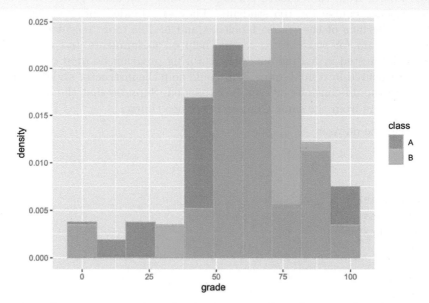

By observation alone, it seems that the final scores of students in the offering where the tutoring program was available (B) is slightly to the right of the distribution corresponding to scores when the program did not exist (A). Could this be chalked up to chance?

As we have done throughout this chapter, we can address this question by means of a hypothesis test. We will state a null and alternative hypothesis that arise from the problem.

Null hypothesis: In the population, the distribution of final exam scores where the tutoring program was available is the same as those when the service did not exist. The difference seen in the sample is because of chance.

Alternative hypothesis: In the population, the distribution of final exam scores when the tutoring program was available are, on average, *higher* than the scores when the program was not.

According to the alternative hypothesis, the average final score in offering B should be *higher* than the average final score in offering A. Therefore, a good test statistic we can use is the difference in the mean between the two groups. That is,

$$\text{test statistic} = \mu_B - \mu_A$$

where μ denotes the mean of the group.

First, we form two vectors `finalsA` and `finalsB` that contain final scores with respect to the course offering.

```
finalsA <- finals |>
  filter(class == 'A') |> pull(grade)
finalsB <- finals |>
  filter(class == 'B') |> pull(grade)
```

The observed value of the statistic can be computed as the following.

```
observed_statistic <- mean(finalsB) - mean(finalsA)
observed_statistic
```

```
## [1] 6.105506
```

We can write a function that computes the statistic for us. We call it `mean_diff`.

```
mean_diff <- function(a, b) {
  return(mean(a) - mean(b))
}
```

Observe how it returns the same value for the observed statistic.

```
mean_diff(finalsB, finalsA)
```

```
## [1] 6.105506
```

To predict the statistic under the null hypothesis, we defer to an idea called the *permutation test*.

6.4.3   A permutation test

Suppose that we are given the following vector of integers.

```
1:10
```

```
##  [1]  1  2  3  4  5  6  7  8  9 10
```

We can interpret these numbers as indices that refer to an element inside a vector. We imagine that the first half of indices belong to a group A, and the second half group B.

Under the assumption of the null hypothesis, there should be no difference between the two distributions A and B with respect to the underlying population. For example, whether a final exam score belongs to the course offering A or B should have no effect on the mean final score. If so, there should be no consequences if we place both groups into a pot, shuffle them around, and compute the mean difference from the result. The resulting value we get from this process is one simulated value of the test statistic under the null hypothesis.

The first bit of machinery we need is a function that shuffles a sequence of integers. We actually already know one: `sample`.

```
shuffled <- sample(1:10)
shuffled
```

```
##  [1]  7 10  5  3  4  1  2  6  9  8
```

In this example, `sample` receives a vector of numbers 1 through 10 and returns the result after shuffling them. We might also call the result a *permutation* of the original sequence – hence, its namesake.

If we again interpret the resulting vector as indices, we take the first half to be the indices of the shuffled group A and the second half the shuffled group B.

```
shuffled[1:5]  # shuffled group A
```

```
## [1]  7 10  5  3  4
```

```
shuffled[6:10] # shuffled group B
```

```
## [1] 1 2 6 9 8
```

The remaining work then is to compute the difference in means between the shuffled groups.

The function one_mean_difference puts everything together. It receives two vectors, a and b, puts them together in a pot, and deals out two shuffled vectors with the same size as a and b, respectively. The function returns the value of the simulated statistic by calling the functional compute_statistic. For this example, we use mean_diff.

```
one_difference <- function(a, b, compute_statistic) {
  pot <- c(a, b)
  sample_indices <- sample(1 : length(pot))
  shuffled_a <- pot[sample_indices[1 : length(a)]]
  shuffled_b <- pot[sample_indices[(length(a) + 1) : length(pot)]]
  return(compute_statistic(shuffled_a, shuffled_b))
}
```

We are now ready to perform a permutation test for the tutoring program example. We would like to simulate the test statistic under the null hypothesis multiple times and collect the values into a vector. As before, we can use replicate. We will simulate 10,000 values.

```
differences <- replicate(n = 10000,
                         one_difference(finalsA, finalsB, mean_diff))
```

6.4.4 Conclusion

Let's visualize the results.

```
ggplot(tibble(differences)) +
  geom_histogram(aes(x = differences, y = after_stat(density)),
                 col="grey", fill = "darkcyan", bins = 20) +
  geom_point(aes(x = observed_statistic, y = 0),
             color = "salmon", size = 3)
```

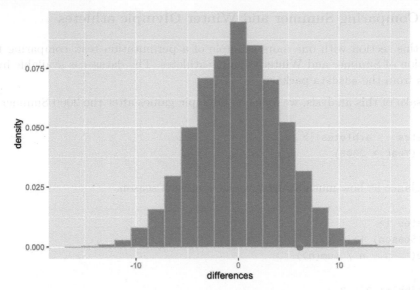

First, observe how the distribution is centered around 0. Under the assumption of the null hypothesis, there is no difference between the final exam averages in the two course offerings and, therefore, the difference clusters around 0.

Also observe that the observed test statistic is quite far from the center. To get a better sense of how far, we compute the p-value.

```
sum(differences >= observed_statistic) / 10000
```

```
## [1] 0.0776
```

This means that the chance of obtaining a mean difference at least as large as 6.10 is around 8%. By standards of the conventional cut-off points we have discussed, we would have enough evidence to refute the null hypothesis at a 90% significance level. Would this be enough to convince us that the tutoring program is indeed effective? Let us consider for a moment what it would mean if it does not.

If we were to demand a higher significance level, say 95%, our observed statistic is no longer significant. The logical next step would be to conclude that the null hypothesis is true, bearing the implication that the tutoring program is ineffective. This would be a statistical fallacy! Even if our results are not significant at the desired level, we do **NOT** take the null hypothesis to be true. Put another way, **we fail to reject the null hypothesis**. That is a mouthful!

The problem here is a lack of evidence. A lack of evidence does not prove that something *does not* exist, e.g., the tutoring program is *not* effective; it very well could be, but our study missed it. Indeed, our permutation test only evaluated one criteria – that is, difference in final exam scores – as a measure for improvement. There are other test statistics or criteria we could have considered, like class participation, which may have benefited from the program. It would be up to the judgment of the department on how to use these results in deciding the merit of the tutoring program.

6.4.5   Comparing Summer and Winter Olympic athletes

We end this section with one more example of a permutation test: comparing the weight information of Summer and Winter Olympic athletes. The dataset is available in the name `athletes` from the `edsdata` package.

For the sake of this analysis, we focus on Olympic games after the 2000 Summer Olympics.

```
my_athletes <- athletes |>
  filter(Year > 2000)
```

We can glance at how much athletes we have in each season.

```
my_athletes |>
  count(Season) |>
  mutate(prop = n / sum(n))
```

```
## # A tibble: 2 x 3
##   Season      n  prop
##   <chr>   <int> <dbl>
## 1 Summer   7964 0.792
## 2 Winter   2088 0.208
```

We observe that Summer athletes make up the bulk of this dataset. Before proceeding any further, we should visualize the weight information with an overlaid histogram.

```
my_athletes |>
  ggplot() +
    geom_histogram(aes(x = Weight, y = after_stat(density),
                       fill = Season),
                   bins = 13, color = "gray",
                   alpha = 0.7, position = "identity")
```

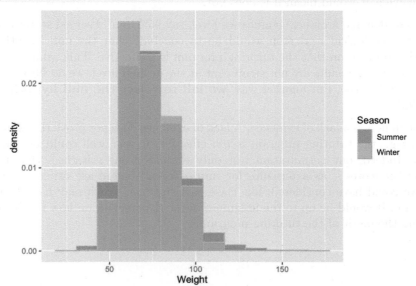

We give our hypothesis statements.

Null hypothesis: In the population, the distribution of weight information in the Summer Olympics is, on average, the same as the Winter Olympics.

Alternative hypothesis: In the population, the distribution of weight information in the Summer Olympics is, on average, different from the Winter Olympics.

Note that the alternative hypothesis, unlike the tutoring program example, does not care whether the weight information for athletes competing in the Winter Olympics is *higher* or *less* than that of athletes in the Summer Olympics. It only states that some *difference* exists. Therefore, the absolute difference in the means would be a good test statistic to use for this problem.

$$\text{test statistic} = |\mu_B - \mu_A|$$

Note how it does not matter which group ends up as A and likewise for B. Let's write a function to compute this statistic; it is a slight variation of the `mean_diff` we saw before.

```
mean_abs_diff <- function(a, b) {
  return(abs(mean(a) - mean(b)))
}
```

6.4.6 The test

We form two vectors `winter_weights` and `summer_weights` that contain the weight information with respect to the season.

```
winter_weights <- my_athletes |>
  filter(Season == "Winter") |>
  pull(Weight)
summer_weights <- my_athletes |>
  filter(Season == "Summer") |>
  pull(Weight)
```

The observed value of the statistic can be computed as the following.

```
observed_statistic <- mean_abs_diff(winter_weights, summer_weights)
observed_statistic
```

```
## [1] 1.298946
```

We are now ready to perform the permutation test. As before, let us simulate the test statistic under the null hypothesis 10,000 times.

```
differences <- replicate(n = 10000,
        one_difference(winter_weights, summer_weights, mean_abs_diff))
```

6.4.7 Conclusion

We are ready to visualize the results.

```
ggplot(tibble(differences)) +
  geom_histogram(aes(x = differences, y = after_stat(density)),
                col="grey", fill = "darkcyan", bins = 15) +
  geom_point(aes(x = observed_statistic, y = 0),
            color = "salmon", size = 3)
```

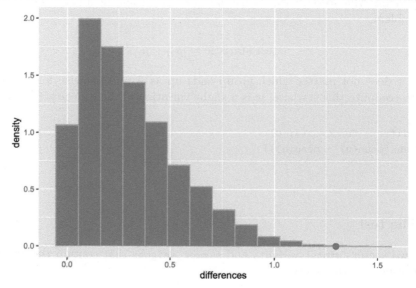

The observed statistic is quite far away from the distribution of simulated test statistics. Let's do a numerical check.

```
sum(differences >= observed_statistic) / 10000
```

```
## [1] 0.001
```

The chance of obtaining a mean absolute difference of 1.29 is roughly 0.1%. We can safely reject the null hypothesis at a significance level over 99%. This confirms that, assuming our dataset is representative of the population of Olympic athletes, the weight information between Summer and Winter Olympic players are likely, on average, to be different.

6.5 Exercises

Be sure to install and load the following packages into your R environment before beginning this exercise set.

```
library(tidyverse)
library(edsdata)
library(gapminder)
```

Question 1 The College of Galaxy makes available applicant acceptance information for different ethnicities: White ("White"), American Indian or Alaska Native ("AI/AN"), Asian ("Asian"), Black ("Black"), Hispanic ("Hispanic"), and Native Hawaiian or Other Pacific Islander ("NH/OPI"). The tibble `galaxy_acceptance` gives the acceptance result from one year.

```
galaxy_acceptance <- tribble(
  ~Ethnicity, ~Applied, ~Accepted,
  "White",   925, 811,
  "NH/OPI", 50, 7,
  "Hispanic", 601, 348,
  "Black", 331, 236,
  "Asian", 237, 101,
  "AI/AN", 84, 30)
galaxy_acceptance
```

```
## # A tibble: 6 x 3
##    Ethnicity Applied Accepted
##    <chr>       <dbl>    <dbl>
## 1 White         925      811
## 2 NH/OPI         50        7
## 3 Hispanic      601      348
## 4 Black         331      236
## 5 Asian         237      101
## 6 AI/AN          84       30
```

- **Question 1.1** What proportion of total accepted applicants are of some ethnicity? Add a new variable named `prop_accepted` that gives the proportion of each ethnicity with respect to the total number of accepted candidates. Assign the resulting tibble to the name `galaxy_distribution`.

  Based on these observations, you may be convinced that the college is biased in favor of enrolling White applicants. Is it justifiable?

  To explore the question, you conduct a *hypothesis test* by comparing the ethnicity distribution at the college to that of degree-granting institutions in the United States. You decide to test the hypothesis that the ethnicity distribution at the College of Galaxy looks like a random sample from the population of accepted applicants in universities across the United States. Using simulation, this is what the data would look like if *the hypothesis were true*. If it doesn't, you reject the hypothesis.

  Thus, you offer the null hypothesis:

  - **Null hypothesis:** "The distribution of ethnicities of accepted applicants at the College of Galaxy was a random sample from the population of accepted applicants at degree-granting institutions in the United States."

- **Question 1.2** With every *null hypothesis* we write down a corresponding **alternative hypothesis**. What is the alternative hypothesis in this case?

We have that there are 1533 accepted applicants at the College of Galaxy. Imagine drawing a random sample of 1533 students from among the admitted students at universities across the United States. This is one student admissible pool we could see if the null hypothesis were true.

The Integrated Postsecondary Education Data System (IPEDS)[5] at the National Center for Education Statistics gives data on U.S. colleges, universities, and technical and vocational institutions. As of Fall 2020, they reported the following ethnicity information about admitted applicants at Title IV degree-granting institutions in the U.S:

```
ipeds2020 <- tribble(~Ethnicity, ~`2020`,
    "White", 9316458,
    "NH/OPI", 46144,
    "Hispanic", 3538778,
    "Black", 2254757,
    "Asian", 1285154,
    "AI/AN", 115951)
```

- **Question 1.3** Repeat **Question 1.1** for `ipeds2020`. Assign the resulting tibble to the name `ipeds2020_dist`.

Under the null hypothesis, we can simulate one "admissible pool" from the population of students in the U.S as follows:

```
total_admitted <- galaxy_distribution |> pull(Accepted) |> sum()
prop_accepted <- ipeds2020_dist |> pull(prop_accepted)
rmultinom(n = 1, size = total_admitted, prob = prop_accepted)
```

The first element in this vector contains the number of White students in this sample pool, the second element the number of Native Hawaiian or Other Pacific Islander students, and so on.

- **Question 1.4** For the ethnicity distribution in our sample, we are interested in the *proportion* of ethnicities that appear in the admissible pool. Write a function `prop_from_sample()` that takes as an argument some distribution (e.g., `ipeds2020_distribution`) and returns a vector containing the proportion of ethnicities that appear in the sample of 1533 people.

- **Question 1.5** Call `prop_from_sample()` to create one vector called `one_sample` that represents one sample of 1533 people from among the admissible students in the United States.

The *total variation distance* (TVD) is a useful test statistic when comparing two distributions. This distance should be small if the null hypothesis is true because samples will have similar proportions of ethnicities as the population from which the sample is drawn.

- **Question 1.6** Write a function called `compute_tvd()`. It takes as an argument a vector of proportions of ethnicities. The first element in the vector is the proportion of White students, the second element the proportion of Native Hawaiian or Other Pacific Islander students, and so on. The function returns the TVD between the given ethnicity distribution and that of the national population.

---

[5] https://nces.ed.gov/ipeds

```
compute_tvd(galaxy_distribution |> pull(prop_accepted)) # example
```

- **Question 1.7** Write a function called `one_simulated_tvd()`. This function takes no arguments. It generates a "sample pool" under the null hypothesis, computes the test statistic, and then return it.

```
one_simulated_tvd()  # an example call
```

- **Question 1.8** Using `replicate()`, run the simulation **10,000** times to produce **10,000** test statistics. Assign the results to a vector called `sample_tvds`.

  The following chunk shows your simulation augmented with an orange dot that shows the TVD between the ethnicity distribution at College of Galaxy and that of the national population.

```
ggplot(tibble(sample_tvds)) +
  geom_histogram(aes(x = sample_tvds, y = after_stat(density)),
              bins = 15,
              fill = "darkcyan", color = 'gray') +
  geom_point(aes(
    x = compute_tvd(galaxy_distribution |> pull(prop_accepted)),
              y = 0), size = 3, color = "salmon")
```

- **Question 1.9** Determine whether the following conclusions can be drawn from these data. Explain your answer.
  - The ethnicity distribution of the admitted applicant pool at the College of Galaxy does not look like that of U.S. universities.
  - The ethnicity distribution of the admitted applicant pool at the College of Galaxy is biased toward white applicants.

Question 2: A strange dice. Your friend Jerry invites you to a game of dice. He asks you to roll a dice 10 times and says that he wins $1 each time a 3 turns up and loses $1 on any other face. Jerry's dice is six-sided, however, the "2" and "4" faces have been replaced with "3"'s. The following code chunk simulates the results after one game:

```
weird_dice_probs <- c(1/6, 0/6, 3/6, 0/6, 1/6, 1/6)
rmultinom(n = 1, size = 10, prob = weird_dice_probs)
```

```
##        [,1]
## [1,]    3
## [2,]    0
## [3,]    4
## [4,]    0
## [5,]    1
## [6,]    2
```

While the game seems like an obvious scam, Jerry claims that his dice is no different than a fair dice in the long run. Can you disprove his claim using a hypothesis test?

- **Question 2.1** Write a function `sample_prop` that receives two arguments `distribution` (e.g., `weird_dice_probs`) and `size` (e.g., 10 rolls). The function simulates the game using a dice where the probability of each face is given by `distribution` and the dice is rolled `size` many times. The proportion of each face that appeared after the simulation is returned.

The following code chunk simulates the result after playing one round of Jerry's game. You record the sample proportions of the faces that appeared in a tibble named `jerry_die_dist`.

```
set.seed(2022)
jerry_die_dist <- tibble(
  face = 1:6,
  prob = sample_prop(weird_dice_probs, 10)
)
jerry_die_dist
```

Let us define the distribution for what we know is a *fair* six-sided die.

```
fair_die_dist <- tibble(
  face = seq(1:6),
  prob = rep(1/6, 6)
)
fair_die_dist
```

Here is what the `jerry_die_dist` distribution looks like when visualized:

```
ggplot(jerry_die_dist) +
  geom_bar(aes(x = as.factor(face), y = prob), stat = "identity")
```

- **Question 2.2** Define a null hypothesis and an alternative hypothesis for this question.

  We saw in Section 5.4 that the mean is equivalent to weighing each face by the proportion of times it appears. The mean of `jerry_die_dist` can be computed as follows:

```
jerry_die_dist |>
  summarize(mean = sum(face * prob))
```

For reference, here is the mean of a *fair* six-sided dice. Observe how close this value is to the mean of Jerry's dice:

```
fair_die_dist |>
  summarize(mean = sum(face * prob))
```

The following function `mystery_test_stat1()` takes a single tibble `dist` (e.g., `jerry_die_dist`) as its argument and computes a test statistic by comparing it to `fair_die_dist`.

```
mystery_test_stat1 <- function(dist) {
  x <- dist |>
    summarize(mean = sum(face * prob)) |>
    pull(mean)
  y <- fair_die_dist |>
```

```
        summarize(mean = sum(face * prob)) |>
        pull(mean)
    return(abs(x-y))
}
```

- **Question 2.3** What test statistic is being used in `mystery_test_stat1`?

- **Question 2.4** Write a function called `one_simulated_stat`. The function receives a single argument `stat_func`. The function generates sample proportions after one round of Jerry's game *under the assumption of the null hypothesis*, computes the test statistic from this sample using the argument `stat_func`, and returns it.

```
one_simulated_stat(mystery_test_stat1) # an example call
```

- **Question 2.5** Complete the following function called `simulate_dice_experiment`. The function receives two arguments, an `observed_dist` (e.g., `jerry_die_dist`) and a `stat_func`. The function computes the observed value of the test statistic using `observed_dist`. It then simulates the game **10,000** times to produce **10,000** different test statistics. The function then prints the p-value and plots a histogram of your simulated test statistics. Also shown is where the observed value falls on this histogram (orange dot) and the cut-off for the 95% significance level.

```
simulate_dice_experiment <- function(observed_dist, stat_func) {

  p_value_cutoff <- 0.05
  print(paste("P-value: ",
          (sum(test_stats >= observed_stat) / length(test_stats))))
  ggplot(tibble(test_stats)) +
    geom_histogram(aes(x = test_stats, y = after_stat(density)),
                   bins=10, color = "gray", fill='darkcyan') +
    geom_vline(aes(xintercept=quantile(test_stats,
                                       1-p_value_cutoff)),
               color='red') +
    geom_point(aes(x=observed_stat,y=0),size=4,color='orange')
}
```

- **Question 2.6** Run the experiment using your function `simulate_dice_experiment` using the observed distribution from `jerry_die_dist` and the mystery test statistic.

  The evidence so far has been unsuccessful in refuting Jerry's claim. Maybe you should stop playing games with Jerry...

  As a desperate final attempt before giving up and agreeing to play Jerry's game, you try using a different test statistic to simulate called `mystery_test_stat2`.

```
mystery_test_stat2 <- function(dist) {
  sum(abs((dist |> pull(prob)) -
          (fair_die_dist |> pull(prob))) /2)
}

mystery_test_stat2(jerry_die_dist) # an example call
```

- **Question 2.7** Repeat **Question 2.6**, this time using `mystery_test_stat2` instead.

- **Question 2.8** At a significance level of 95%, what do we conclude from the first experiment? How about the second experiment?

- **Question 2.9** Examine the difference between the test statistics in `mystery_test_stat1` and `mystery_test_stat2`. Why is it that the conclusion of the test is different depending on the test statistic selected?

- **Question 2.10** Which of the following statements are **FALSE**? Indicate them by including its number in the following vector `pvalue_answers`.

  - The p-value printed is the probability that the die is fair.
  - The p-value printed is the probability that the die is NOT fair.
  - The p-value cutoff (5%) is the probability that the die is NOT fair.
  - The p-value cutoff (5%) is the probability of seeing a test statistic as extreme or more extreme than this one if the null hypothesis were true.

- **Question 2.11** For the statements you selected to be FALSE, explain why they are wrong.

Question 3 This question is a continuation of **Question 2**. The following incomplete function `experiment_rejects_null` receives four arguments: a tibble describing the probability distribution of a dice, a function to compute a test statistic, a p-value cutoff, and a number of repetitions to use. The function simulates 10 rolls of the given dice, and tests the null hypothesis about that dice using the test statistic given by `stat_func`. The function returns a Boolean: `TRUE` if the experiment *rejects* the null hypothesis at `p_value_cutoff`, and `FALSE` otherwise.

```
experiment_rejects_null <- function(die_probs,
                      stat_func, p_value_cutoff, num_repetitions) {
  observed_dist <- tibble(
    face = 1:6,
    prob = sample_prop(die_probs, 10)
  )



  p_value <- sum(test_stats >= observed_stat) / num_repetitions
  return(p_value < p_value_cutoff)
}
```

- **Question 3.1** Read and understand the above function. Then complete the missing portion that computes the observed value of the test statistic and simulates `num_repetitions` many test statistics under the null hypothesis.

  The following code chunk simulates the result after testing Jerry's dice with `mystery_test_stat1` at the P-value cut-off of 5%. Run it a few times to get a rough sense of the results.

- **Question 3.2** Repeat the experiment `experiment_rejects_null(weird_dice_probs, mystery_test_stat1, 0.05, 250)` 300 times using `replicate`. Assign `experiment_results` to a vector that stores the result of each experiment.

  Note: This code chunk will need some time to finish (approximately a few minutes). This will be a little slow. 300 repetitions of the simulation should require a minute or so of computation, and should be enough to get an answer that is roughly correct.

- **Question 3.3** Compute the proportion of times the function returned TRUE in experiment_results. Assign your answer to prop_reject.

- **Question 3.4** Does your answer to **Question 3.3** make sense? What value did you expect to get? Put another way, what is the probability that the null hypothesis is *rejected* when the dice is actually fair?

- **Question 3.5** What does it mean for the function to return TRUE when weird_dice_probs is passed as an argument? From the perspective of finding the truth about Jerry's (phony) claim, is the experiment successful? What if the function returned TRUE when fair_die_dist is passed as an argument instead?

Question 4. The United States House of Representatives in the 116th Congress (2019-2021) had 435 members. According to the Center for American Women and Politics (CAWP)[6], 101 were women and 334 men. The following tibble house gives the head counts:

```
house <- tribble(~gender, ~num_members,
                 "Female", 101,
                 "Male",   334)
house
```

```
## # A tibble: 2 x 2
##   gender num_members
##   <chr>        <dbl>
## 1 Female         101
## 2 Male           334
```

In this question, we will examine whether women are underrepresented in the chamber.

- **Question 4.1** If men and women are equally represented in the chamber, then the chance of either gender occupying any seat should be like that of a fair coin flip. For instance, if the chamber consisted of just 10 seats, then one "House of Representatives" might look like:

```
sample(c("Female", "Male"), size = 10,
       replace = TRUE, prob = c(0.5, 0.5))
```

Using this, write a *null* and *alternative* hypothesis for this problem.

- **Question 4.2** Using **Question 4.1**, write a function called one_sample_house that simulates one "House" under the null hypothesis. The function receives two arguments, gender_prop and house_size. The function samples "Female" or "Male" house_size many times where the chance of either gender appearing is given by gender_prop. The function then returns a tibble with the gender head counts in the simulated sample. Following is one possible returned tibble:

Gender	num_members
Female	207
Male	228

---

[6]https://cawp.rutgers.edu/facts/levels-office/congress/history-women-us-congress

```
one_sample_house <- function(gender_prop, house_size) {



}

total_seats <- house |> pull(num_members) |> sum()
one_sample_house(c(0.5, 0.5), total_seats) # an example call
```

- **Question 4.3** A good test statistic for this problem is the *difference* in the head count of males from the head count of females. Write a function that takes a tibble `head_count_tib` as an argument (that has the format as in **Question 4.2**). The function computes and returns the test statistic from this tibble.

- **Question 4.4** Compute the *observed value* of the test statistic using your `one_diff_stat()`. Assign the resulting value to the name `observed_value_house`.

- **Question 4.5** Write a function called `simulate_one_stat` that simulates one test statistic. The function receives two arguments, the gender proportions `prop` and the total seats (`total_seats`) to fill in the simulated "House". The function simulates a sample under the null hypothesis, and computes and returns the test statistic from the sample.

```
simulate_one_stat(c(0.5, 0.5), 100) # an example call
```

- **Question 4.6** Simulate 10,000 different test statistics under the null hypothesis. Store the results to a vector named `test_stats`.

  The following `ggplot2` code visualizes your results:

```
ggplot(tibble(test_stats)) +
  geom_histogram(aes(x = test_stats), bins=18, color="gray") +
  geom_point(aes(x = observed_value_house, y = 0),
             size=2, color="red")
```

- **Question 4.7** Based on the experiment, what can you say about the representation of women in the House?

  Let us now approach the analysis another way. Instead of assuming equal representation, let us base the comparison by using the representation of women candidates in the preceding 2018 U.S. House Primary elections. The tibble `house_primary` from the `edsdata` package compiles primary election results for Democratic and Republican U.S. House candidates running in elections from 2012 to 2018. The data is prepared by the Michael G. Miller Dataverse[7] part of the Harvard Dataverse[8].

```
library(edsdata)
house_primary

## # A tibble: 5,716 x 25
##    raceid  year stcd  state  seat party redist    fr   law
##    <chr>  <dbl> <chr> <chr> <dbl> <dbl>  <dbl> <dbl> <dbl>
```

---

[7]https://dataverse.harvard.edu/dataset.xhtml?persistentId=doi:10.7910/DVN/CXVMSY
[8]https://dataverse.harvard.edu/

```
##  1 20182~  2018 2305  minn~     2      1      0      0      0
##  2 20181~  2018 1006  geor~     0      1      0      2      0
##  3 20181~  2018 1006  geor~     0      1      0      2      0
##  4 20183~  2018 3401  nort~     3      0      0      0      0
##  5 20143~  2014 3204  new ~     2      1      0      9      1
##  6 20182~  2018 2210  mich~     0      1      0      1      0
##  7 20162~  2016 2210  mich~     3      1      0      9      0
##  8 20183~  2018 3205  new ~     1      1      0      0      1
##  9 20121~  2012 1705  kent~     0      1      1      0      1
## 10 20180~  2018 0927  flor~     3      0      0      9      1
## # ... with 5,706 more rows, and 16 more variables:
## #    candnumber <dbl>, prez <dbl>, votep <dbl>,
## #    type <dbl>, incname <chr>, candidate <chr>,
## #    candvotes <dbl>, tvotes <dbl>, candpct <dbl>,
## #    winner <dbl>, inc <dbl>, definc <lgl>, gender <dbl>,
## #    qual <dbl>, office <dbl>, runoff <dbl>
```

- **Question 4.8** Form a two-element vector named `primary_prop` that gives the proportion of female and male candidates, respectively, in the 2018 U.S. House Primary elections. This can be accomplished as follows:
    - Filter the data to the year `2018`. The data should not contain the results for any elections that resulted in a runoff (where `runoff = 1`).
    - Summarize and count each gender that appears in `gender` in the resulting tibble.
    - Add a variable that computes the proportions from these counts.
    - Pull the proportions as a vector and assign it to `primary_prop`.

- **Question 4.9** Repeat **Question 4.6** this time using the proportions given by `primary_prop`.

  The following code visualizes the revised result:

```
ggplot(tibble(test_stats)) +
  geom_histogram(aes(x = test_stats), bins=18, color="gray") +
  geom_point(aes(x = observed_value_house, y = 0),
             size=2, color="red")
```

- **Question 4.10** Compute the p-value using the `test_stats` you generated by comparing it with `observed_value_house`. Assign your answer to `p_value`.

- **Question 4.11** Why is it that in the first histogram the simulated test statistics cluster around 0 and in the second histogram the simulated values cluster around a value much greater? Is the statement of the null hypothesis the same in both cases?

- **Question 4.12** Now that we have analyzed the data in two ways, are women equally represented in the House? Why or why not?

Question 5. Cathy recently received from a friend a replica dollar coin which appears to be slightly biased towards "Heads". Cathy tosses the coin 20 times in a row counts how many times "Heads" turns up. She repeats this for 10 trials. Her results are summarized in the following tibble:

```
cathy_heads_game <- tibble(
  trial = 1:10,
```

```
    num_heads = c(12, 13, 13, 11, 17, 10, 10, 14, 9, 15),
    num_tails = 20 - num_heads
)
cathy_heads_game
```

```
## # A tibble: 10 x 3
##    trial num_heads num_tails
##    <int>     <dbl>     <dbl>
## 1      1        12         8
## 2      2        13         7
## 3      3        13         7
## 4      4        11         9
## 5      5        17         3
## 6      6        10        10
## 7      7        10        10
## 8      8        14         6
## 9      9         9        11
## 10    10        15         5
```

- **Question 5.1** What is the total heads observed? Assign your result to a double named `total_heads_observed`.

  Let us write an experiment and check how plausible it is for this coin to be fair.

- **Question 5.2** Given the outcome of 20 trials, which of the following test statistics would be reasonable for this hypothesis test?

    - The total number of heads.
    - The total number of heads minus the total number of tails.
    - Whether there is at least one head.
    - Whether there is at least one tail.
    - The total variation distance between the probability distribution of a fair coin and the observed distribution of heads and tails.
    - The trial with the minimum number of heads.

  Assign the name `good_test_stats` to a vector of integers corresponding to these test statistics.

- **Question 5.3** Let us write a code that simulates tossing a fair coin. Write a function called `one_test_stat` that receives a parameter `num_trials`. The function simulates a fair coin toss 20 times, records the number of heads, and repeats this procedure `num_trials` many times. The function returns the total number of heads over the given number of trials.

```
one_test_stat(10) # an example call after cathy's 10 trials
```

- **Question 5.4** Repeat Cathy's experiment 10,000 times. Store the results in to-tal_head_stats.

- **Question 5.5** Compute a p-value using `total_head_stats`. Assign the result to `p_value`.

- **Question 5.6** From the experiment how plausible do you say Cathy's coin is fair?

Question 6. A popular course in the College of Groundhog is an undergraduate programming course CSC1234. In the spring semester of 2022, the course had three sections, A, B, and C. The sections were taught by different instructors. The course had the same textbook, the same assignments, and the same exams. One same formula was applied to determine the final grade. At the end of a semester, some students in Sections A and C came to their instructors and asked if the instructors had been harsher than the instructor for Section B, because several buddies of theirs in Section B did better in the course. Time for a hypothesis test!

The section and score information for the semester is available in the tibble csc1234 from the edsdata package.

```
library(edsdata)
csc1234
```

```
## # A tibble: 388 x 2
##    Section Score
##    <chr>   <dbl>
##  1 A         100
##  2 A         100
##  3 A         100
##  4 A         100
##  5 A         100
##  6 A         100
##  7 A         100
##  8 A         100
##  9 A         100
## 10 A         100
## # ... with 378 more rows
```

We will use a permutation test to see if the scores for Sections A and C are indeed significantly lower than the scores for Section B. That is, we will compare three groups: Section A with B, Section A with C, and Section B with C.

- **Question 6.1** Compute the group-wise mean for each section of the course. The tibble should contain two variables: the section name and the mean of that section. Assign the resulting tibble to the name section_means.

- **Question 6.2** Visualize a histogram of the scores in csc1234. Use a facet wrap on Section so that you can view the three distributions together separately. We suggest using 10 bins.

  We can develop a chance model by hypothesizing that any section's scores looks like a random sample out of all of the student scores across all three sections. We can then see the difference in mean scores for each of the three pairs of randomly drawn "sections". This is a specified chance model we can use to simulate and, therefore, is the **null hypothesis**.

- **Question 6.3** Define a good **alternative** hypothesis for this problem.

- **Question 6.4** Write a function called mean_differences that takes a tibble as its single argument. It then summarizes this tibble by computing the average mean score (in Score) for each section (in Section). The function returns a *three-element vector* of mean differences for each pair: the difference in mean scores between A and B ("A-B"), C and B ("C-B"), and C and A ("C-A").

- **Question 6.5** Compute the observed differences in the means of the three sections using mean_differences. Store the results in observed_differences.

The following code chunk puts your observed values into a tibble named observed_diff_tibble.

```
observed_diff_tibble <- tibble(
  pair = c("A-B", "C-B", "C-A"),
  test_stat = observed_differences
)
observed_diff_tibble
```

- **Question 6.6** Write a function scores_permute_test that does the following:
  - From csc1234 form a new variable that shuffles the values in Score using sample. Overwrite the variable Score with the shuffled values.
  - Call mean_differences on this shuffled tibble.
  - Return the vector of differences.

```
scores_permute_test() # an example call
```

- **Question 6.7** Use replicate on the scores_permute_test function you wrote to generate 1,000 sample differences.

The following code chunk creates a tibble named differences_tibble from the simulated test statistics you generated above.

```
differences_tibble <- tibble(
            `A-B` = test_stat_differences[1,],
            `C-B` = test_stat_differences[2,],
            `C-A` = test_stat_differences[3,])
differences_tibble
```

- **Question 6.8** Generate three histograms using the results in differences_tibble. As with **Question 6.2**, use a facet wrap on each pairing (i.e., A-B, C-B, and C-A). Then attach a red point to each histogram indicating the observed value of the test statistic (use observed_diff_tibble). We suggest using 20 bins for the histograms.

- **Question 6.9** The bulk of the distribution in each of the three histograms is centered around 0. Using what you know about the stated null hypothesis, why do the distributions turn out this way?

- **Question 6.10** By examining the above three histograms and where the observed value of the test statistic falls, which difference among the three do you think is the most egregious?

- **Question 6.11** Based on your answer to **Question 6.10**, can we say that the hypothesis test brings enough evidence to show that the drop in student scores was deliberate and that the instructor was unfair in grading?

7

Quantifying Uncertainty

So far we have developed ways to show how to use data to decide between competing questions about the world. For instance, does Harvard enroll proportionately less Asian Americans than other private universities in the United States; are the exam grades in one lab section of a course too low when compared to other lab sections; can an experimental drug bring an improvement to patients recovering from brain trauma? We evaluated questions like these by means of an *hypothesis test* where we put forward two hypotheses: a *null* hypothesis and an *alternative* hypothesis.

Often we are just interested in what a value looks like. For instance, airlines might be interested in the median flight delay of their flights to preserve customer satisfaction; political candidates may look to the percentage of voters favoring them to gauge how aggressive their campaigning should be.

Put in the language of statistics, we are interested in *estimating* some unknown *parameter* about a population. If all the data have been made available to us, we could compute the parameter directly with ease. However, we often do not have access to the full population (as is the case with polling voters) or there may be too much data to work with that it becomes computationally prohibitive (as is the case with flights).

We have seen before that sampling distributions can provide reliable approximations to the true (and usually unknown) distribution and that, likewise, a statistic computed from it can provide a reliable estimate of the parameter in question. However, the value of a statistic can turn out *differently* depending on the random samples that are drawn to compose a sampling distribution. How much can the value of a statistic vary? Could we quantify this uncertainty?

This chapter develops a way to answer this question using an important technique in data science called *resampling*. We begin by introducing order statistics and percentiles. This will provide us the tools needed to develop the resampling method to produce distributions from a sample, in which we apply order statistics to the generated distributions to obtain something called the *confidence interval*.

7.1 Order Statistics

The minimum, maximum, and the median are part of what we call *order statistics*. Order statistics are values at certain positions in numerical data after reordering the data in ascending order.

DOI: 10.1201/9781003320845-7

7.1.1   Prerequisites

This section will make use of data for all flights that departed New York City in 2013. The dataset is made available by the Bureau of Transportation Statistics[1] in the United States. Let's also load in the tidyverse as usual.

```
library(tidyverse)
library(nycflights13)
```

7.1.2   The `flights` data frame

In our prior exploration of this data frame, we generated empirical distributions of departure delays. Let's revisit this study and visualize the departure delays again.

```
ggplot(flights) +
  geom_histogram(aes(x = dep_delay, y = after_stat(density)),
                 color="grey", bins = 30)
```

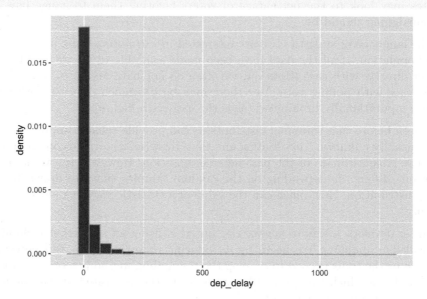

As before, we are interested in the bulk of the data here, so we can ignore the 1.83% of flights with delays of more than 150 minutes.

```
flights150 <- flights |>
  filter(dep_delay <= 150)
```

```
ggplot(flights150) +
  geom_histogram(aes(x = dep_delay, y = after_stat(density)),
                 color="grey", bins = 30)
```

---

[1] https://www.transtats.bts.gov/

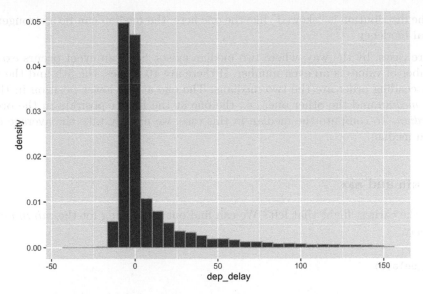

Let's extract the departure delay column as a vector.

```
dep_delays <- flights150 |> pull(dep_delay)
```

7.1.3 `median`

The *median* is a popular order statistic that gives us a sense of the central tendency of the data. It is the value at the middle position in the data after reordering of the values.

```
median(dep_delays)
```

```
## [1] -2
```

This tells us that half of the flights had early departures – not bad! Recall that we also used the mean to understand the central tendency. Note that the *mean* (or *average*) is a statistic, but it is not an order statistic. Let's compare with the mean of departure delays.

```
mean(dep_delays)
```

```
## [1] 8.716037
```

There is quite a bit of discrepancy between the two. Observe that the histogram above has a very long right tail; the mean is pulled upward by flights with long departure delays. In general:

If a distribution has a long tail, the mean will be pulled away from the median in the direction of the tail. Otherwise, if the distribution is symmetrical, the mean and the median will equal.

When the distribution is "skewed" like the one here, the median can be a stronger indicator of central tendency.

There are cases, by the way, where two median exists. Such an event occurs exactly when the number of values is an even number. If there are 10 values, the 5th and the 6th values in the ascending order are the two medians. The one at the lower position in the order is the *odd median* and the other one, i.e., the one at the higher position in the order, is the *even median*. To compute the median in this case, we usually take the average of the odd and even median.

7.1.4 `min` and `max`

What is the earliest flight that left? We can find out by looking for the *minimum* departed flight delay.

```
min(dep_delays)
```

```
## [1] -43
```

The authors admit that this flight might have left a little too early for their liking. What about the latest flight?

```
max(dep_delays)
```

```
## [1] 150
```

Recall that this maximum is actually artificial because we filtered all rows whose departure delay was more than 150. To recover the true maximum, we need to refer to the original `flights` data.

```
flights |>
  pull(dep_delay) |>
  max(na.rm = TRUE)
```

```
## [1] 1301
```

That is almost a *22 hour* delay – better get a sleeping bag!

7.2 Percentiles

Now that we have an understanding of order statistics, we can use it to develop the notion of a *percentile*. We will also explore a closely related concept called the *quartile*.

You are probably already familiar with the concept of percentiles from sports or standardized testing like the SAT[2]. Organizations like the College Board talk so much about percentiles – to the extent of writing full guides on how to interpret them – because they are indicators of how students perform relative to other exam-takers. Indeed, the percentile is another

---

[2]https://collegereadiness.collegeboard.org/pdf/understanding-sat-scores.pdf

order statistic that tells us something about the *rank* of a data point after reordering the elements in a dataset. Now that we have an understanding of order statistics, we can use it to develop the notion of a *percentile*. We will also explore a closely related concept called the *quartile*.

7.2.1   Prerequisites

Before starting, let's load the tidyverse as usual.

```
library(tidyverse)
library(edsdata)
```

7.2.2   The `finals` tibble

In the spirit of the College Board, we will examine exam scores to develop an understanding of percentiles. Recall that the `finals` data frame contains hypothetical final exam scores from two offerings of an undergraduate computer science course. Let's load it in.

```
finals
```

```
## # A tibble: 102 x 2
##      grade class
##      <dbl> <chr>
## 1       89 A
## 2       17 A
## 3       94 A
## 4       51 A
## 5       49 A
## 6       93 A
## 7       52 A
## 8       54 A
## 9       57 A
## 10      65 A
## # ... with 92 more rows
```

The dataset contains final scores from a total of 105 students.

```
nrow(finals)
```

```
## [1] 102
```

We will not concern ourselves with the individual offerings of the course this time. Since the scores are of interest for this study, let us extract a vector of scores from the tibble.

```
scores <- finals |>
  pull(grade)
```

To orient ourselves to the data, we can look at the maximum and minimum scores.

```
max(scores)
```

```
## [1] 98
```

```
min(scores)
```

```
## [1] 0
```

We may be alarmed to see that the minimum score of 0. Some insight into the course would reveal that there were a few students who did not appear for the final exam (don't be one of them!).

Finally, let us visualize the distribution of scores.

```
ggplot(finals) +
  geom_histogram(aes(x = grade, y = after_stat(density)),
              col="grey", fill = "darkcyan", bins = 10)
```

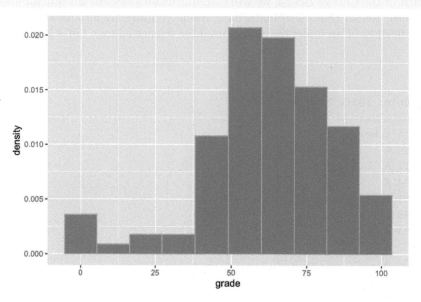

7.2.3 The `quantile` function

The percentile is an order statistic where the position of the data is not the rank but a percentage that specifies relative position in the data. For instance, the 50th percentile is the smallest value that is at least as large as 50% of the elements in `scores`; it must be a value on the list of scores. We can compute this simply with the `quantile` function in R.

```
quantile(scores, c(0.5), type = 1)
```

```
## 50%
##  65
```

The value at the 50th percentile is something we already know: the median score!

```
median(scores)
```

```
## [1] 65
```

The `quantile` function gives the value that cuts off the first n percent of the data values when it is sorted in ascending order. There are many ways to compute percentiles (see the help page for a sneak peek, with `?quantile`). The one that matches the definition used here corresponds to `type = 1`.

The additional argument passed in is a vector of desired percentages. These must be between 0 and 1. This is how `quantile` gets its name: *quantiles* are *percentiles* scaled to have a value between 0 and 1, e.g. `0.5` rather than `50`.

Let us look at some more percentile values in the vector of scores.

```
quantile(scores, c(0.05, 0.2, 0.9, 0.95, 0.99, 1), type = 1)
```

```
##   5%  20%  90%  95%  99% 100%
##   17   49   89   93   97   98
```

Let's pick apart some of these values. We see that the value at the 5th percentile is the lowest exam score that is at least as large as 5% of the scores. We can confirm this easily by summing the number of scores less than 17 and dividing by the total number of scores.

```
sum(scores < 17) / length(scores)
```

```
## [1] 0.04901961
```

Moving on up, we see that the 95th and 99th percentiles are quite close together. We also observe that the 100th percentile is 98, which corresponds to the maximum score obtained on the final. That is, a 98 is at least as large as 100% of the scores, which is the entire class.

The 0th percentile is simply the smallest value in the dataset, as 0% of the data is at least as large as it. In other words, there is no exam score in the class lower than a 0.

```
quantile(scores, c(0), type = 1)
```

```
## 0%
##  0
```

If the College Board says a student is in the "top 10 percentile", this would be a misnomer. What they really mean to say is that the student is in the $1 -$ top X percentile, or 90th percentile.

7.2.4 Quartiles

In addition to the medians, common percentiles are the 1/4th and 3/4th, which we often call the bottom quarter and the top quarter. Basically, we chop the data in quarters and use the boundaries between the neighboring quarters. Since these percentiles partition the data into quarters, these are given a special name: *quartiles*.

```
quantile(scores, c(0/4, 1/4, 2/4, 3/4, 4/4), type = 1)
```

```
##   0%  25%  50%  75% 100%
##    0   52   65   77   98
```

Observe what happens when we omit the vector of percentages.

```
quantile(scores, type = 1)
```

```
##   0%  25%  50%  75% 100%
##    0   52   65   77   98
```

The corresponding 0th, 25th, 50th, 75th, and 100th percentiles of the vector are returned.

7.2.5 Combining two percentiles

By combining two percentiles, we can get a rough sense of the distribution. For example, the combination of 25th and 75th percentiles represents the "middle" 50%. Similarly, the 2.5th and 97.5th percentiles represent the middle 95% of the data. That is,

```
quantile(scores, c(0.025, 0.975), type = 1)
```

```
##  2.5% 97.5%
##     0    95
```

95% of the scores is between 0 and 95. We could find this more directly by realizing that the middle 95% corresponds to going up and down from the 50th percentile by half of that amount, which is 47.5%.

```
middle_area <- 0.95
quantile(scores, 0.5 + (middle_area / 2) * c(-1, 1), type = 1)
```

```
##  2.5% 97.5%
##     0    95
```

As one more example, here is the middle 90% of scores.

```
middle_area <- 0.90
quantile(scores, 0.5 + (middle_area / 2) * c(-1, 1), type = 1)
```

```
##  5% 95%
##  17  93
```

7.2.6 Advantages of percentiles

Percentile is a useful concept because it eliminates the use of population size in specifying the position; that is, the position specification does not directly take into account the size of the data. What do we mean by that?

Let's return to the example of final exam scores. Suppose that one offering of the class contained 50 students while another had 200 students. Consider the "top 10 students" in each class.

Since top 10 is 20% of 50 students, there is a 20% chance for a student to be among the top 10, while the chances decrease to 5% for the class of 200. That is, if we specify a top group with its size, the significance being in the top group varies depending on the size of the population and so we must to specify the size of the underlying group, e.g., "top 10 in a group of 4000 students". Percentiles are nice in that they are not sensitive to these changes.

7.3 Resampling

It is usually the case that a data scientist will receive a sample from an underlying population to which she has no access. If she had access to the underlying population, she could calculate the parameter value directly. Since that is impossible, is there a way for her to use the sample at hand to generate a range of values for the statistic?

Yes! This is a technique we call *resampling*, which is also known as the *bootstrap*. In bootstrapping, we treat the dataset at hand as the "population" and generate "new" samples from it. But there is a catch. Each sample data set that we generate should be equal in size to the original. This necessarily means that our sampling plan be done *with* replacement.

Since the samples have the same size as the original with the use of replacement, duplicates and omissions can arise. That is, there are items that will appear multiple times as well as items that are missing. Because randomness is involved, the discrepancy varies.

7.3.1 Prerequisites

This section will defer again to the New York City flights in 2013 from the Bureau of Transportation Statistics[3]. Let's also load in the tidyverse as usual.

```
library(tidyverse)
library(nycflights13)
```

7.3.2 Population parameter: the median time spent in the air

When studying this dataset, we have spent a lot of time examining flight departure delays. This time we will turn our attention to another variable in the tibble which tracks the amount of time a flight spent in air, in minutes. The variable is called air_time.

---

[3]https://www.transtats.bts.gov/

Let's visualize the distribution of air time in `flights`.

Recall the distribution of departure delays in `flights150`.

```
ggplot(flights) +
  geom_histogram(aes(x = air_time, y = after_stat(density)),
                 col="grey", fill = "darkcyan", bins = 20)
```

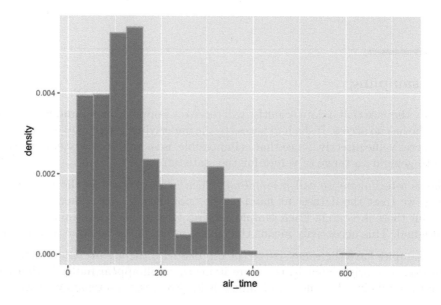

As before, let's concentrate on the bulk of the data and filter out any flights that flew for more than 400 minutes.

```
flights400 <- flights |>
  filter(air_time < 400) |>
  drop_na()
```

We plot this distribution one more time.

```
ggplot(flights400) +
  geom_histogram(aes(x = air_time, y = after_stat(density)),
                 col="grey", fill = "darkcyan", bins = 20)
```

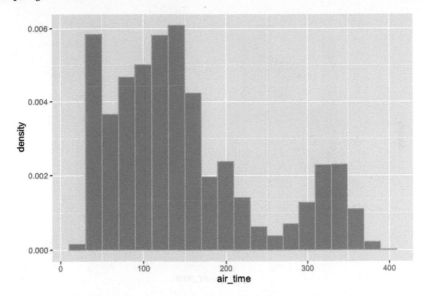

The parameter we will select for this study is the mean air time.

```
pop_mean <- flights400 |>
  pull(air_time) |>
  mean()
pop_mean
```

```
## [1] 149.6463
```

Let us see how well we can estimate this value based on a sample of the flights. We will study two such samples: an artificial sample and a random sample.

7.3.3 First try: A mechanical sample

For our mechanical sample, we will assume that we have been given only a cross-section of the flights data and try to estimate the population median based on this sample. Let us suppose we have been given the flight data for only the months of September and October.

```
flights_sample <- flights400 |>
  filter(month == 9 | month == 10)
```

There are 55,522 flights appearing in the subset. Let's visualize the distribution of air time from our sample.

```
ggplot(flights_sample) +
  geom_histogram(aes(x = air_time, y = after_stat(density)),
                 col="grey", fill = "darkcyan", bins = 20)
```

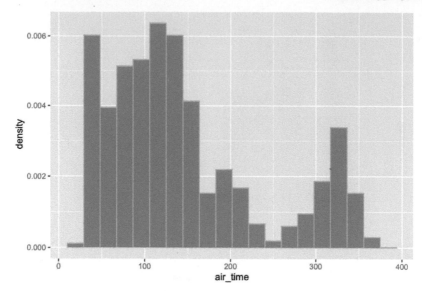

It appears close to the population of flights, though there are notable differences: flights that have longer air times (between 300 and 400 minutes) appear exaggerated in this dataset. Let's compute the mean from this sample.

```
sample_mean <- flights_sample |>
  pull(air_time) |>
  mean()
sample_mean
```

```
## [1] 145.378
```

It is quite different from the population median. Nevertheless, this subset of flights will serve as the dataset from which we will bootstrap our samples. Put another way, **we will treat this sample as if it were the population.**

7.3.4 Resampling the sample mean

To perform a bootstrap, we will draw from the sample, at random **with replacement**, the same number of times as the size of the sample dataset.

To simplify the work, let us extract the column of air times as a vector.

```
air_times <- flights_sample |>
  pull(air_time)
```

We know already how to sample at random with replacement from a vector using `sample`. Computing the sample mean is also straightforward: just pipe the returned vector into `mean`.

```
sample_mean <- air_times |>
  sample(replace = TRUE) |>
  mean()
sample_mean
```

```
## [1] 145.3686
```

Let us move this work into a function we can call.

```
one_sample_mean <- function() {
  sample_mean <- flights_sample |>
    pull(air_time) |>
    sample(replace = TRUE) |>
    mean()
  return(sample_mean)
}
```

Give it a run!

```
one_sample_mean()
```

```
## [1] 144.6964
```

This function is actually quite useful. Let's generalize the function so that we may call it with other datasets we will work with. The modified function will receive three parameters: (1) a tibble to sample from, (2) the column to work on, and (3) the statistic to compute.

```
one_sample_value <- function(df, label, statistic) {
  sample_value <- df |>
    pull({{ label }}) |>
    sample(replace = TRUE) |>
    statistic()
  return(sample_value)
}
```

We can now call it as follows.

```
one_sample_value(flights_sample, air_time, mean)
```

```
## [1] 145.398
```

Q: What's the deal with those (ugly) double curly braces ({{) ? To make R programming more enjoyable, the tidyverse allows us to write out column names, e.g. air_time, just like we would variable names. The catch is that when we try to use such syntax sugar from inside a function, R has no idea what we mean. In other words, when we say pull(label) R thinks that we want to extract a vector from a column called label, despite the fact we passed in air_time as an argument. To lead R in the right direction, we surround label with {{ so that R knows to interpret label as, indeed, air_time.

7.3.5   Distribution of the sample mean

We now have all the pieces in place to perform the bootstrap. We will replicate this process many times so that we can compose an empirical distribution of all the bootstrapped sample means. Let's repeat the process 10,000 times.

```r
bstrap_means <- replicate(n = 10000,
                one_sample_value(flights_sample, air_time, mean))
```

Let us visualize the bootstrapped sample means using a histogram.

```r
df <- tibble(bstrap_means)
ggplot(df, aes(x = bstrap_means, y = after_stat(density))) +
  geom_histogram(col="grey", fill = "darkcyan", bins = 8)
```

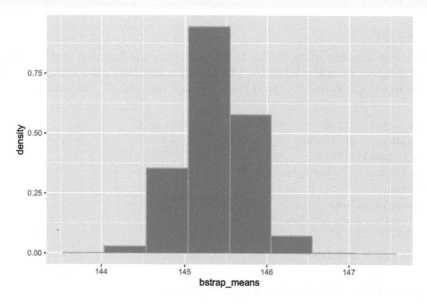

7.3.6   Did it capture the parameter?

How often does the population mean fall somewhere in the empirical histogram? Does it reside "somewhere at the center" or at the fringes where the tails are? Let us be more specific by what we mean when we say "somewhere at the center": the middle 95% of bootstrapped means containing the population mean.

We can identify the "middle 95%" using the percentiles we learned from the last section. Here they are:

```r
desired_area <- 0.95
middle95 <- quantile(bstrap_means,
                0.5 + (desired_area / 2) * c(-1, 1), type = 1)
middle95
```

```
##      2.5%     97.5%
## 144.6075 146.1229
```

Let us annotate this interval on the histogram.

```
df <- tibble(bstrap_means)

ggplot(df, aes(x = bstrap_means, y = after_stat(density))) +
  geom_histogram(col="grey", fill = "darkcyan", bins = 8) +
  geom_segment(aes(x = middle95[1], y = 0,
                xend = middle95[2], yend = 0),
              size = 2, color = "salmon")
```

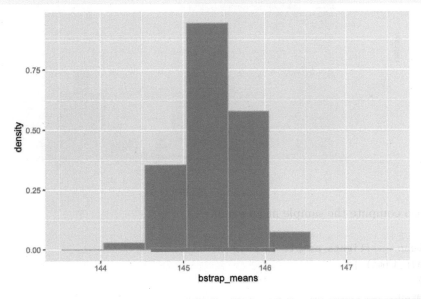

```
pop_mean
```

```
## [1] 149.6463
```

Our population mean is 149.6 minutes – that is nowhere to be seen in this interval or even in the histogram! It would seem then that in all of the 10,000 replications of the bootstrap, not even one was able to capture the population mean. What happened?

Recall the subset selection we used: all flights in September or October. This was a very artificial selection that is prone to bias. We learned before when we discussed sampling plans that bias in the sample can mislead the statistic computed from it, especially when using a convenience sample such as the one here.

7.3.7 Second try: A random sample

We will now try to estimate the population mean using a random sample of flights. Let us select at random without replacement 10,000 flights from the data.

```
flights_sample <- flights400 |>
  slice_sample(n = 10000, replace = FALSE)
```

We will visualize what our random sample looks like.

```
ggplot(flights_sample) +
  geom_histogram(aes(x = air_time, y = after_stat(density)),
             col="grey", fill = "darkcyan", bins = 20)
```

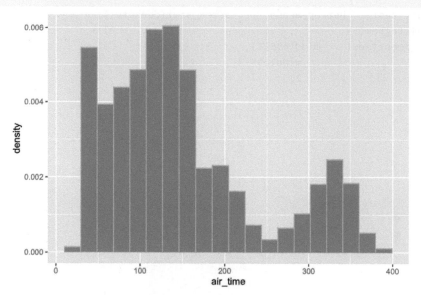

Let us also compute the sample mean again.

```
sample_mean <- flights_sample |>
  pull(air_time) |>
  mean()
sample_mean
```

```
## [1] 150.7474
```

We observe that the sample mean is also much closer to the population mean, unlike our mechanical selection attempt. This is confirmation of the Law of Averages (finally) at work: when we sample at random and the sample size is large, the distribution of the sample closely follows that of the flight population.

Let us now repeat the bootstrap. Recall that we will treat this sample as if it were the population.

7.3.8 Distribution of the sample mean (revisited)

We have done all the hard work already in setting up the bootstrap. To redo the process, we need only to pass in the random sample contained in `flights_sample`. As before, let us repeat the process 10,000 times.

```
bstrap_means <- replicate(n = 10000,
             one_sample_value(flights_sample, air_time, mean))
```

We will identify the "middle 95%". Here is the interval:

```
desired_area <- 0.95
middle95 <- quantile(bstrap_means,
                     0.5 + (desired_area / 2) * c(-1, 1), type = 1)
middle95
```

```
##      2.5%     97.5%
## 148.9982 152.5637
```

Let us annotate this interval on the histogram. We will also plot the population mean as a red dot.

```
df <- tibble(bstrap_means)

ggplot(df, aes(x = bstrap_means, y = after_stat(density))) +
  geom_histogram(col="grey", fill = "darkcyan", bins = 8) +
  geom_segment(aes(x = middle95[1], y = 0,
                   xend = middle95[2], yend = 0),
                   size = 2, color = "salmon") +
  geom_point(aes(x = pop_mean, y = 0), color = "red", size = 3)
```

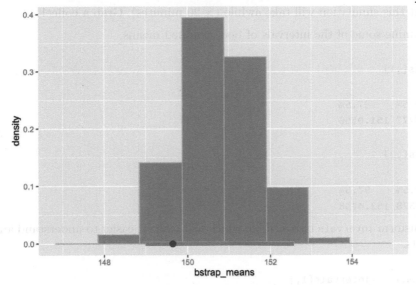

The population mean of 149.6 minutes falls in this interval. We conclude that the "middle 95%" interval of bootstrapped means successfully captured the parameter.

7.3.9 Lucky try?

Our interval of bootstrapped means captured the parameter in the air time data. But were we just lucky? We can test it out.

We would like to see how often the "middle 95%" interval captures the parameter. We will need to redo the entire process many times to find an answer. More specifically, we will follow the recipe:

- Collect a fresh sample of size 10,000 from the population. For the sampling plan, sample at random without replacement.

- Do 10,000 replications of the bootstrap process and find the "middle 95%" interval of bootstrapped means. We will repeat this process 100 times so that we end up with 100 intervals; we will count how many of them contain the population mean.

```
all_the_bootstraps <- function() {
  desired_area <- 0.95

  flights_sample <- flights400 |>
    slice_sample(n = 10000, replace = FALSE)

  bstrap_means <- replicate(n = 10000,
                    one_sample_value(flights_sample, air_time, mean))
  middle95 <- quantile(bstrap_means,
                    0.5 + (desired_area / 2) * c(-1, 1), type = 1)
  return(middle95)
}
```

```
intervals <- replicate(n = 100, all_the_bootstraps())
```

Note that this simulation will take awhile (> 20 minutes). Grab a coffee!

Let's examine some of the intervals of bootstrapped means.

```
intervals[,1]
```

```
##     2.5%     97.5%
## 147.4277 151.0550
```

```
intervals[,2]
```

```
##     2.5%     97.5%
## 148.7328 152.4258
```

Let's transform intervals into a tibble which will make it easier to understand and visualize the results.

```
left_column <- intervals[1,]
right_column <- intervals[2,]
```

```
interval_df <- tibble(
  replication = 1:100,
  left = left_column,
  right = right_column
)
interval_df
```

```
## # A tibble: 100 x 3
##    replication  left right
##          <int> <dbl> <dbl>
```

```
##  1        1  147.   151.
##  2        2  149.   152.
##  3        3  148.   152.
##  4        4  149.   152.
##  5        5  149.   153.
##  6        6  147.   150.
##  7        7  148.   151.
##  8        8  148.   151.
##  9        9  148.   152.
## 10       10  149.   152.
## # ... with 90 more rows
```

How many of these contain the population mean? We can count the number of intervals where the population mean is between the left and right endpoints.

```
interval_df |>
  filter(left <= pop_mean & right >= pop_mean) |>
  nrow()
```

```
## [1] 94
```

We can visualize these intervals by stacking them on top of each other vertically. The vertical red line shows where the population mean lies. Under real-life circumstances, we do not know where it is.

```
ggplot(interval_df) +
  geom_segment(aes(x = left, y = replication,
                   xend = right, yend = replication),
               color = "salmon") +
  geom_vline(xintercept = pop_mean, color = "red") +
  labs(x = "Air time (minutes)")
```

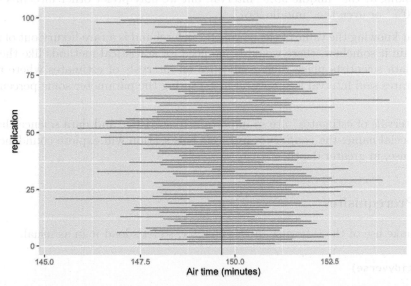

We expect about 95 of the 100 intervals to cross the vertical line; meaning, it contains the parameter. We would label such intervals as "good". If an interval does not, oh well – that's the nature of chance. Fortunately, these do not occur often. In fact, they should occur about 5 times among 100 trials, or 95%. The strength of statistics is not clairvoyance, but the ability to quantify uncertainty.

7.3.10   Resampling round-up

Before we close this section, we end with a quick summary on how to perform a bootstrap.

Goal: To estimate some population parameter we do not know about, e.g., the mean air time of New York City flights.

- Select a sampling plan. A safe bet is to sample at *random* without replacement from the population. Be sure the sample drawn is *large* in size and remember that in reality sampling is an expensive process. It is likely you will get only one chance to draw a sample from the population.
- Bootstrap the random sample (this time, *with* replacement) and compute the desired statistic from it.
- Replicate this process a great number of times to obtain many bootstrapped samples.
- Find the "middle 95%" interval of the bootstrapped samples.

7.4   Confidence Intervals

The previous section developed a way to estimate the value of a parameter we do not know. Because chance is an inevitable part of drawing a random sample, we cannot be precise and offer a single value for this estimate, e.g., we can determine that the mean height of all individuals in the United States is *exactly* 5.3 feet. Instead, we provide an *interval* of estimates by looking at a bulk of values that are "somewhere in the center". Typically this entails looking at the "middle 95%" interval, but we may prefer other intervals such as the "middle 90%" or even the "middle 99%".

Recall that knowing the value of the parameter beforehand is a rare luxury out of reach; if we could obtain it somehow, there would be no need for statistical methods like the bootstrap. Instead, data scientists place their confidence on intervals of estimates where the process that generates said interval is successful in capturing the parameter some percentage of the time.

These "intervals of estimates" are so important to statistics and data science that they are given a special name: the *confidence interval*. This section will explore confidence intervals, and their use, in greater depth.

7.4.1   Prerequisites

We will make use of the tidyverse in this chapter, so let's load it in as usual.

```
library(tidyverse)
```

We will also bring forward the `one_sample_value` function we wrote in the previous section.

```
one_sample_value <- function(df, label, statistic) {
  sample_value <- df |>
    pull({{ label }}) |>
    sample(replace = TRUE) |>
    statistic()
  return(sample_value)
}
```

For the running example in this section, we turn to survey data collected by the US National Center for Health Statistics (NCHS) on nutrition and health information. This data is available in the tibble NHANES from the NHANES package. In accordance to the documentation (see ?NHANES), the dataset can be treated as if it were a simple random sample from the American population. We use this dataset as an example where we *do not know* the population parameter.

```
library(NHANES)
NHANES
```

```
## # A tibble: 10,000 x 76
##        ID SurveyYr Gender   Age AgeDe~1 AgeMo~2 Race1 Race3
##     <int> <fct>    <fct>  <int> <fct>     <int> <fct> <fct>
##  1 51624 2009_10  male      34 " 30-3~     409 White <NA>
##  2 51624 2009_10  male      34 " 30-3~     409 White <NA>
##  3 51624 2009_10  male      34 " 30-3~     409 White <NA>
##  4 51625 2009_10  male       4 " 0-9"       49 Other <NA>
##  5 51630 2009_10  female    49 " 40-4~     596 White <NA>
##  6 51638 2009_10  male       9 " 0-9"      115 White <NA>
##  7 51646 2009_10  male       8 " 0-9"      101 White <NA>
##  8 51647 2009_10  female    45 " 40-4~     541 White <NA>
##  9 51647 2009_10  female    45 " 40-4~     541 White <NA>
## 10 51647 2009_10  female    45 " 40-4~     541 White <NA>
## # ... with 9,990 more rows, 68 more variables:
## #   Education <fct>, MaritalStatus <fct>, HHIncome <fct>,
## #   HHIncomeMid <int>, Poverty <dbl>, HomeRooms <int>,
## #   HomeOwn <fct>, Work <fct>, Weight <dbl>,
## #   Length <dbl>, HeadCirc <dbl>, Height <dbl>,
## #   BMI <dbl>, BMICatUnder20yrs <fct>, BMI_WHO <fct>,
## #   Pulse <int>, BPSysAve <int>, BPDiaAve <int>, ...
```

7.4.2 Estimating a population proportion

Let us use this dataset to estimate the proportion of healthy sleepers in the American population. A "healthy amount of sleep" is defined by the American Academy of Sleep Medicine[4] as 7 to 9 hours per night for adults between the ages of 18 and 60.

---

[4]https://www.ncbi.nlm.nih.gov/pmc/articles/PMC4434546/#:~:text=Adults%20should%20sleep%207%20or,and%20increased%20risk%20of%20death

With this information, we perform some basic preprocessing of the data:

- Drop any observations that contain a missing value in the column SleepHrsNight.
- Filter the data to contain observations for adults between the ages of 18 and 60.
- Create a new Boolean variable healthy_sleep that indicates whether a participant gets a healthy amount of sleep.

```
# BEGIN SOLUTION
NHANES_relevant <- NHANES |>
  drop_na(c(SleepHrsNight)) |>
  filter(between(Age, 18, 60)) |>
  mutate(healthy_sleep = between(SleepHrsNight, 7, 9)) |>
  relocate(healthy_sleep, .before = SurveyYr)
NHANES_relevant
```

```
## # A tibble: 5,748 x 77
##         ID healthy_sl~1 Surve~2 Gender   Age AgeDe~3 AgeMo~4
##      <int> <lgl>        <fct>   <fct>  <int> <fct>     <int>
##  1 51624 FALSE        2009_10 male      34 " 30-3~     409
##  2 51624 FALSE        2009_10 male      34 " 30-3~     409
##  3 51624 FALSE        2009_10 male      34 " 30-3~     409
##  4 51630 TRUE         2009_10 female    49 " 40-4~     596
##  5 51647 TRUE         2009_10 female    45 " 40-4~     541
##  6 51647 TRUE         2009_10 female    45 " 40-4~     541
##  7 51647 TRUE         2009_10 female    45 " 40-4~     541
##  8 51656 FALSE        2009_10 male      58 " 50-5~     707
##  9 51657 FALSE        2009_10 male      54 " 50-5~     654
## 10 51666 FALSE        2009_10 female    58 " 50-5~     700
## # ... with 5,738 more rows, 70 more variables:
## #   Race1 <fct>, Race3 <fct>, Education <fct>,
## #   MaritalStatus <fct>, HHIncome <fct>,
## #   HHIncomeMid <int>, Poverty <dbl>, HomeRooms <int>,
## #   HomeOwn <fct>, Work <fct>, Weight <dbl>,
## #   Length <dbl>, HeadCirc <dbl>, Height <dbl>,
## #   BMI <dbl>, BMICatUnder20yrs <fct>, BMI_WHO <fct>, ...
```

```
# END SOLUTION
```

We can inspect the resulting table. Note that there are 5,748 observations in the tibble.

```
NHANES_relevant
```

```
## # A tibble: 5,748 x 77
##         ID healthy_sl~1 Surve~2 Gender   Age AgeDe~3 AgeMo~4
##      <int> <lgl>        <fct>   <fct>  <int> <fct>     <int>
##  1 51624 FALSE        2009_10 male      34 " 30-3~     409
##  2 51624 FALSE        2009_10 male      34 " 30-3~     409
##  3 51624 FALSE        2009_10 male      34 " 30-3~     409
##  4 51630 TRUE         2009_10 female    49 " 40-4~     596
##  5 51647 TRUE         2009_10 female    45 " 40-4~     541
##  6 51647 TRUE         2009_10 female    45 " 40-4~     541
```

```
##  7 51647 TRUE           2009_10 female    45 " 40-4~      541
##  8 51656 FALSE          2009_10 male      58 " 50-5~      707
##  9 51657 FALSE          2009_10 male      54 " 50-5~      654
## 10 51666 FALSE          2009_10 female    58 " 50-5~      700
## # ... with 5,738 more rows, 70 more variables:
## #    Race1 <fct>, Race3 <fct>, Education <fct>,
## #    MaritalStatus <fct>, HHIncome <fct>,
## #    HHIncomeMid <int>, Poverty <dbl>, HomeRooms <int>,
## #    HomeOwn <fct>, Work <fct>, Weight <dbl>,
## #    Length <dbl>, HeadCirc <dbl>, Height <dbl>,
## #    BMI <dbl>, BMICatUnder20yrs <fct>, BMI_WHO <fct>, ...
```

We will apply bootstrapping to the NHANES_relevant tibble to estimate an unknown parameter: the proportion of healthy sleepers in the American population.

Let us visualize the distribution of healthy sleepers using a bar chart.

```
ggplot(NHANES_relevant) +
  geom_bar(aes(x = healthy_sleep),
           col="grey", fill = "darkcyan", bins = 20)
```

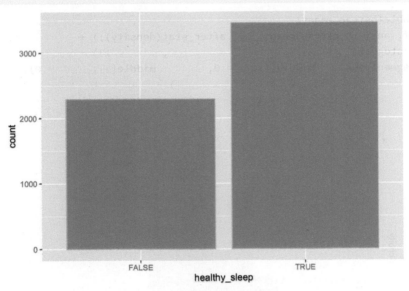

The proportion of healthy sleepers is the fraction of TRUE's in the healthy_sleep column. Recall that Boolean variables are just 1's and 0's. Thus, we can sum the number of TRUE's and divide by the total number of subjects. This is equivalent to computing the mean for the healthy_sleep column.

```
NHANES_relevant |>
  summarize(prop = mean(healthy_sleep))
```

```
## # A tibble: 1 x 1
##      prop
##     <dbl>
## 1   0.601
```

We are now ready to bootstrap from this random sample. Recall that `one_sample_value` will perform the bootstrap for us. We will replicate the bootstrap process a large number of times, say 10,000, so that we can plot a sampling histogram of the bootstrapped medians.

```
# Do the bootstrap!
bstrap_means <- replicate(n = 10000,
                one_sample_value(NHANES_relevant, healthy_sleep, mean))
```

As before, we will identify the 95% confidence interval. Here is the interval:

```
desired_area <- 0.95
middle <- quantile(bstrap_means,
                0.5 + (desired_area / 2) * c(-1, 1), type = 1)
middle
```

```
##      2.5%     97.5%
## 0.5887265 0.6139527
```

Let us plot the sampling histogram and annotate the interval on this histogram.

```
df <- tibble(bstrap_means)
ggplot(df, aes(x = bstrap_means, y = after_stat(density))) +
  geom_histogram(col="grey", fill = "darkcyan", bins = 13) +
  geom_segment(aes(x = middle[1], y = 0, xend = middle[2], yend = 0),
                size = 2, color = "salmon") +
  labs(x = "Proportion of healthy sleepers")
```

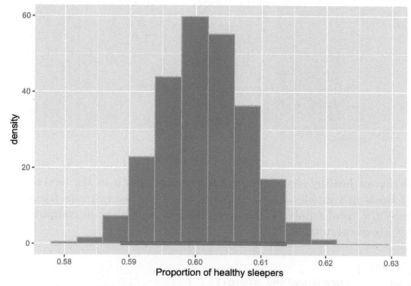

This looks a lot like what we saw in the previous section, with one key difference: there is no dot indicating where the parameter is! We do not know where the dot will fall or if it is even on this interval.

Statistics does not promise clairvoyance. It is a tool for quantifying uncertainty. What we have obtained is a 95% confidence interval of estimates. Meaning, this bootstrap process will be successful in capturing the parameter about 95% of the time. But that also leaves a 5% chance where we are totally off. Can we control the level of uncertainty?

7.4.3 Levels of uncertainty: 80% and 99% confidence intervals

So far we have examined the 95% confidence interval. Let us see what happens to the interval of estimates when we increase our level of confidence. We will examine a 99% confidence interval.

```
desired_area <- 0.99
middle <- quantile(bstrap_means,
                   0.5 + (desired_area / 2) * c(-1, 1), type = 1)
middle
```

```
##       0.5%      99.5%
## 0.5847251  0.6179541
```

```
df <- tibble(bstrap_means)
ggplot(df, aes(x = bstrap_means, y = after_stat(density))) +
  geom_histogram(col="grey", fill = "darkcyan", bins = 13) +
  geom_segment(aes(x = middle[1], y = 0, xend = middle[2], yend = 0),
               size = 2, color = "salmon") +
  labs(x = "Proportion of healthy sleepers")
```

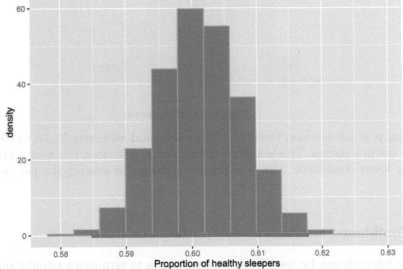

The interval is much wider! The proportion of healthy sleepers in the population goes from about 58.4% to 61.7%. This points to a trade-off: as we increase our confidence in the interval of estimates, this is compensated by making the interval *wider*. That is, a confidence interval generated by this resampling process has a chance of missing the parameter only 1% of the time. That probability does not correspond to the specific interval we found, but to the *process* that generated said interval. For the [0.584, 0.617] interval we found, the parameter either sits on the interval or not.

Let us move in the other direction and try a 80% confidence interval.

```
desired_area <- 0.80
middle <- quantile(bstrap_means,
                   0.5 + (desired_area / 2) * c(-1, 1), type = 1)
middle
```

```
##       10%       90%
## 0.5929019 0.6096033
```

```
df <- tibble(bstrap_means)
ggplot(df, aes(x = bstrap_means, y = after_stat(density))) +
  geom_histogram(col="grey", fill = "darkcyan", bins = 13) +
  geom_segment(aes(x = middle[1], y = 0, xend = middle[2], yend = 0),
               size = 2, color = "salmon") +
  labs(x = "Proportion of healthy sleepers")
```

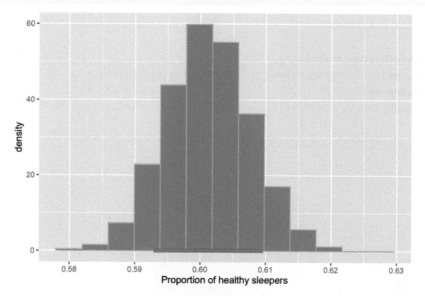

This interval is *much* narrower than the 99% interval and estimates 59.3% to 60.9% healthy sleepers in the population. This is a much tighter set of estimates, but we traded a narrower interval for lower confidence. This interval has a chance of missing the parameter 20% of the time.

7.4.4   Confidence intervals as a hypothesis test

Confidence intervals can be used for more than trying to estimate a population parameter. One popular use case for the confidence interval is something we saw in the previous chapter: the hypothesis test.

Let us reconsider the 95% confidence interval we obtained. The proportion of healthy sleepers in the population goes from 58.8% to 61.4%. Suppose that a researcher is interested in testing the following hypothesis:

Null hypothesis. The proportion of healthy sleepers in the population is 61%.

Alternative hypothesis. The proportion of healthy sleepers in the population is *not* 61%.

If we were testing this hypothesis at the 95% significance level, we would *fail* to reject the null hypothesis. Why?

The value supplied by the null (61%) sits on our 95% confidence interval for the population proportion. Therefore, at this level of significance, this value is plausible.

If we were to lower our confidence (to say 90% or 80%), the conclusion could have been different. This raises an important point about cut-offs: some fields demand a high level of significance for a result to be accepted by its scientific community; other fields may require much less convincing. For instance, experimental studies[5] in Physics demand significance levels at 99.9% or even 99.99%[6] for a result to be even considered publishable. It is not hard to imagine why: findings in Physics are usually axiomatic and rejecting a null hypothesis implies the discovery of phenomena in nature. A 99.99% confidence interval would guarantee that such a discovery is a fluke only 0.01% of the time.

The basis for using confidence intervals as a hypothesis test is rooted in statistical theory. In practice, we simply check whether the value supplied by the null hypothesis sits on the confidence interval or not.

7.4.5   Final remarks: resampling with care

We end this section with some points to keep in mind when applying resampling.

- Avoid introducing bias into the sample that is used as input for resampling. The sampling plan of simple random sampling will usually work best. And, even with simple random samples, it is possible to draw a "weird" original sample such that the confidence interval generated using it fails to capture the parameter.
- When the size of a random sample is moderately sized enough, the chance of the boot-strapped sample being identical to it is extremely rare. Therefore, you should aim to work with large random samples.
- Resampling does not work well when estimating extreme values, for instance, estimating the minimum or maximum value of a population.
- The distribution of the statistic should look roughly "bell" shaped. The histogram of the resampled statistics will be a hint.

7.5   Exercises

Be sure to install and load the following packages into your R environment before beginning this exercise set.

Question 1. The following vector `lucky_numbers` contains several numbers:

```
lucky_numbers <- c(5, 10, 17, 25, 31, 36, 43)
lucky_numbers
```

---

[5]https://arxiv.org/pdf/0706.3283.pdf
[6]https://journals.aps.org/prd/pdf/10.1103/PhysRevD.93.122002?casa_token=LX4dEzE
BwoEAAAAA%3ATEemb6ulsnfx89acCPP-GoEzMsjLDng26HX5YJ0CMZczkdOrPpmiHTLaFy
UR3c6jZcuDYVgJys92eRU

```
## [1]   5 10 17 25 31 36 43
```

Using the function `quantile` as shown in the textbook, determine the lucky number that results from the order statistics: (1) min, (2) max, and (3) median.

Question 2 The University of Lost World has conducted a staff and faculty survey regarding their most favorite rock bands. The university received 200 votes, which are summarized as follows:

- Pink Floyd (35%)
- Led Zeppelin (22%)
- Allman Brothers Band (20%)
- Yes (12%)
- Uncertain (11%)

In the following, we will use `"P"`, `"L"`, `"A"`, `"Y"`, and `"U"` to refer to the artists. The following tibble `rock_bands` summarizes the information:

```
rock_bands <- tibble(
  band_initial = c("P", "L", "A", "Y", "U"),
  proportion = c(0.35, 0.22, 0.20, 0.12, 0.11),
  votes = proportion * 200
)
rock_bands
```

```
## # A tibble: 5 x 3
##    band_initial proportion votes
##    <chr>            <dbl> <dbl>
## 1 P                 0.35    70
## 2 L                 0.22    44
## 3 A                 0.2     40
## 4 Y                 0.12    24
## 5 U                 0.11    22
```

These proportions represent just a *sample* of the population of University of Lost World. We will attempt to estimate the corresponding population parameters - the *proportion* of listening preference for each rock band in the population of University of Lost World staff and faculty. We will use confidence intervals to compute a range of values that reflects the uncertainty of our estimate.

- **Question 2.1** Using `rock_bands`, generate a tibble `votes` containing 200 rows corresponding to the votes. You can group by `band_initial` and repeat each band's row `votes` number of times by using `rep(1, each = votes)` within a `slice()` call (remember computing *within groups?*). Then form a tibble with a single column named `vote`.

  Here is what the first few rows of this tibble should look like:

vote
A
A
A
A
A
...

We will conduct bootstrapping using the tibble votes.

- **Question 2.2** Write a function `one_resampled_statistic(num_resamples)` that receives the number of samples to sample *with replacement* (why not without?) from votes. The function resamples from the tibble votes `num_resamples` number of times and then computes the proportion of votes for each of the 5 rock bands. It returns the result as a *tibble* in the same form as `rock_bands`, but containing the resampled votes and proportions from the bootstrap.

  Here is one possible tibble after running `one_resampled_statistic(100)`. The answer will be different each time you run this!

vote	votes	proportion
A	23	0.23
L	19	0.19
P	40	0.40
U	7	0.07
Y	11	0.11

  ```
  one_resampled_statistic <- function(num_resamples) {

  }

  one_resampled_statistic(100) # a sample call
  ```

- **Question 2.3** Let us set two names, `num_resamples` and `trials`, to use when conducting the bootstrapping. `trials` is the desired number of resampled proportions to simulate for each of the bands. This can be set to some large value; let us say 1,000 for this experiment. But what value should `num_resamples` be set to, which will be the argument passed to `one_resampled_statistic(num_resamples)` in the next step?

  The following code chunk conducts the bootstrapping using your `one_resampled_statistic()` function and the names `trials` and `num_resamples` you created above. It stores the results in a vector `bstrap_props_tibble`.

  ```
  bstrap_props_tibble <- replicate(n = trials,
                      one_resampled_statistic(num_resamples),
                      simplify = FALSE) |>
                  bind_rows()
  bstrap_props_tibble
  ```

- **Question 2.4** Generate an overlaid histogram using `bstrap_props_tibble`, showing the five distributions for each band. Be sure to use a positional adjustment to avoid stacking in the bars. You may also wish to set an alpha to see each distribution better. Use 20 for the number of bins.

  We can see significant difference in the popularity between some bands. For instance, we see that the bootstrapped proportions for P is significantly higher than Y's by virtue of no overlap between their two distributions; conversely, U and Y overlap each other completely showing no significant preference for U over Y and vice versa. Let us formalize this intuition for these three bands using an approximate 95% confidence interval.

- **Question 2.5** Define a function `cf95` that receives a *vector* `vec` and returns the approximate "middle 95%" using `quantile`.

Let us examine the 95% confidence intervals of the bands P, Y, and U, respectively.

```
bstrap_props_tibble |>
  filter(vote %in% c("P", "Y", "U")) |>
  group_by(vote) |>
  summarize(ci = list(cf95(proportion))) |>
  unnest_wider(ci)
```

- **Question 2.6** By looking at the upper and lower endpoints of each interval, and the overlap between intervals (if any), can you say whether P is more popular than Y or U? How about for Y, is Y more popular than U?

- **Question 2.7** Suppose you computed the following approximate 95% confidence interval for the proportion of band P votes.

$$[.285, .42]$$

Is it true that 95% of the population of faculty lies in the range $[.285, .42]$? Explain your answer.

- **Question 2.8** Can we say that there is a 95% probability that the interval $[.285, .42]$ contains the *true* proportion of the population who listens to band P? Explain your answer.

- **Question 2.9** Suppose that you created 80%, 90%, and 99% confidence intervals from one sample for the popularity of band P, but forgot to label which confidence interval represented which percentages. Match the following intervals to the percent of confidence the interval represents.

  - $[0.265, 0.440]$
  - $[0.305, 0.395]$
  - $[0.285, 0.420]$

Question 3. Recall the tibble `penguins` from the package `palmerpenguins` includes measurements for 344 penguins in the Palmer Archipelago. Let us try using the method of resampling to estimate using confidence intervals some useful parameters of the population.

```
library(palmerpenguins)
penguins
```

```
## # A tibble: 344 x 8
##    species island    bill_~1 bill_~2 flipp~3 body_~4 sex
##    <fct>   <fct>        <dbl>   <dbl>   <int>   <int> <fct>
## 1 Adelie  Torgersen     39.1    18.7     181    3750 male
## 2 Adelie  Torgersen     39.5    17.4     186    3800 fema~
## 3 Adelie  Torgersen     40.3    18       195    3250 fema~
## 4 Adelie  Torgersen     NA      NA       NA      NA  <NA>
## 5 Adelie  Torgersen     36.7    19.3     193    3450 fema~
## 6 Adelie  Torgersen     39.3    20.6     190    3650 male
## 7 Adelie  Torgersen     38.9    17.8     181    3625 fema~
## 8 Adelie  Torgersen     39.2    19.6     195    4675 male
```

```
##  9 Adelie  Torgersen    34.1    18.1    193   3475 <NA>
## 10 Adelie  Torgersen    42      20.2    190   4250 <NA>
## # ... with 334 more rows, 1 more variable: year <int>,
## #   and abbreviated variable names 1: bill_length_mm,
## #   2: bill_depth_mm, 3: flipper_length_mm,
## #   4: body_mass_g
```

- **Question 3.1** First, let us focus on estimating the *mean* body mass of the penguins, available in the variable `body_mass_g`. Form a tibble named `penguins_pop_df` that is identical to `penguins` but does not contain any missing values in the variable `body_mass_g`.

  We will imagine the 342 penguins in `penguins_pop_df` to be the *population* of penguins of interest. Of course, direct access to the population is almost never possible in a real-world setting. However, for the purposes of this question, we will claim clairvoyance and see how close the method of resampling approximates some population *parameter*, i.e., the mean body mass of penguins in the Palmer Archipelago.

- **Question 3.2** What is the mean body mass of penguins in `penguins_pop_df`? Store it in `pop_mean`.

- **Question 3.3** Draw a sample *without replacement* from the population in `penguins_pop_df`. Because samples can be expensive to collect in real settings, set the sample size to 50.

  The sample in `one_sample` is what we will use to *resample* from a large number of times. We saw in the textbook a function that resamples from a tibble, computes a statistic from it, and returns it. Following is the function:

```
one_sample_value <- function(df, label, statistic) {
  sample_value <- df |>
    pull({{label}}) |>
    sample(replace = TRUE) |>
    statistic()
  return(sample_value)
}
```

- **Question 3.4** What is the size of the resampled tibble when `one_sample` is passed as an argument? Assign your answer to the name `resampled_size_answer`.

    1. 342
    2. 684
    3. 50
    4. 100
    5. 1

- **Question 3.5** Using `replicate`, create 1,000 resampled mean statistics from `one_sample` using the variable `body_mass_g`. Assign your answer to the name `resampled_means`.

- **Question 3.6** Let us combine the steps from **Question 3.4** and **Question 3.5** into a function. Write a function `resample_mean_procedure` that takes no arguments. The function draws a sample of size 50 from the population (**Question 3.4**), and then generates 1,000 resampled means from it (**Question 3.5**) which are then returned.

- **Question 3.7** Write a function `get_mean_quantile` that takes a single argument `desired_area`. The function performs the resampling procedure using resample_mean_procedure and returns the middle desired_area interval (e.g., 90% or 95%) as a vector.

Here is an example call that obtains an approximate 90% confidence interval. Also shown is the population mean. Does your computed interval capture the parameter? Try running the cell a few times. The interval printed should be different each time you run the code chunk.

```
print(get_mean_quantile(0.9))
print(pop_mean)
```

- **Question 3.8** Repeat the get_mean_quantile procedure to obtain 100 different approximate 90% confidence intervals. Assign the intervals to the name mean_intervals.

  The following code chunk organizes your results into a tibble named interval_df.

```
interval_df <- tibble(
  replication = 1:100,
  left = mean_intervals[1,],
  right = mean_intervals[2,]
)
interval_df
```

- **Question 3.9** Under an approximate 90% confidence interval, how many of the above 100 intervals do you expect captures the population mean? Use what you know about confidence intervals to answer this; do not write any code to determine the answer.

  The following code chunk visualizes your intervals with a vertical line showing the parameter:

```
ggplot(interval_df) +
  geom_segment(aes(x = left, y = replication,
                   xend = right, yend = replication),
               color = "magenta") +
  geom_vline(xintercept = pop_mean, color = "red")
```

- **Question 3.10** Now feed the tibble interval_df to a filter that keeps only those rows whose approximate 90% confidence interval includes pop_mean. How many of those intervals actually captured the parameter? Store the number in number_captured.

Question 4. This problem is a continuation of *Question 3*. We will now streamline the previous analysis by generalizing the functions we wrote. This way we can try estimating different parameters and compare the results.

- **Question 4.1** Let us first generalize the resample_mean_procedure from **Question 3.6**. Call the new function resample_procedure. The function should receive the following arguments:

  - pop_df, a tibble
  - label, the variable under examination. Recall the use of {{ to refer to it properly.
  - initial_sample_size, the sample size to use for the initial draw from the population
  - n_resamples, the number of resampled statistics to generate
  - stat, the statistic function

  The function returns a vector containing the resampled statistics.

```
resample_procedure <- function(pop_df,
                               label,
                               initial_sample_size,
                               n_resamples,
                               stat) {

}
```

- **Question 4.2** Generalize the function `get_mean_quantile` from **Question 3.7**. Call the new function `get_quantile`. This function receives the same arguments as `resample_mean_procedure` with the addition of one more argument, `desired_area`, the interval width. The function then calls `resample_procedure` to obtain the resampled statistics. The function returns the middle quantile range of these statistics according to `desired_area`, e.g., the "middle 90%" if `desired_area` is 0.9.

```
get_quantile <- function(pop_df,
                         label,
                         initial_sample_size,
                         n_resamples,
                         stat,
                         desired_area) {

}
```

- **Question 4.3** We can now package all the actions into one function. Call the function `conf_interval_test`. The function receives the same arguments as `get_quantile` with one new argument, `num_intervals`, the number of confidence intervals to generate. The function performs the following actions (in order):
  - Compute the population parameter from `pop_df` (assuming access to the population is possible in `pop_df`) by running the function `stat_func` on the variable `label`. Recall the use of `{{` to refer to `label` properly. Assign this number to the name `pop_stat`.
  - Obtain `num_intervals` many confidence intervals by repeated calls to the function `get_quantile`. Assign the resulting intervals to the name `intervals`.
  - Arrange the results in `intervals` into a tibble named `interval_df` with three variables: `replication`, `left`, and `right`.
  - Print the number of confidence intervals that capture the parameter `pop_stat`.
  - Visualize the intervals with a vertical red line showing where the parameter is.

NOTE: If writing this function seems daunting, don't worry! All of the code you need is already written. You should be able to simply copy your work from this question and from the steps in **Question 3**.

```
conf_interval_test <- function(pop_df,
                               label,
                               init_samp_size,
                               n_resamples, stat_func,
                               desired_area, num_intervals) {

}
```

Let us now try some experiments.

- **Question 4.4** Run `conf_interval_test` on `penguins_pop_df` to estimate the mean body mass in the population using the variable `body_mass_g`. Set the initial draw size to 50 and number of resampled statistics to 1000. Generate 100 different approximate 90% confidence intervals.

- **Question 4.5** Repeat **Question 4.4**, this time estimating the *max* body mass in the population instead of the *mean*.

- **Question 4.6** Repeat **Question 4.4**, this time increasing the initial draw size. First try 100, then 200, and 300.

- **Question 4.7** For the max-based estimates, why is it that so many of the 90% confidence intervals are unsuccessful in capturing the parameter?

- **Question 4.8** For the mean-based estimates, at some point when increasing the initial draw size from 50 to 300, all of the 100 differently generated confidence intervals capture the parameter. Given what we know about 90% confidence intervals, how can this be possible?

Question 5 Let's return to the College of Groundhog CSC1234 simulation from **Question 6** in Chapter 6. We evaluated the claim that the final scores of students from Section B were significantly lower than those from Sections A and C by means of a *permutation test*. Permutation analysis seeks to quantify what the null distribution looks like. For this reason, it tries to *break* whatever structure may be present in the dataset and quantify the patterns we would expect to see under a chance model.

Recall the tibble `csc1234` from the `edsdata` package:

```
library(edsdata)
csc1234
```

```
## # A tibble: 388 x 2
##      Section Score
##      <chr>   <dbl>
##   1 A          100
##   2 A          100
##   3 A          100
##   4 A          100
##   5 A          100
##   6 A          100
##   7 A          100
##   8 A          100
##   9 A          100
## 10 A          100
## # ... with 378 more rows
```

- **Question 5.1** How many students are in each section? Form a tibble that gives an answer and assign the resulting tibble to the name `section_counts`.

There is another way we can approach the analysis. We can quantify the uncertainty in the mean score difference between two sections by estimating a *confidence interval* with the *resampling* technique. Under this scheme, we assume that each section performs identically and that the student scores available in each section (116 from A, 128 from B, and 144 from C) is a sample from some larger population of student scores for the CSC1234 course, which we do not have access to.

Thus, we will sample *with* replacement from each section. Then, as with the permutation exercise, we can compute the mean difference in scores for each pair of sections ("A-B", "C-B", "C-A") using the bootstrapped sample. The interval we obtain from this process can be used to test the hypothesis that the average score difference is different from chance.

- **Question 5.2** Recall the work from **Question 6** in Chapter 6. Copy over your work for creating the function `mean_differences` and the observed group mean differences in `observed_differences`.

- **Question 5.3** Generate an overlaid histogram for Score from `csc1234` showing three distributions in the same plot, the scores for Section A, for Section B, and for Section C. Use 10 bins and a `dodge` positional adjustment this time to compare the distributions.

Resampling calls for sampling *with replacement*. Suppose that we are to resample scores with replacement from the "Section A" group, then likewise for the "Section B" group, and finally, the "Section C" group. Then we compute the difference in means between the groups (A-B, C-B, C-A). Would the bulk of this distribution be centered around 0? Let's find out!

- **Question 5.4** State a null and alternative hypothesis for this problem.

Let us use resampling to build a confidence interval and address the hypothesis.

- **Question 5.5** Write a function `resample_tibble` that takes a tibble as its single argument, e.g., `csc1234`. The function samples Score **with replacement WITHIN each group** in Section. It overwrites the variable Score with the result of the sampling. The resampled tibble is returned.

```
resample_tibble(csc1234)   # an example call
```

- **Question 5.6** Write a function `csc1234_one_resample` that takes no arguments. The function resamples from `csc1234` using the function `resample_tibble`. It then computes the mean difference in scores using the `mean_differences` function you wrote from the permutation test. The function returns a one-element *list* containing a vector with the computed differences.

```
csc1234_one_resample()   # an example call
```

- **Question 5.7** Using `replicate`, generate 10,000 resampled mean differences. Store the resulting vector in the name `resampled_differences`.

The following code chunk organizes your results into a tibble `differences_tibble`:

```
differences_tibble <- tibble(
        `A-B` = map_dbl(resampled_differences, function(x) x[1]),
        `C-B` = map_dbl(resampled_differences, function(x) x[2]),
        `C-A` = map_dbl(resampled_differences, function(x) x[3])) |>
    pivot_longer(`A-B`:`C-A`,
                 names_to = "Section Pair",
                 values_to = "Statistic")  |>
    mutate(`Section Pair` =
```

```
            factor(`Section Pair`, levels=c("A-B", "C-B", "C-A")))
differences_tibble
```

- **Question 5.8** Form a tibble named `section_intervals` that gives an approximate 95% confidence interval for each pair of course sections in `resampled_differences`. The resulting tibble should look like:

Section Pair	left	right
A-B	...	...
C-A	...	...
C-B	...	...

To accomplish this, use `quantile` to summarize a grouped tibble and then a pivot function. Don't forget to `ungroup`!

The following plots a histogram of your results for each course section pair. It then annotates each histogram with the approximate 95% confidence interval you found.

```
print(observed_differences)
differences_tibble |>
  ggplot() +
  geom_histogram(aes(x = Statistic, y = after_stat(density)),
                 color = "gray", fill = "darkcyan", bins = 20) +
  geom_segment(data = section_intervals,
               aes(x = left, y = 0, xend = right, yend = 0),
               size = 2, color = "salmon") +
  facet_wrap(~`Section Pair`)
```

Note how the observed mean score differences in `observed_differences` fall squarely in its respective interval (if you like, plot the points on your visualization!).

- **Question 5.9** Draw the conclusion of the hypothesis test for each of the three confidence intervals. Do we reject the null hypothesis? If not, what conclusion can we make?

- **Question 5.10** Suppose that the 95% confidence interval you found for "A-B" is $[-9.35, -1.95]$. Does this mean that 95% of the student scores were between $[-9.35, -1.95]$? Why or why not?

8

Towards Normality

We have studied several statistics in this text such as the mean and order statistics like the median. We have also drawn sampling distributions to simulate a statistic to see if it approximates well some parameter value of interest. For some statistics, like the total variation distance, we saw that the distribution skewed in some direction. But the sampling distribution for the sample mean has consistently shown something that resembles a bell shape, regardless of the underlying population.

Recall that a goal of this text is to make inferences about data in a population for which we know very little about. If we can extract properties of random samples that are true *regardless of the underlying population*, we have at hand a powerful tool for doing inference. The distribution of the sample mean is one such property, which we will develop more in depth in this chapter.

When we refer to these bell-shaped curves, we are talking about a distribution called the *normal distribution*. To develop an intuition for this, we will first examine a new (but important) statistic called the *standard deviation*, which measures generally how much the data points are away from the mean. It is important because the shape of a normal distribution is completely determinable from its mean and standard deviation. We will then learn that by taking a large number of samples from a population, where each sample has no relation to another, the resulting distribution will look like – you guessed it – a normal distribution.

8.1 Standard Deviation

The *standard deviation* (or *SD* for short) measures how much the data points are away from the mean. Put another way, it is a measure of the *spread* in the data.

8.1.1 Prerequisites

We will continue to make use of the tidyverse in this section so let us load it in.

```
library(tidyverse)
```

8.1.2 Definition of standard deviation

Suppose we have some number, say N, of samples and for each sample, we have its measurement. Let us say $x_1, \ldots, x_N$ are the measurements. The mean of the samples is the total

DOI: 10.1201/9781003320845-8

of $x_1, \ldots, x_N$ divided by the number of samples N, that is

$$\frac{x_1 + \cdots + x_N}{N}$$

Let μ denote the average. The *variance* of the samples is the average of the square amounts of these samples from the mean.

$$\frac{(x_1 - \mu)^2 + \cdots + (x_N - \mu)^2}{N}$$

We use symbol σ^2 to represent the variance. Because of the subtraction of the mean from the individual values, the variance measures the spread of the data around the mean. Since each term in the total is a square, the unit of the variance is the square of the unit of the samples. For example, if the original measure is meters then the unit of the variance is square meters. To make the two units comparable to each other, we take the square root of the variance, which we denote by σ.

$$\sigma = \sqrt{\frac{(x_1 - \mu)^2 + \cdots + (x_N - \mu)^2}{N}}$$

We call the quantity σ the *standard deviation*.

8.1.3 Example: exam scores

Suppose we have drawn ten sample exam scores.

```
sample_scores <- c(78, 89, 98, 90, 96, 90, 84, 91, 98, 76)
sample_scores
```

```
## [1] 78 89 98 90 96 90 84 91 98 76
```

Let us first compute the mean and the squared element-wise differences from the mean.

```
mu <- mean(sample_scores)
diffs_from_mu_squared <- (sample_scores - mean(sample_scores)) ** 2
mu
```

```
## [1] 89
```

```
diffs_from_mu_squared
```

```
## [1] 121   0  81   1  49   1  25   4  81 169
```

Computing the variance is straightforward. We need only to sum up `diffs_from_mu_squared` and divide the resulting quantity by the number of scores.

```
variance <- sum(diffs_from_mu_squared) / length(sample_scores)
variance
```

```
## [1] 53.2
```

Finding the standard deviation is easy – just take the square root!

```
sdev <- sqrt(variance)
sdev
```

```
## [1] 7.293833
```

The mean is 89 points and the standard deviation is about 7.3 points. Let us visualize the distribution of scores.

```
tibble(sample_scores) |>
  ggplot() +
  geom_histogram(aes(x = sample_scores, y = after_stat(density)),
                 color = "gray", fill = "darkcyan", binwidth = 3) +
  scale_x_continuous(breaks = seq(70, 100, 4))
```

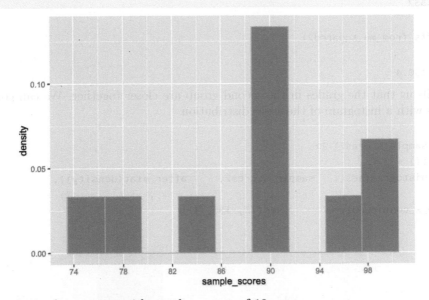

Let us compare these scores with another group of 10 scores.

```
sample_scores2 <- c(88, 89, 93, 90, 86, 90, 84, 91, 93, 80)
sample_scores2
```

```
##  [1] 88 89 93 90 86 90 84 91 93 80
```

We will repeat the above steps.

```
mu2 <- mean(sample_scores2)
diffs_from_mu_squared2 <- (sample_scores2 - mean(sample_scores2)) ** 2
variance2 <- sum(diffs_from_mu_squared2) / length(sample_scores2)
sdev2 <- sqrt(variance2)
```

Let us examine the mean and standard deviation of this group.

```
mu2
```

```
## [1] 88.4
```

```
sdev2
```

```
## [1] 3.878144
```

Interestingly, the group has the same mean but the standard deviation is much smaller. Notice that the sum of the squared difference values are much larger in the first vector than in the second.

```
sum(diffs_from_mu_squared)
```

```
## [1] 532
```

```
sum(diffs_from_mu_squared2)
```

```
## [1] 150.4
```

This tells us that the grades in the second group are closer together. We can confirm our findings with a histogram of the score distribution.

```
tibble(sample_scores2) |>
  ggplot() +
  geom_histogram(aes(x = sample_scores2, y = after_stat(density)),
                 color = "gray", fill = "darkcyan", binwidth = 3) +
  scale_x_continuous(breaks = seq(70, 100, 4))
```

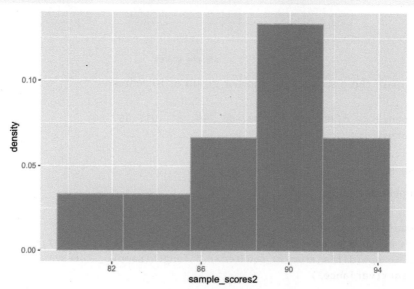

Observe that the scores are much closer together compared to the first histogram.

8.1.4 Sample standard deviation

Often in the calculation of standard deviation, instead of dividing by N, we divide by $N-1$. We call the versions of σ^2 and σ with division by N the *population variance* and the *population standard deviation*, respectively. The versions that use division by $N-1$ are the *sample variance* and *sample standard deviation*.

There is a subtle consideration in why we want to use $N-1$ for the variance. To see why, suppose you have made up your mind on the value of the mean, say it is 6, and determine all the values in your data **except for the last one**. Here is an example of the situation.

$$\mu = 6 = \frac{2+8+4+\ ?}{4}$$

Once you have finished determining three out of the four values, you no longer have anymore freedom in choosing the fourth value. This is because of a constraint composed by the mean formula. That is, if we isolate the sum of the numbers from above...

$$2+8+4+\ ? = 6*4$$

Thus, we can figure quite easily the value of the ?.

$$? = 6*4 - (2+8+4) = 10$$

The phenomena we observed is representative of a general fact. In the formula that involves μ and $x_1, \dots, x_N$, like the one for variance, there are no longer N independent values, there are only $N-1$ of them.

The difference in quantity by using $N-1$ instead of N may not be large, specifically, when N is large and the differences are small. Still, for an accurate understanding of the data spread, the choice can be highly critical. For this reason, statisticians prefer to work with the sample standard deviation when dealing with samples. We will do so as well for the remainder of this text.

8.1.5 The sd function

Fortunately R comes with a function sd that computes the *sample* standard deviation. We compare them side-by-side with the values we computed earlier.

```
c(sdev,sd(sample_scores))
```

```
## [1] 7.293833 7.688375
```

```
c(sdev2,sd(sample_scores2))
```

```
## [1] 3.878144 4.087923
```

Because of the use of $N-1$ in place of N, the values in the right column are greater than those on the left.

8.2   More on Standard Deviation

The previous section introduced the notion of standard deviation (SD) and how it is a
measure of *spread* in the data. We build upon our discussion of SD in this section.

8.2.1   Prerequisites

We will continue to make use of the tidyverse. To demonstrate an example of the use (as well
as misuses) of SD, we also load in two synthetically generated datasets from the package
datasauRus.

```
library(tidyverse)
library(datasauRus)
```

We retrieve the two datasets here, and will explore them shortly.

```
star <- datasaurus_dozen |>
  filter(dataset == "star")
bullseye <- datasaurus_dozen |>
  filter(dataset == "bullseye")
```

8.2.2   Working with SD

To see what we can glean from the SD, let us turn to a more interesting dataset. The tibble
starwars from the dplyr package contains data about all characters in the Star Wars canon.
The table records a wealth of information about each character, e.g., name, height (cm),
weight (kg), home world, etc. Pull up the help (?starwars) for more information about the
dataset.

```
starwars
```

```
## # A tibble: 87 x 14
##      name         height   mass hair_~1  skin_~2  eye_c~3  birth~4
##      <chr>        <int>   <dbl> <chr>    <chr>    <chr>    <dbl>
##  1 Luke Skyw~      172      77 blond    fair     blue       19
##  2 C-3PO          167      75 <NA>     gold     yellow    112
##  3 R2-D2           96      32 <NA>     white,~  red        33
##  4 Darth Vad~     202     136 none     white    yellow    41.9
##  5 Leia Orga~     150      49 brown    light    brown      19
##  6 Owen Lars      178     120 brown,~  light    blue       52
##  7 Beru Whit~     165      75 brown    light    blue       47
##  8 R5-D4           97      32 <NA>     white,~  red        NA
##  9 Biggs Dar~     183      84 black    light    brown      24
## 10 Obi-Wan K~     182      77 auburn~  fair     blue-g~    57
## # ... with 77 more rows, 7 more variables: sex <chr>,
## #   gender <chr>, homeworld <chr>, species <chr>,
```

```
## #    films <list>, vehicles <list>, starships <list>, and
## #    abbreviated variable names 1: hair_color,
## #    2: skin_color, 3: eye_color, 4: birth_year
```

Let us focus on the characters' heights. It turns out that the height information is missing for some of them; let us remove these entries from the dataset.

```
starwars_clean <- starwars |>
  drop_na(height)
```

Here is a histogram of the characters' heights.

```
ggplot(starwars) +
  geom_histogram(aes(x = height, y = after_stat(density)),
                 color = "gray", fill = "darkcyan", bins = 20)
```

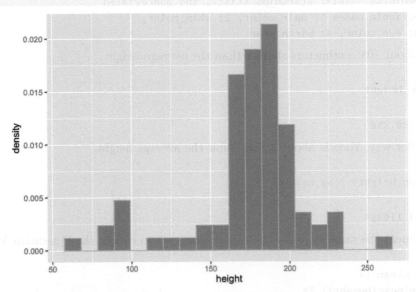

The average height of Star Wars characters is just over 174 centimeters (or 5'8"), which is about 1.4 centimeters shorter than the average height of men in the United States (5'9").

```
mean_height <- starwars_clean |>
  summarize(mean(height)) |>
  pull()
mean_height
```

```
## [1] 174.358
```

The SD tells us how far off a character's height is from the average, which is about 34.77 centimeters.

```
sd_height <- starwars_clean |>
  summarize(sd(height)) |>
  pull()
sd_height
```

```
## [1] 34.77043
```

The shortest character in cannon is the legendary Jedi Master Yoda, registering a height of just 66 centimeters!

```
starwars_clean |>
  arrange(height) |>
  head(1)
```

```
## # A tibble: 1 x 14
##    name   height  mass hair_~1 skin_~2 eye_c~3 birth~4 sex
##    <chr>   <int> <dbl> <chr>   <chr>   <chr>     <dbl> <chr>
## 1 Yoda       66    17 white   green   brown       896 male
## # ... with 6 more variables: gender <chr>,
## #   homeworld <chr>, species <chr>, films <list>,
## #   vehicles <list>, starships <list>, and abbreviated
## #   variable names 1: hair_color, 2: skin_color,
## #   3: eye_color, 4: birth_year
```

Yoda is about 108 centimeters shorter than the average height.

```
66 - mean_height
```

```
## [1] -108.358
```

Put another way, Yoda is about 3 SDs *below* the average height.

```
(66 - mean_height) / sd_height
```

```
## [1] -3.116384
```

We can repeat the same steps for the tallest character in canon: the Quermian Yarael Poof.

```
starwars_clean |>
  arrange(desc(height)) |>
  head(1)
```

```
## # A tibble: 1 x 14
##    name   height  mass hair_~1 skin_~2 eye_c~3 birth~4 sex
##    <chr>   <int> <dbl> <chr>   <chr>   <chr>     <dbl> <chr>
## 1 Yara~     264    NA none    white   yellow       NA male
## # ... with 6 more variables: gender <chr>,
## #   homeworld <chr>, species <chr>, films <list>,
## #   vehicles <list>, starships <list>, and abbreviated
## #   variable names 1: hair_color, 2: skin_color,
## #   3: eye_color, 4: birth_year
```

Yarael Poof's height is about 2.58 SDs *above* the average height.

```
(264 - mean_height) / sd_height
```

```
## [1] 2.57811
```

It seems then that the tallest and shortest characters are only a few SDs away from the mean – no more than 3. The mean and SD is useful in this way: all the heights of Star Wars characters can be found within 3 SDs of the mean. This gives us a good sense of the spread in the data.

8.2.3 Standard units

The number of standard deviations a value is away from the mean can be calculated as follows:

$$z = \frac{\text{value} - \text{mean}}{\text{SD}}$$

The resulting quantity z measures *standard units* and is sometimes called the *z-score*. We saw two such examples of standard units when studying the tallest and shortest Star Wars characters.

R provides a function `scale` that converts values into standard units. Such a transformation is called *scaling*. For instance, we can add a new column with all of the heights in standard units.

```
starwars_clean |>
  mutate(su = scale(height)) |>
  select(name, height, su)
```

```
## # A tibble: 81 x 3
##    name                 height   su[,1]
##    <chr>                 <int>    <dbl>
##  1 Luke Skywalker          172  -0.0678
##  2 C-3PO                   167  -0.212
##  3 R2-D2                    96  -2.25
##  4 Darth Vader             202   0.795
##  5 Leia Organa             150  -0.701
##  6 Owen Lars               178   0.105
##  7 Beru Whitesun lars      165  -0.269
##  8 R5-D4                    97  -2.22
##  9 Biggs Darklighter       183   0.249
## 10 Obi-Wan Kenobi          182   0.220
## # ... with 71 more rows
```

As before, note that the standard units are much less than 3 SDs (in general, they need not be).

Scaling is often used in data analysis as it allows us to put data in a comparable format. Let us visit another example to see why.

8.2.4 Example: judging a contest

Suppose three eccentric judges evaluated eight contestants at a contest. They evaluate each contestant on the scale of 1 to 10. Enter the judges.

- **Mrs. Sweet.** Her cakes are loved by everyone. She also has a reputation for giving high scores. She has never given a score below 5.
- **Mr. Coolblood.** He has a reputation of being ruthless with a strong penchant for corn dogs. He has never given a high score to anyone in the past.
- **MS Hodgepodge.** Her salsa dip is quite spicy. Yet she tends to spread her scores fairly. If there are enough contestants, she gives 1 or 2 to at least one contestant and 9 or 10 to at least one contestant.

Here are the results.

```
contest <- tribble(~name, ~sweet, ~coolblood, ~hodgepodge,
                    "Ashley",   9, 5, 9,
                    "Bruce",    8, 6, 4,
                    "Cathryn",  7, 5, 5,
                    "Drew",     8, 2, 1,
                    "Emily",    9, 7, 7,
                    "Frank",    6, 1, 1,
                    "Gail",     8, 4, 4,
                    "Harry",    5, 3, 3
                    )
contest
```

```
## # A tibble: 8 x 4
##   name    sweet coolblood hodgepodge
##   <chr>   <dbl>     <dbl>      <dbl>
## 1 Ashley      9         5          9
## 2 Bruce       8         6          4
## 3 Cathryn     7         5          5
## 4 Drew        8         2          1
## 5 Emily       9         7          7
## 6 Frank       6         1          1
## 7 Gail        8         4          4
## 8 Harry       5         3          3
```

Would it be enough to simply total their scores to determine the winner? Or should we take into account their eccentricities and *scale* the scores somehow?

That is where standard units come to play. We adjust each judge's scores by subtracting his/her mean and then dividing it by his/her standard deviation. After adjustment, each score represents relative to the standard deviation how much away the original score is from the mean.

Let us compute the raw total of the three scores, and compare this with the scaled scores.

```
contest |> mutate(
  sweet_su = scale(sweet),
  hodge_su = scale(hodgepodge),
  cool_su = scale(coolblood),
  raw_sum = sweet + coolblood + hodgepodge,
  scaled_sum = sweet_su + hodge_su + cool_su
) |>
  select(name, raw_sum, scaled_sum)
```

```
## # A tibble: 8 x 3
##   name    raw_sum scaled_sum[,1]
##   <chr>     <dbl>         <dbl>
## 1 Ashley       23          3.21
## 2 Bruce        18          1.19
## 3 Cathryn      17          0.349
## 4 Drew         11         -1.87
## 5 Emily        23          3.47
## 6 Frank         8         -3.77
## 7 Gail         16          0.202
## 8 Harry        11         -2.77
```

Based on the raw scores, we identify a two-way tie between Ashely and Emily. The scaled scores allow us to break the scores and we can announce, with confidence, Emily as the winner!

8.2.5 Be careful with summary statistics!

The SD is part of what we call *summary statistics* as they are useful for *summarizing* information about a distribution. The SD tells us about the *spread* of the data and where a histogram might sit on a number line.

Indeed, SD is an important tool and we will continue our study of it in the following sections. However, SD – as with all summary statistics – must be used with caution. We end our discussion in this section with an instructive example as to why.

Recall that we have two datasets star and bullseye. Each contains some x and y coordinate pairs. Let us examine some of the x coordinates in each.

```
star_x <- star |>
  pull(x)
head(star_x)
```

```
## [1] 58.21361 58.19605 58.71823 57.27837 58.08202 57.48945
```

```
bullseye_x <- bullseye |>
  pull(x)
head(bullseye_x)
```

```
## [1] 51.20389 58.97447 51.87207 48.17993 41.68320 37.89042
```

The values seem close, but look different enough. Let us compute the SD.

```
sd(star_x)
```

```
## [1] 16.76896
```

```
sd(bullseye_x)
```

```
## [1] 16.76924
```

How about the mean?

```
mean(star_x)
```

```
## [1] 54.26734
```

```
mean(bullseye_x)
```

```
## [1] 54.26873
```

The result points to a clear answer: both distributions represented by star_x and bulls-eye_x have the *exact* same SD and mean. We may be tempted then to conclude that both distributions are *equivalent*. We would be mistaken!

Whenever in doubt, turn to visualization. We plot a histogram of the distributions overlaid on top of each other. If the distributions are actually equal, we expect this to be reflected in the histogram.

```
datasaurus_dozen |>
  filter(dataset == "star" | dataset == "bullseye") |>
  ggplot() +
  geom_histogram(aes(x = x, y = after_stat(density), fill = dataset),
                 color = "gray", position = "identity",
                 alpha = 0.7, bins = 10) +
  scale_x_continuous(breaks = seq(15, 90, 5))
```

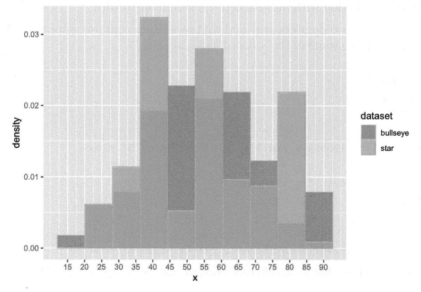

The histogram confirms that, contrary to what we might expect, these distributions are very much different – despite having the same SD and mean! Thus, the moral of this lesson: be careful with summary statistics and **always visualize your data.**

The star and bullseye datasets have an additional y coordinate which we have ignored in this study. We leave it as an exercise to the reader to determine if the distributions represented by y are also different yet have identical summary statistics.

8.3   The Normal Curve

The mean and SD are key pieces of information in determining the shape of some distributions. The most famous of them is the *normal distribution*, which we turn to in this section.

8.3.1   Prerequisites

We will continue to make use of the tidyverse so let us load it in. We will also work with some datasets from the `edsdata` package so let us load that in as well.

```
library(tidyverse)
library(edsdata)
```

8.3.2   The standard normal curve

The standard normal curve has a rather complicated formula:

$$\phi(z) = \frac{1}{\sqrt{2\pi}} e^{-\frac{1}{2}z^2}, \quad -\infty < z < \infty$$

where π is the constant $3.141592\ldots$ and e is *Euler's number* $2.71828\ldots$. It is best to think of this visually as in the following plot.

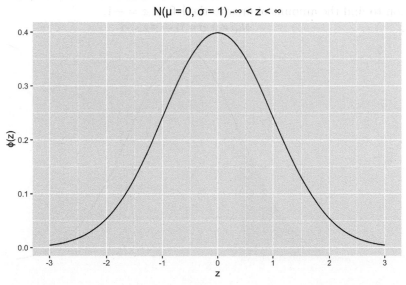

The values on the x-axis are in *standard units* (or z-scored values). We observe that the curve is symmetric around 0 where the "bulk" of the data is close to the mean. Following are some properties about the curve:

- The total area under the curve is 1.
- The curve is symmetric so it inherits a property we know about symmetric distributions: the mean and median are both 0.

- The SD is 1 which, fortunately for us, is clearly identifiable on the x-axis.
- The curve has two *points of inflection* at -1 and +1, which are annotated on the following plot.

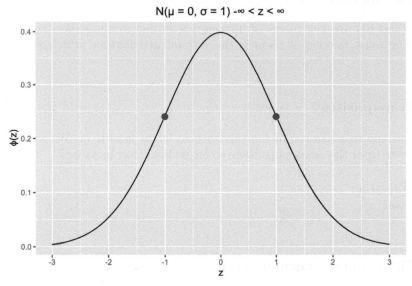

Observe how the curve looks like a salad bowl in the regions $(-\infty, -1)$, and $(1, \infty)$ and in the region $(-1, 1)$ the bowl is flipped upside down!

8.3.3   Area under the curve

We can find the area under the standard normal curve with the function pnorm. Let us use this function to find the amount of area to the left of $z = -1$.

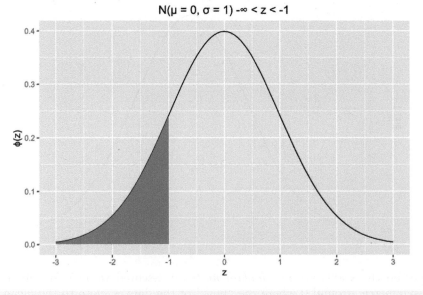

```
pnorm(-1)
```

```
## [1] 0.1586553
```

So about 15.9% of the data lies to the left of $z = -1$. Recall from our properties that the curve is symmetric and that the total area must sum to 1. We can take advantage of these two to calculate other areas, e.g., the area to the *right* of -1.

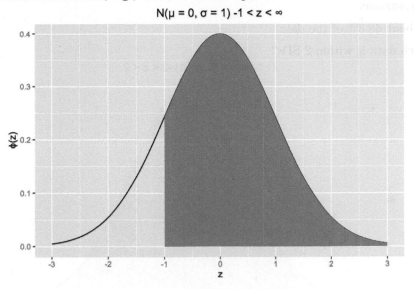

```
1 - pnorm(-1)
```

```
## [1] 0.8413447
```

That's about 84% of the data.

Here is a trickier problem: how much area is within 1 SD? Put another way, what is the area between $z = -1$ and $z = 1$? That's the orange-shaded area in the following plot.

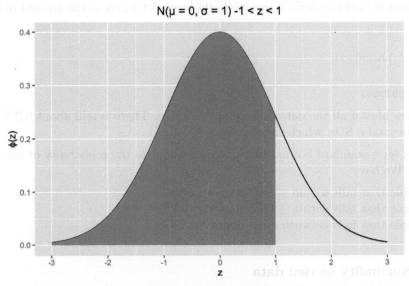

You might be able to guess at a few ways to answer this. One way to do it is to find the area to left of $z = -1$ (shaded in dark cyan) and subtract it from the area to the left of $z = 1$. This resulting subtraction is the area in orange, between $z = -1$ and $z = 1$.

```
pnorm(1) - pnorm(-1)
```

```
## [1] 0.6826895
```

That's about 68.3% of the data.

How much data is within 2 SDs?

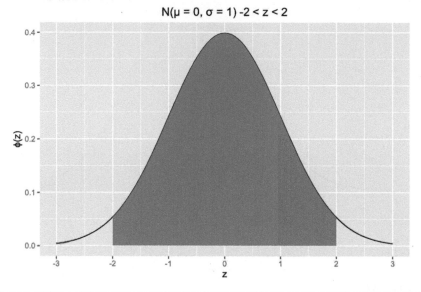

```
pnorm(2) - pnorm(-2)
```

```
## [1] 0.9544997
```

That's about 95% of the data. To complete the story, let's look at the amount of data within 3 SDs.

```
pnorm(3) - pnorm(-3)
```

```
## [1] 0.9973002
```

That covers *almost* all the data, but mind the italics. There is still about 0.3% of the data that lies beyond 3 SDs, which can happen.

Therefore, for a standard normal curve, we can calculate the probability of where a sample might lie. We have:

- a sample that falls within ± 1 SD is about **68%**.
- a sample that falls within ± 2 SDs is about **95%**.
- a sample that falls between ± 3 is about **99.73%**.

8.3.4 Normality in real data

Normality frequently occurs in real datasets. Let us have a look at a few attributes from the Olympic athletes dataset in `athletes` from the `edsdata` package.

We focus specifically on the heights, weights, and age variables.

```
my_athletes <- athletes |>
  drop_na() |>
  select(c(Height, Weight, Age))
```

Here is the histogram of these variables.

```
my_athletes |>
  pivot_longer(everything()) |>
  ggplot(aes(x = value, y = after_stat(density), fill = name)) +
  geom_histogram(bins=15, color = "gray") +
  facet_wrap(~name, scales = "free") +
  guides(fill = "none")
```

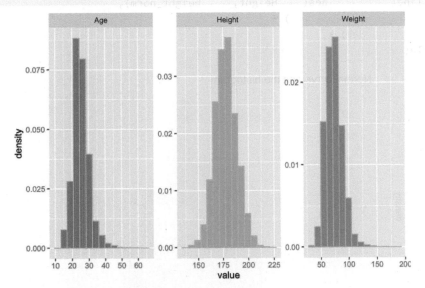

The heights most closely resemble the bell curve. We can also spot normality in the weight and age distributions, though they appear to be somewhat "lopsided" when compared to the heights.

Let us compute the mean and standard deviation of the athletes' heights.

```
summarized <- athletes |>
  summarize(mean(Height), sd(Height))
summarized
```

```
## # A tibble: 1 x 2
##   `mean(Height)` `sd(Height)`
##            <dbl>        <dbl>
## 1           178.         10.9
```

```
height_mean <- summarized[[1]]
height_sd <- summarized[[2]]
```

8.3.5 Athlete heights and the `dnorm` function

We can create a normal curve using the function dnorm. It takes as arguments a vector of x values, the mean, and SD; in terms of the plot, it returns the "y coordinate" for each corresponding x value. Let us construct a normal curve for the Olympic athlete heights using the mean and SD we have just obtained, and overlay it atop the histogram.

```
height_norm <- dnorm(athletes |> pull(Height),
                     mean = height_mean, sd = height_sd)
```

```
ggplot(athletes, aes(x = Height, y = after_stat(density))) +
  geom_histogram(col="grey", fill = "darkcyan", bins=14) +
  geom_line(mapping = aes(x = Height, y = height_norm),
            color = "salmon") +
  labs(x = "Height") +
  ggtitle("Normal curve and the histogram")
```

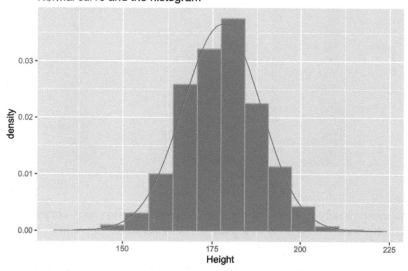

We observe the following.

- The histogram shows strong resemblance to the normal curve.
- The tails of the normal curve extend towards infinity, but there are no athletes shorter than 136 cm and taller than 223 cm.

How close is the curve to the histogram? Let us compare the proportion of athletes whose height is at most 200 cm with respect to the two distributions.

```
pnorm(200, mean = height_mean, sd = height_sd)
```

```
## [1] 0.9796522
```

```
num_rows <- athletes |>
  filter(Height <= 200) |>
```

```
  nrow()
num_rows / nrow(athletes)
```

```
## [1] 0.9812465
```

The normal distribution says 97.96% while the actual record is 98.12%. Pretty close. How about 181?

```
pnorm(181, mean = height_mean, sd = height_sd)
```

```
## [1] 0.6207147
```

```
num_rows <- athletes |>
  filter(Height <= 181) |>
  nrow()
num_rows / nrow(athletes)
```

```
## [1] 0.634704
```

The two values are 62.07% versus 63.47%.

8.4 Central Limit Theorem

As noted at the outset of this chapter, the bell-shaped distribution has been a running motif through most of our examples. While most of the data histograms we studied have not turned out bell-shaped, the sampling distributions representing some simulated statistic has reliably turned out that way.

This is no coincidence. In fact, it is the consequence of an impressive theory in statistics called the *Central Limit Theorem*. We will study the theorem in this section, but let us first see a situation where a bell-shaped distribution results to set up the context.

8.4.1 Prerequisites

We will make use of the tidyverse in this chapter, so let's load it in as usual.

```
library(tidyverse)
library(gapminder)
```

We will also work with GDP per capita data from the gapminder package, so let us load that in as well.

8.4.2   Example: Net allotments from a clumsy clerk

Recall that in an earlier section, we simulated the story of a minister and the doubling grains. In that story, the amount of grains a minister receives doubles each day; by the end of the 64th day, he expects to receive an impressive total of $2^{64} - 1$ grains.

The king has put a clerk in charge to assign the correct amount of grains each day. However, she is human and tabulates the grains by hand, so is prone to error: she may double count the number of grains on some days or forget to count a day altogether. But for the most part she gets it right.

This yields three events on a given day: the counting was *as expected*, it was *double counted*, or the clerk *forgot*. We said that the probabilities for these events are 2/3, 1/3, and 1/3, respectively. Here is the distribution.

```
df <- tribble(~event, ~probability,
              "as expected", 2/4,
              "double counted", 1/4,
              "forgot", 1/4
              )
ggplot(df) +
  geom_bar(aes(x = event, y = probability),
           stat = "identity", fill = "darkcyan")
```

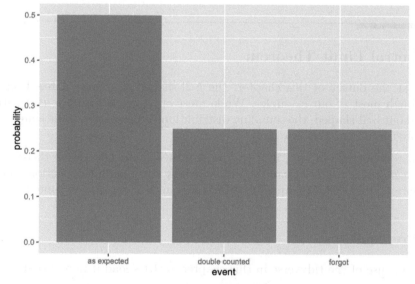

The number of grain allotments received after 64 days is the *sum* of draws made at random with replacement from this distribution.

Using `sample` we can see the result of the grain allotment on any given day.

```
allotment <- sample(c(0, 1, 2), prob = c(1/3, 2/3, 1/3),
                    replace = TRUE, size = 1)
allotment
```

```
## [1] 1
```

If `allotment` turns out 1, the minister received the correct amount of grains that day; 0 indicates that no grains were received and 2 means he received double the amount. The minister hopes that the allotments will total to be 64 and, to the king's dismay, perhaps even more.

We are now ready to simulate one value of the statistic. Let us put our work into a function we can use called `one_simulation`.

```
one_simulation <- function() {
  allotments <- sample(c(0, 1, 2),
                    prob = c(1/4, 2/4, 1/4),
                    replace = TRUE, size = 64)
  return(sum(allotments))
}
one_simulation()
```

```
## [1] 70
```

The following code simulates 10,000 times the net allotments the minister received at the end of the 64th day.

```
num_repetitions <- 10000
net_allotments <- replicate(n = num_repetitions, one_simulation())

results <- tibble(
  repetition = 1:num_repetitions,
  net_allotments = net_allotments
)
results
```

```
## # A tibble: 10,000 x 2
##     repetition net_allotments
##          <int>          <dbl>
## 1            1             77
## 2            2             67
## 3            3             63
## 4            4             66
## 5            5             66
## 6            6             63
## 7            7             60
## 8            8             65
## 9            9             57
## 10          10             64
## # ... with 9,990 more rows
```

```
ggplot(results) +
  geom_histogram(aes(x = net_allotments, y = after_stat(density)),
                 color = "gray", fill = "darkcyan", bins = 15)
```

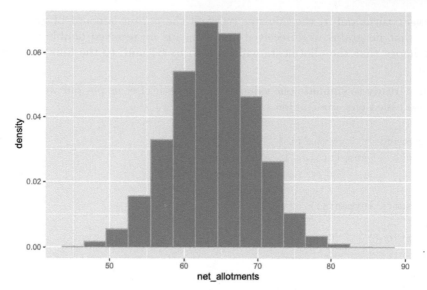

We observe a bell-shaped curve, even though the distribution we drew from does not look
bell-shaped. Note the the distribution is **centered** around 64 days, as expected. With kudos
to the clerk for a job well done, we note that accomplishing such a feat in reality would be
impossible.

To understand the **spread**, look for the inflection point in this histogram. That point occurs
at around 70 days, which means the SD is the distance from the center to this point – that
looks to be about 5.6 days.

```
results |>
  pull(net_allotments) |>
  sd()
```

```
## [1] 5.662408
```

To confirm the bell-shaped curve we are seeing, we can create a normal distribution from
this mean and SD and overlay it atop the sampling histogram. We observe that it provides
a good fit.

```
curve <- dnorm(net_allotments, mean = 64, sd = 5.66)
```

```
ggplot(results) +
  geom_histogram(aes(x = net_allotments, y = after_stat(density)),
                 color = "gray", fill = "darkcyan", bins = 15) +
  geom_line(mapping = aes(x = net_allotments, y = curve),
            color = "salmon")
```

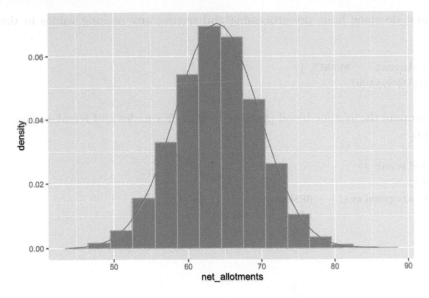

8.4.3 Central Limit Theorem

The Central Limit Theorem is a formalization of the phenomena we observed from the doubling grains story. Formally, it states the following.

The sum or average of a large random sample that is identically and independently distributed will resemble a normal distribution, regardless of the underlying distribution from which the sample is drawn.

The "identically and independently distributed" condition, often abbreviated simply as "IID", is a mouthful. It means that the random samples we draw must have no relation to each other, i.e., the drawing of one sample (say, the event *forgot*) does not make another event more or less likely of occcuring. Hence, we prefer a sampling plan of sampling with replacement.

The Central Limit Theorem is a powerful concept because it becomes possible to make inferences about unknown populations with very little knowledge. And, the larger the sample size becomes, the stronger the resemblance.

8.4.4 Comparing average systolic blood pressure

The Central Limit Theorem says that the sum of IID samples follows a distribution that resembles a normal distribution. Let us return to the NHANES package and compare average systolic blood pressure for males and females. If we take a large sample of participants, what is the sample average systolic blood pressure as given by BPSysAve? According to the theorem, we expect the distribution to be roughly normal.

Let us first do some basic pre-processing and remove any missing values in the BPSysAve column.

```
NHANES_relevant <- NHANES |>
  drop_na(BPSysAve)
```

Let us compare the systolic blood pressure readings for males and females using an overlaid histogram.

```
NHANES_relevant |>
  ggplot() +
  geom_histogram(aes(x = BPSysAve, fill = Gender,
                 y = after_stat(density)),
             color = "gray", alpha = 0.7,
             position = "identity", bins=15)
```

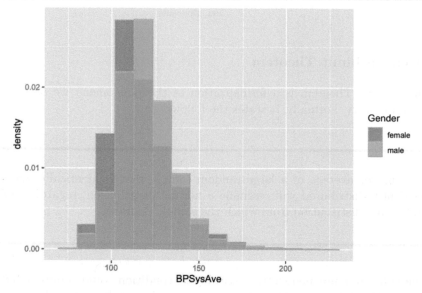

Observe the skew in these data histograms, as shown by the long right-tail in each. Neither closely resembles a normal distribution.

Let us also note the standard deviation and mean for these distributions.

```
summary_stats <- NHANES_relevant |>
  group_by(Gender) |>
  summarize(mean = mean(BPSysAve),
            sd = sd(BPSysAve))
summary_stats
```

```
## # A tibble: 2 x 3
##   Gender  mean    sd
##   <fct>  <dbl> <dbl>
## 1 female  116.  17.8
## 2 male    120.  16.4
```

Let us deal out the participants into two separate datasets according to the gender status.

```
NHANES_female <- NHANES_relevant |>
  filter(Gender == "female")
NHANES_male <- NHANES_relevant |>
  filter(Gender == "male")
```

Let us simulate the average systolic blood pressure in a sample of 100 participants for both males and females. We will write a function to simulate one value for us. Observe that the sampling is done with replacement, in accordance to the "IID" precondition needed by the Central Limit Theorem.

```
one_simulation <- function(df, label, sample_size) {
  df |>
    slice_sample(n = sample_size, replace = TRUE) |>
    summarize(mean({{ label }})) |>
    pull()
}
```

The function `one_simulation` takes as arguments the tibble `df`, the column `label` to use for computing the statistic, and the sample size `sample_size`.

Here is one run of the function with the female data.

```
one_simulation(NHANES_female, BPSysAve, 100)
```

```
## [1] 116.2
```

As before, we will simulate the statistic for female data 10,000 times.

```
num_repetitions <- 10000
sample_means <- replicate(n = num_repetitions,
                          one_simulation(NHANES_female, BPSysAve, 100))

female_results <- tibble(
  repetition = 1:num_repetitions,
  sample_mean = sample_means,
  gender = "female"
)
female_results
```

```
## # A tibble: 10,000 x 3
##    repetition sample_mean gender
##         <int>       <dbl> <chr>
## 1           1        116. female
## 2           2        118. female
## 3           3        114. female
## 4           4        115. female
## 5           5        115. female
## 6           6        117. female
## 7           7        115. female
## 8           8        118. female
## 9           9        117. female
```

```
## 10            10           115. female
## # ... with 9,990 more rows
```

Let us repeat the simulation for the male data.

```
sample_means <- replicate(n = num_repetitions,
                          one_simulation(NHANES_male, BPSysAve, 100))

male_results <- tibble(
  repetition = 1:num_repetitions,
  sample_mean = sample_means,
  gender = "male"
)
male_results
```

```
## # A tibble: 10,000 x 3
##     repetition sample_mean gender
##          <int>       <dbl> <chr>
## 1            1        118. male
## 2            2        120. male
## 3            3        120. male
## 4            4        120. male
## 5            5        119. male
## 6            6        122. male
## 7            7        118. male
## 8            8        119  male
## 9            9        121. male
## 10          10        118. male
## # ... with 9,990 more rows
```

Let us merge the two tibbles together so that we can plot the sampling distributions together.

```
results <- bind_rows(female_results, male_results)
```

We plot an overlaid histogram in density scale showing the two distributions.

```
ggplot(results) +
  geom_histogram(aes(x = sample_mean, y = after_stat(density)),
                 bins = 20, color = "gray",
                 position = "identity", alpha = 0.8) +
  facet_wrap(~gender, scales = "free")
```

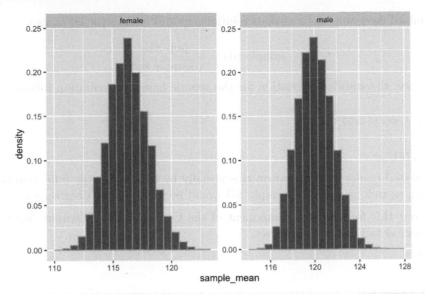

Indeed, we see that both distributions are approximately normal, where each centers a different mean. To confirm the shape, we overlay a normal curve atop each histogram.

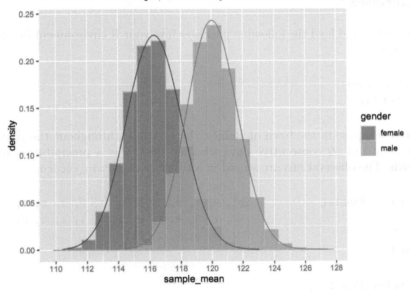

```
summary_stats
```

```
## # A tibble: 2 x 3
##   Gender   mean    sd
##   <fct>   <dbl> <dbl>
## 1 female   116.  17.8
## 2 male     120.  16.4
```

Let us briefly compare the sampling histograms to the summary statistics computed earlier in `summary_stats`, which gives the mean and standard deviation for this dataset, which we are treating as the "population." We observe that both sampling histograms are centered at this *population* mean.

The standard deviation has a curious relationship and follows the equation:

$$\text{sample SD} = \frac{\text{pop SD}}{\sqrt{\text{sample size}}}$$

For instance, the standard deviation for the sample female distribution follows:

```
17.81539 / sqrt(100)
```

```
## [1] 1.781539
```

The interested reader should confirm this visually by looking for the inflection point in the histogram, as well as what the sample SD would be for the male distribution.

It turns out this formula is a component of the Central Limit Theorem. We will explore this in more detail in the exercise set.

8.5   Exercises

Be sure to install and load the following packages into your R environment before beginning this exercise set.

```
library(tidyverse)
library(edsdata)
```

Question 1 Suppose in the town of Raines, the rain falls throughout the year. A student created a record of 30 consecutive days whether there was a precipitation of at least a quarter inch. The observations are stored in a tibble named rain_record.

```
rain_record <- tibble(had_rain = c(1, 0, 1, 0, 1, 0, 0, 1, 0, 1, 0, 1,
                                   1, 1, 0, 0, 0, 0, 1, 0, 1, 1, 0, 0,
                                   0, 0, 0, 0, 1, 0))
rain_record
```

```
## # A tibble: 30 x 1
##      had_rain
##         <dbl>
## 1          1
## 2          0
## 3          1
## 4          0
## 5          1
## 6          0
## 7          0
## 8          1
## 9          0
## 10         1
## # ... with 20 more rows
```

In the variable `had_rain`, `1` represents the day in which there was enough precipitation and `0` represents the day in which there was not enough precipitation.

The Central Limit Theorem (CLT) tells us that the probability distribution of the *sum* or *average* of a large random sample drawn **with replacement** will look roughly normal (i.e., bell-shaped), *regardless of the distribution of the population from which the sample is drawn.*

- **Question 1.1** Let us visualize the precipitation distribution in Raines (given by `had_rain`) using a histogram. Plot the histogram under density scale so that the y-axis shows the chance of the event. Use only 2 bins.

  It looks like there is about a 40% chance of rain and a 60% chance of no rain, which definitely does not look like a normal distribution. The proportion of rainy days in a month is equal to the average number of `1`s that appear, so the CLT should apply if we compute the sample proportion of rainy days in a month many times.

- **Question 1.2** Let us examine the Central Limit Theorem using `rain_record`. Write a function called `rainy_proportion_after_period` that takes a `period` number of days to simulate as input. The function simulates `period` number of days by sampling from the variable `had_rain` in `rain_record` *with replacement*. It then returns the *proportion* of rainy days (i.e., `1`) in this sample as a double.

```
rainy_proportion_after_period(5) # an example call
```

- **Question 1.3** Write a function `rainy_experiment()` that receives two arguments, `days` and `sample_size`, where `days` is the number of days in Raines to simulate and `sample_size` is the number of times to repeat the experiment. It executes the function `rainy_proportion_after_period(days)` you just wrote `sample_size` number of times. The function returns a tibble with the following variables:

  - `iteration`, for the rounds `1:sample_size`
  - `sample_proportion`, which gives the sample proportion of rainy days in each experiment

- **Question 1.4** Here is one example call of your function. We simulate 30 days following the same regimen the student did in the town of Raines, and repeat the experiment for 5 times.

```
rainy_experiment(30, 5)
```

The CLT only applies when sample sizes are "sufficiently large." Let us try a simulation to develop a sense for how the distribution of the sample proportion changes as the sample size is increased.

The following function `draw_ggplot_for_rainy_experiment()` takes a single argument `sample_size`. It calls your `rainy_experiment()` with the argument `sample_size` and then plots a histogram from the sample proportions generated.

```
draw_ggplot_for_rainy_experiment <- function(sample_size) {
  g <- rainy_experiment(30,sample_size) |>
    ggplot(aes(x = sample_proportion)) +
    geom_histogram(aes(y = after_stat(density)),
                   fill = "darkcyan",
                   color="gray",
```

```
                    bins=50)
  return(g)
}
draw_ggplot_for_rainy_experiment(10)
```

Play with different calls to `draw_ggplot_for_rainy_experiment()` by passing in different sample size values. For what value of `sample_size` do you begin to observe the application of the CLT? What does the shape of the histogram look like for the value you found?

Question 2 The CLT states that the standard deviation of a *normal distribution* is given by:

$$\frac{\text{SD of distribution}}{\sqrt{\text{sample size}}}$$

Let us test that the *SD of the sample mean* follows the above formula using flight delays from the tibble `flights` in the `nycflights13` package.

```
nycflights_sd <- flights |>
  summarize(sd = sd(dep_delay, na.rm = TRUE)) |>
  pull(sd)
nycflights_sd
```

- **Question 2.1** Write a function called `theory_sd` that takes a sample size `sample_size` as its single argument. It returns the theoretical standard deviation of the mean for samples of size `sample_size` from the flight delays according to the Central Limit Theorem.

```
theory_sd(10) # an example call
```

- **Question 2.2** The following function `one_sample_mean()` simulates one sample mean of size `sample_size` from the `flights` data.

```
one_sample_mean <- function(sample_size) {
  one_sample_mean <- flights |>
    slice_sample(n = sample_size, replace = FALSE) |>
    summarize(mean = mean(dep_delay, na.rm = TRUE)) |>
    pull(mean)
  return(one_sample_mean)
}

one_sample_mean(10) # an example call
```

Write a function named `sample_sd` that receives a single argument: a sample size `sample_size`. The function simulates 200 samples of size `sample_size` from `flights`. The function returns the standard deviation of the 200 sample means. This function should make repeated use of the `one_sample_mean()` function above.

- **Question 2.3** The chunk below puts together the theoretical and sample SDs for flight delays for various sample sizes into a tibble called `sd_tibble`.

```
sd_tibble <- tibble(
    sample_size = seq(10, 100, 5),
    theory_sds = map_dbl(sample_size, theory_sd),
    sample_sds = map_dbl(sample_size, sample_sd)
    ) |>
  pivot_longer(c(theory_sds,sample_sds),
               names_to = "category",
               values_to = "sd")

sd_tibble
```

Plot these theoretical and sample SD quantities from `sd_tibble` using either a line plot or scatter plot with `ggplot2`. A line plot may be easier to spot differences between the two quantities, but feel free to use whichever visualization makes most sense to you. Regardless, your visualization should show both quantities in a single plot.

- **Question 2.4** As the sample size increases, do the theory and sample SDs change in a way that is consistent with what we know about the Central Limit Theorem?

Question 3. In the textbook, we examined judges' evaluations of contestants. We have another example here with a slightly bigger dataset. The data is from the evaluation of applicants to a scholarship program by four judges. Each judge evaluated each applicant on a scale of 5 points. The applicants have already gone through a tough screening process and, in general, they are already high achievers.

To begin, let us load the data into `applications` from the `edsdata` package.

```
library(edsdata)
applications
```

```
## # A tibble: 22 x 5
##    Last       Mary Nancy Olivia Paula
##    <chr>     <dbl> <dbl>  <dbl> <dbl>
## 1  Arnold    4.9   4.95   4.75  4
## 2  Baxter    4.05  3.55   4.35  3.6
## 3  Chromovich 4.3  3.55   4.1   3.25
## 4  Dempsey   4     3.75   4.35  3.8
## 5  Engels    3.85  4.3    3.8   3.9
## 6  Franks    4.1   4.74   3.89  4.6
## 7  Greene    4.55  4.4    3.7   4.7
## 8  Hanks     3.55  3.55   3.9   3.7
## 9  Ingels    4.35  4.35   3.95  3.5
## 10 Jules     4.7   4.6    4.4   4.5
## # ... with 12 more rows
```

- **Question 3.1** If the observational unit is defined as the score an applicant received by some judge, then the current presentation of the data are not tidy (why?). Apply a pivot transformation to `applications` so that three variables are apparent in the dataset: `Last` (last name of the student), `Judge`(the judge who scored the student), and `Score` (the score the student received). Assign the resulting tibble to the name `app_tidy`.

- **Question 3.2** Use the `scale` function to obtain the scaled version of the scores with respect to each judge. Add a new variable to `app_tidy` called `Scaled` that contains these scaled scores. Assign the resulting tibble to the name `with_scaled`.

- **Question 3.3** Which of the following statements, if any, are accurate based on the tibble `with_scaled`?

    - Arnold's scaled Judge Mary score is about 1.9 standard deviations higher than the mean score Arnold received from the four different judges.
    - The score Judge Mary gave to Baxter is roughly 0.02 standard deviations below Judge Mary's mean score over all applicants.
    - Based on the raw score alone, we can tell the score Arnold received from Judge Mary is higher than the average score Judy Mary gave.
    - The standard deviation of Judge Mary's scaled scores is higher than the standard deviation of Judge Olivia's scaled scores.

- **Question 3.4** Add two new variables to `with_scaled`, `Total Raw` and `Total Scaled`, that gives the total raw score and total scaled score, respectively, for each applicant. Save the resulting tibble to the name `total_scores`.

- **Question 3.5** Sort the rows of the data in `total_scores` in descending order of `Total Raw` and create a variable `Raw Rank` that gives the index of the sorted rows. Store the resulting tibble in the name `app_raw_sorted`. Then use `total_scores` to sort the rows in descending order of `Total Scaled` and, likewise, create a variable `Scaled Rank` that gives the index of the row when sorted this way. Store the result in `app_scaled_sorted`.

- **Question 3.6** Join the results from `app_raw_sorted` with `app_scaled_sorted` using an appropriate join function. The resulting tibble should contain five variables: the applicant's last name, their total raw score, total scaled score, the raw "rank", and the scaled "rank". Assign the resulting tibble to the name `ranks_together`.

- **Question 3.7** Create a new variable `Rank difference` that gives the difference between the raw rank and the scaled rank. Assign the resulting tibble back to the name `ranks_together`.

- **Question 3.8** For which applicant does the first rank difference occur? For which applicant does the largest rank difference occur?

- **Question 3.9** Why do you think there are differences in the rankings given by the raw and scaled scores? Does the use of scaling bring any benefit when making a decision about which applicant should be granted admission? Which one would you choose?

- **Question 3.10** How does the ranking change when removing applicants with large discrepancies in ranking? Remove Baxter from the dataset and then repeat all above steps. What differences do you observe in the ranking? Do the new findings change your answer to **Question 3.9**?

Question 4. Recall that the Central Limit Theorem states that the distribution of the average of independent and identically distributed samples gets closer to a normal distribution as we increase the number of samples. We have used the sample mean for examining the phenomenon, but let us try a different statistic – the sample variance – and see if the phenomenon holds. Moreover, we will compute this statistic using some quantity we will make up that is not normally distributed, and see if the Central Limit Theorem still applies regardless of the underlying quantity we are using.

For this exploration, we will continue our examination of departure delays in the `flights` tibble from `nycflights13`.

Note: Be careful when dealing with missing values for this problem! For this problem, it is enough to simply eliminate missing values in the `arr_delay` and `dep_delay` variables.

```
library(nycflights13)
flights <- flights |>
  drop_na()
```

- **Question 4.1** The quantity we will examine here is the *absolute difference* in departure delay (`dep_delay`) and arrive delay (`arr_delay`). Add a new variable called `dep_arr_abs_diff` that gives this new quantity. Assign the resulting tibble to the name `flights_with_diff`.

- **Question 4.2** Assuming that we can treat the `flights_with_diff` tibble as the *population* of all flights that departed NYC, compute the population variance of the `dep_arr_abs_diff` variable. Assign the resulting double to the name `pop_var_abs_diff`.

- **Question 4.3** What is the max of the absolute differences in `dep_arr_abs_diff`? What is the mean of them? Store the answers in the names `max_diff` and `mean_diff`, respectively. How about the quantile values at 0.5, 0.15, 0.35, 0.50, 0.65, 0.85, 0.95, and 0.99? Store the quantile values in `quantile_values`.

- **Question 4.4** Plot a histogram of `dep_arr_abs_diff` from `flights_with_diff` in density scale with 30 bins. Add to the histogram the point on the x-axis indicating the max, the mean, the 0.5 quantile, and the 0.99 quantile. Use `"black"` for the 0.99 quantile and `"red"` for the max. Use two other different colors for the mean and the median.

- **Question 4.5** Write a function `var_from_sample` that receives a single argument `n_sample`. The function samples `n_samples` from the the tibble `flights_with_diff` without replacement, and computes the *sample variance* in the new variable `dep_arr_abs_diff` we created. The sample variance is then returned.

```
# example calls
var_from_sample(10)
var_from_sample(100)
var_from_sample(1000)
```

- **Question 4.6** Write a function called `hist_for_sample` that receives a single argument `sample_size`. This function should accomplish the following:

  - Call repeatedly the function `var_from_sample` with the given `sample_size`, say, 1,000 times.
  - Generate a histogram of the simulated sample statistics under density scale.
  - Annotate this histogram with (1) a vertical blue line showing where the population parameter is (`pop_var_abs_diff`) and (2) a red point indicating the *mean* of the generated sample statistics.

```
hist_for_sample <- function(sample_size) {

}
```

The following code calls your function with different sample sizes:

```
map(c(10, 20, 50, 100, 1000, 10000), hist_for_sample)
```

- **Question 4.7** As the sample size is increased, does the variance of the simulated statistics increase or decrease? How can you tell?

- **Question 4.8** As the sample size is increased, the red point moves closer and closer to the vertical blue line. Is this observation a coincidence due to the data we used? If not, what does it suggest about the mean computed from our sample and the population parameter?

- **Question 4.9** Alex, Bob, and Jeffrey are grumbling about whether we can use the Central Limit Theorem (CLT) to help think about what the histograms should look like in the above parts.

  - Alex believes we cannot use the CLT since we are looking at the sampling histogram of the test statistic and we do not know what the probability distribution looks like.
  - Bob believes the CLT does not apply because the distribution of `dep_arr_abs_diff` is not normally distributed.
  - Jeffrey believes that both of these concerns are invalid, and the CLT is helpful.

  Who is right? Are they all wrong? Explain your reasoning.

Modeling

Modeling

9

Regression

In this chapter we turn to making guesses, or *predictions*, about the future. There are many examples in which you, your supervisor, or an employer would want to make claims about the future – that are accurate! For instance:

- Can we predict the miles per gallon of a brand new car model using aspects of its design and performance?
- Can we predict the price of suburban homes in a city using information collected from towns in the city, e.g. crime levels, demographics, median income, and distance to closest employment center?
- Can we predict the magnitude of an earthquake given temporal and geological characteristics of the area, e.g. fault length and recurrence interval?

To answer these questions, we need a framework or *model* for understanding how the world works. Fortunately, data science has offered up many such models for addressing the above proposed questions. This chapter turns to an important one known as *regression*, which is actually a family of methods designed for modeling the value of a *response* variable given some number of *predictor* variables.

We have already seen something that resembles regression in the introduction to visualization, where we guessed by examination the line explaining the highway mileage using what we know about a car model's displacement. We build upon that intuition to examine more formally what regression is and how to use it appropriately in data science tasks.

9.1 Correlation

Linear regression is closely related to a statistic called the *correlation*, which refers to how tightly clustered points are about a straight line with respect to two variables. If we observe such clustering, we say that the two variables are *correlated*; otherwise, we say that there is no correlation or, put differently, that there is no association between the variables.

In this section, we build an understanding of what correlation is.

9.1.1 Prerequisites

We will make use of the tidyverse and the `edsdata` package in this chapter, so let us load these in as usual. We will also return to some datasets from the `datasauRus` package, so let us load that as well.

```
library(tidyverse)
library(datasauRus)
library(edsdata)
```

To build some intuition, we composed a tibble called `corr_tibble` (available in `edsdata`) that contains several artificial datasets. Don't worry about the internal details of how the data is generated; in fact, it would be better to think of it as real data!

```
corr_tibble
```

```
## # A tibble: 2,166 x 3
##            x          y dataset
##        <dbl>      <dbl> <chr>
## 1    0.695     2.26     cone
## 2    0.0824    0.0426   cone
## 3    0.0254   -0.136    cone
## 4    0.898     3.44     cone
## 5    0.413     1.15     cone
## 6   -0.730    -1.57     cone
## 7    0.508     1.26     cone
## 8   -0.0462   -0.0184   cone
## 9   -0.666    -0.143    cone
## 10  -0.723     0.159    cone
## # ... with 2,156 more rows
```

9.1.2 Visualizing correlation with a scatter plot

Let us begin by examining the relationship between the two variables `y` versus `x` in the dataset `linear` within `corr_tibble`. As we have done in the past, we will use a scatter plot for such a visualization.

```
corr_tibble |>
  filter(dataset == "linear") |>
  ggplot() +
  geom_point(aes(x = x, y = y), color = "darkcyan")
```

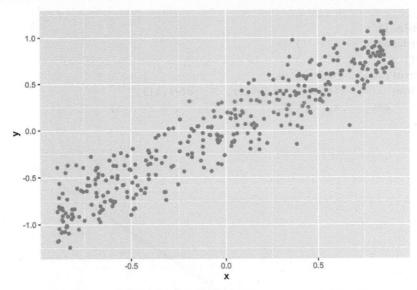

We noted earlier that two variables are correlated if they appear to cluster around a straight line. It appears that there is a strong correlation present here, but let us confirm what we are seeing by drawing a reference line at $y = x$.

```
corr_tibble |>
  filter(dataset == "linear") |>
  ggplot() +
  geom_point(aes(x = x, y = y), color = "darkcyan") +
  geom_line(data = data.frame(x = c(-1,1), y = c(-1,1)),
            aes(x = x, y = y), color = "blue", size = 1)
```

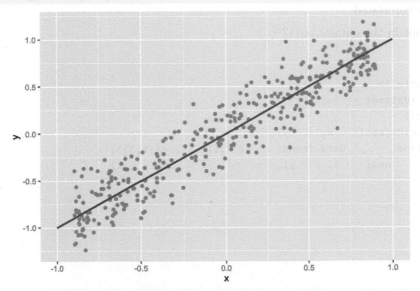

Indeed, we can confirm a "strong" correlation between these two variables. We will define how strong momentarily. Let us turn to another dataset, `perf`.

```
corr_tibble |>
  filter(dataset == "perf") |>
  ggplot() +
  geom_point(aes(x = x, y = y), color = "darkcyan") +
  geom_line(data = tibble(x = c(-1,1), y = c(-1,1)),
            aes(x = x, y = y), color = "blue", size = 1)
```

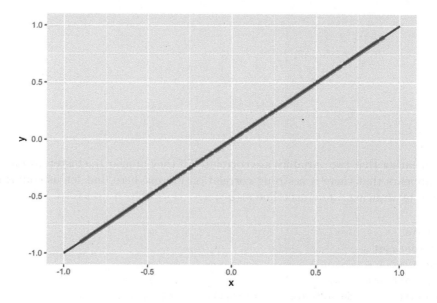

Neat! The points fall *exactly* on the line. We can confidently say that y_perf and x are perfectly correlated.

How about in the dataset null?

```
corr_tibble |>
  filter(dataset == "null") |>
  ggplot() +
  geom_point(aes(x = x, y = y), color = "darkcyan") +
  geom_line(data = data.frame(x = c(-1,1), y = c(-1,1)),
            aes(x = x, y = y), color = "blue", size = 1)
```

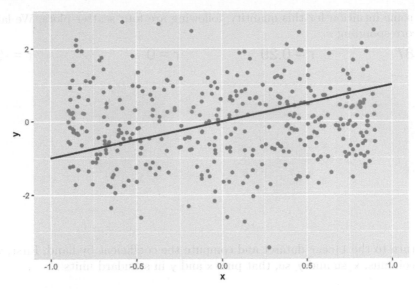

This one, unlike the others, does not cluster well around the line $y = x$ nor does it show a trend whatsoever; in fact, it seems the points are drawn at random, much like static noise on a television. We call a plot that shows such phenomena a *null plot*.

Let us now quantify what correlation is.

9.1.3 The correlation coefficient r

We are now ready to present a formal definition for correlation, which is usually referred to as the correlation coefficient r.

---

r is the mean of the products of two variables that are scaled to standard units.

Here are some mathematical facts about r. Proving them is beyond the scope of this text, so we only state them as properties.

- r can take on a value between 1 and -1.
- $r = 1$ means perfect positive correlation; $r = -1$ means perfect negative correlation.
- Two variables with $r = 0$ means that they are not related by a line, i.e., there is no *linear association* among them. However, they can be related by something else, which we will see an example of soon.

- Because r is scaled to standard units, it is a dimensionless quantity, i.e., it has no units.

- **Association does not imply causation!** Just because two variables are strongly correlated, positively or negatively, does not mean that x *causes* y.

To build some intuition for this quantity, following are four scatter plots. We labeled each with its corresponding r.

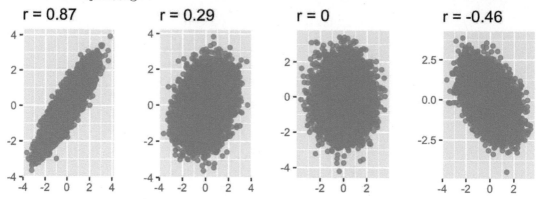

Let us return to the linear dataset and compute the coefficient by hand. First, we append two new columns, x_su and y_su, that puts x and y in standard units.

```
corr_tibble |>
  filter(dataset == 'linear') |>
  transmute(x = x,
            y = y,
            x_su = scale(x),
            y_su = scale(y))
```

```
## # A tibble: 361 x 4
##            x        y x_su[,1] y_su[,1]
##        <dbl>    <dbl>    <dbl>    <dbl>
## 1   0.695    0.757     1.20     1.24
## 2   0.0824   0.326     0.112    0.511
## 3   0.0254   0.128     0.0114   0.175
## 4   0.898    0.708     1.56     1.16
## 5   0.413    0.0253    0.698    0.00109
## 6  -0.730   -0.505    -1.33    -0.900
## 7   0.508    0.701     0.866    1.15
## 8  -0.0462  -0.101    -0.116   -0.213
## 9  -0.666   -0.971    -1.21    -1.69
## 10 -0.723   -0.583    -1.31    -1.03
## # ... with 351 more rows
```

We weill modify the above code to add one more column, prod, that takes the product of the columns x_su and y_su. We will also save the resulting tibble to a variable called linear_df_standard.

```
linear_df_standard <- corr_tibble |>
  filter(dataset == 'linear') |>
  transmute(x = x,
            y = y,
            x_su = scale(x),
            y_su = scale(y),
            prod = x_su * y_su)
linear_df_standard
```

```
## # A tibble: 361 x 5
##             x       y x_su[,1] y_su[,1] prod[,1]
##         <dbl>   <dbl>    <dbl>    <dbl>    <dbl>
##  1  0.695    0.757    1.20     1.24     1.49
##  2  0.0824   0.326    0.112    0.511    0.0574
##  3  0.0254   0.128    0.0114   0.175    0.00199
##  4  0.898    0.708    1.56     1.16     1.81
##  5  0.413    0.0253   0.698    0.00109  0.000761
##  6 -0.730   -0.505   -1.33    -0.900    1.19
##  7  0.508    0.701    0.866    1.15     0.995
##  8 -0.0462  -0.101   -0.116   -0.213    0.0246
##  9 -0.666   -0.971   -1.21    -1.69     2.05
## 10 -0.723   -0.583   -1.31    -1.03     1.36
## # ... with 351 more rows
```

According to the definition, we need only to calculate the mean of the products to find r.

```
r <- linear_df_standard |>
  pull(prod) |>
  mean()
r
```

```
## [1] 0.9366707
```

So the correlation of y and x is about 0.94 which implies a strong positive correlation, as expected.

Thankfully, R comes with a function for computing the correlation so we need not repeat this work every time we wish to explore the correlation between two variables. The function is called cor and receives two vectors as input.

```
linear_df_standard |>
  summarize(r = cor(x, y))
```

```
## # A tibble: 1 x 1
##       r
##   <dbl>
## 1 0.939
```

Note that there may be some discrepancy between the two values. This is due to the $n - 1$ correction that R provides when calculating quantities like sd.

9.1.4 Technical considerations

There are a few technical points to be aware of when using correlation in your analysis. We reflect on these here.

Switching axes does not affect the correlation coefficient. Let us swap x and y in the "linear" dataset in corr_tibble and then plot the result.

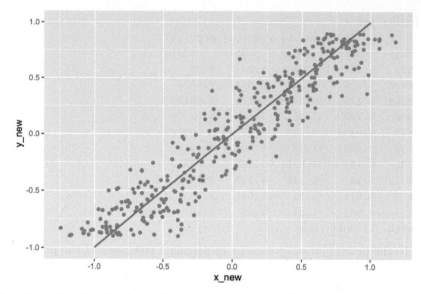

Observe how this visualization is a reflection over the $y = x$ line. We compute the correlation.

```
swapped |>
  summarize(r = cor(x_new, y_new))
```

```
## # A tibble: 1 x 1
##       r
##    <dbl>
## 1 0.939
```

This value is equivalent to the value we found earlier for this dataset.

Correlation is sensitive to "outliers". Consider the following toy dataset and its corresponding scatter.

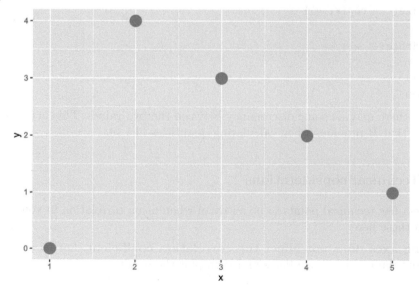

Besides the first point, we would expect this dataset to have a near perfect *negative corre-lation* between x and y. However, computing the correlation coefficient for this dataset tells a different story.

```
weird %>%
  summarize(r = cor(x, y))
```

```
## # A tibble: 1 x 1
##        r
##    <dbl>
## 1      0
```

We call such points that do not follow the overall trend "outliers". It is important to look out for these as said observations have the potential to dramatically affect the signal in the analysis.

$r = 0$ **is a special case.** Let us turn to the dataset curvy to see why.

```
curvy <- corr_tibble |>
  filter(dataset == "curvy")
```

We can compute the correlation coefficient as before.

```
curvy %>%
  summarize(cor(x, y))
```

```
## # A tibble: 1 x 1
##    `cor(x, y)`
##          <dbl>
## 1      -0.0368
```

The small value may suggest that there is no correlation and, therefore, the scatter diagram should resemble a null plot. Let us visualize the data.

```
ggplot(curvy) +
  geom_point(aes(x = x, y = y), color = "darkcyan")
```

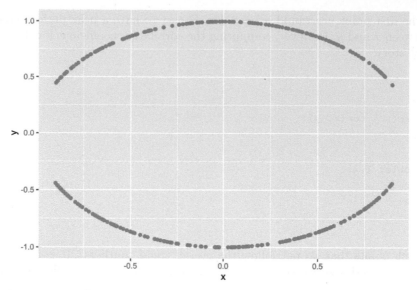

A clear pattern emerges! However, the association is very much not *linear*, as indicated by the obtained r value.

r **is agnostic to units of the input variables.** This is due to its calculation being based on standard units. So, for instance, we can look at the correlation between a quantity in miles per gallon and another quantity in liters.

9.1.5   Be careful with summary statistics! (revisited)

We saw before the danger of using summary statistics like mean and standard deviation without first visualizing data. Correlation is another one to watch out for. Let us see why using the same `bullseye` and `star` datasets we examined before. We will compute the correlation coefficient for each dataset.

```
datasaurus_dozen |>
  filter(dataset == "bullseye" | dataset == "star") |>
  group_by(dataset) |>
  summarize(r = cor(x, y))
```

```
## # A tibble: 2 x 2
##   dataset        r
##   <chr>      <dbl>
## 1 bullseye -0.0686
## 2 star     -0.0630
```

We observe that both datasets have almost identical coefficient values so we may be suspect that both also have seemingly identical associations as well. We may also claim there is some evidence that suggests there is a weak negative correlation between the variables.

As before, the test of any such claim is visualization.

```
datasaurus_dozen |>
  filter(dataset == "bullseye" | dataset == "star") |>
```

```
ggplot() +
  geom_point(aes(x = x, y = y, color = dataset)) +
  facet_wrap( ~ dataset, ncol = 2)
```

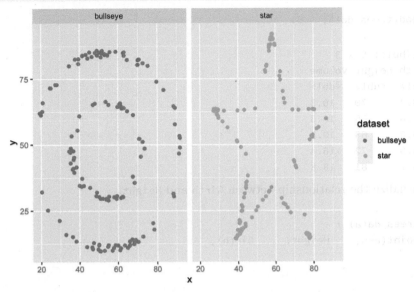

Indeed, we would be mistaken! The variables in each dataset are not identically associated nor do they bear any kind of linear association. The lesson here, then, remains the same: **always visualize your data!**

9.2 Linear Regression

Having a better grasp of what correlation is, we are now ready to develop an understanding of linear regression.

9.2.1 Prerequisites

We will make use of the tidyverse in this chapter, so let us load it in as usual.

```
library(tidyverse)
```

9.2.2 The `trees` data frame

The `trees` data frame contains data on the diameter, height, and volume of 31 felled black cherry trees. Let us first convert the data to a tibble.

```
trees_data <- tibble(datasets::trees)
```

Note that the diameter is erroneously labeled as Girth.

```
slice_head(trees_data, n=5)
```

```
## # A tibble: 5 x 3
##    Girth Height Volume
##    <dbl>  <dbl>  <dbl>
## 1    8.3     70   10.3
## 2    8.6     65   10.3
## 3    8.8     63   10.2
## 4   10.5     72   16.4
## 5   10.7     81   18.8
```

Let us visualize the relationship between Girth and Height.

```
ggplot(trees_data) +
  geom_point(aes(x = Height, y = Girth), color = "darkcyan")
```

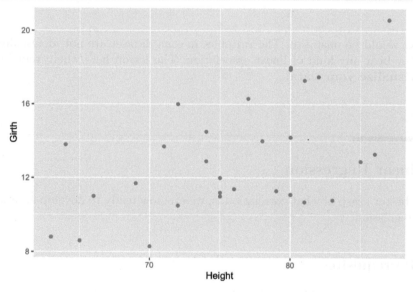

There seems to be some correlation between the two – taller trees tend to have larger diameters. Confident in the trend we are seeing, we propose the question: can we *predict* the average diameter of a black cherry tree from its height?

9.3   First Approach: Nearest Neighbors Regression

One way to make a prediction about the outcome of some individual is to first find others who are similar to that individual and whose outcome you do know. We can then use those outcomes to guide the prediction.

Say we have a new black cherry tree whose height is 65 ft. We can look at trees that are "close" to 65 ft, say, within 1 feet. We can find these individuals using `dplyr`:

```
trees_data |>
  filter(between(Height, 75 - 1, 75 + 1))
```

```
## # A tibble: 7 x 3
##   Girth Height Volume
##   <dbl>  <dbl>  <dbl>
## 1 11        75   18.2
## 2 11.2      75   19.9
## 3 11.4      76   21
## 4 11.4      76   21.4
## 5 12        75   19.1
## 6 12.9      74   22.2
## 7 14.5      74   36.3
```

Here is a scatter with those "close" observations annotated using the color cyan.

```
trees_data |>
  mutate(close_to_75 = between(Height, 75 - 1, 75 + 1)) |>
  ggplot() +
    geom_point(aes(x = Height, y = Girth, color = close_to_75))
```

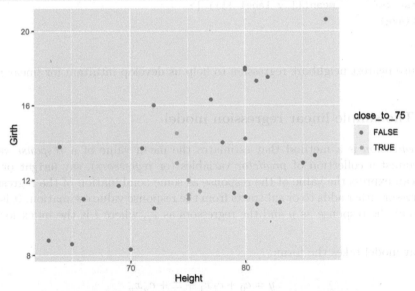

We can take the mean of the outcomes for those observations to obtain a *prediction* for the *average* diameter of a black cherry tree whose height is 65 ft. Let us amend our `dplyr` code to compute this value.

```
trees_data |>
  filter(between(Height, 75 - 1, 75 + 1)) |>
  summarize(prediction = mean(Girth))
```

```
## # A tibble: 1 x 1
##    prediction
##         <dbl>
## 1        12.1
```

This method is called *nearest neighbors regression* because of the use of "neighbors" to make informed predictions about individuals whose outcomes we do not know. Its procedure is as follows:

- Define a threshold t
- Filter each x in the dataset to contain only rows where its x value is within $x \pm t$. This defines a group, or "neighborhood", for each x.
- Take the mean of corresponding y values in each group
- For each x, the prediction is the *average* of the y values in the corresponding group

We write a function called `nn_predict` to obtain a prediction for each height in the dataset. The function is designed so that it can be used to make predictions for any dataset with a given threshold amount.

```
nn_predict <- function(x, tib, x_label, y_label,
                       threshold) {
  tib |>
    filter(between({{ x_label }},
                   x - threshold,
                   x + threshold)) |>
    summarize(avg = mean({{ y_label }})) |>
    pull(avg)
}
```

We will use nearest neighbors regression to help us develop intuition for linear regression.

9.3.1   The simple linear regression model

Linear regression is a method that estimates the mean value of a *response* variable, say `Girth`, against a collection of *predictor* variables (or *regressors*), say `Height` or `Volume`, so that we can express the value of the response as some combination of the regressors, where each regressor either adds to or subtracts from the response value estimation. It is customary to represent the response as y and the regressors as x_i, where i is the index for one of the regressors.

The linear model takes the form:

$$y = c_0 + c_1 x_1 + \ldots + c_n x_n$$

where c_0 is the intercept and the other c_i's are coefficients. This form is hard to digest, so we will begin with using only one regressor. Our model, then, reduces to:

$$y = c_0 + c_1 x$$

which is the equation of a line, just as you have seen it in high school math classes. There are some important assumptions we are making when using this model:

- The model is of the form $y = c_0 + c_1 x$, that is, the data can be modeled linearly.
- The variance of the "errors" is more or less constant. This notion is sometimes referred to as *homoskedasticity*.

We will return to what "errors" mean in just a moment. For now just keep in mind that the linear model may not, and usually will not, pass through all of the points in the dataset and, consequently, some amount of error is produced by the predictions it makes.

You may be wondering how we can obtain the intercept (c_0) and slope (c_1) for this line. For that, let us return to the tree data.

9.3.2 The regression line in standard units

We noted before that there exists a close relationship between correlation and linear regression. Let us see if we can distill that connection here.

We begin our analysis by noting the correlation between the two variables.

```
trees_data |>
  summarize(r = cor(Girth, Height))
```

```
## # A tibble: 1 x 1
##       r
##   <dbl>
## 1 0.519
```

This confirms the positive linear trend we saw earlier. Recall that the correlation coefficient is a dimensionless quantity. Let us standardize `Girth` and `Height` so that they are in standard units and are also dimensionless.

```
girth_height_su <- trees_data |>
  transmute(Girth_su = scale(Girth),
            Height_su = scale(Height))
girth_height_su
```

```
## # A tibble: 31 x 2
##    Girth_su[,1] Height_su[,1]
##           <dbl>         <dbl>
## 1         -1.58        -0.942
## 2         -1.48        -1.73
## 3         -1.42        -2.04
## 4        -0.876        -0.628
## 5        -0.812         0.785
## 6        -0.780         1.10
## 7        -0.716        -1.57
## 8        -0.716        -0.157
## 9        -0.685         0.628
## 10       -0.653        -0.157
## # ... with 21 more rows
```

Let us plot the transformed data using a scatter plot again. Note how the axes of this plot have changed and that we can clearly identify how many SDs each point is away from the mean along each axis.

```
ggplot(girth_height_su) +
  geom_point(aes(x = Height_su, y = Girth_su), color = "darkcyan")
```

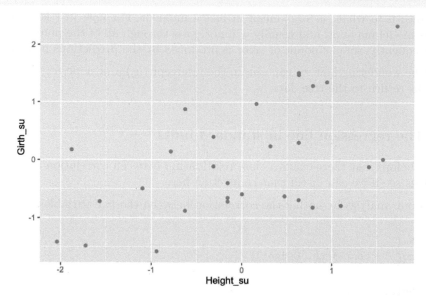

How can we find an equation of a line that "best" passes through this collection of points? We can start with some trial and error – say, the line $y = x$.

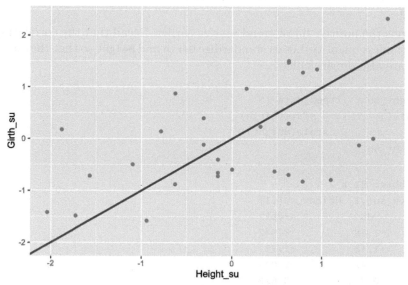

Let us compare this with the nearest neighbors regression line. We use the function nn_predict we developed earlier in combination with a purrr map and set the threshold amount to $\pm$ 1 ft.

```
with_nn_predictions <- girth_height_su |>
  mutate(prediction =
           map_dbl(Height_su,
                   nn_predict, girth_height_su, Height_su, Girth_su, 1))
with_nn_predictions
```

```
## # A tibble: 31 x 3
##    Girth_su[,1] Height_su[,1] prediction
##         <dbl>        <dbl>       <dbl>
## 1      -1.58       -0.942      -0.440
## 2      -1.48       -1.73       -0.767
## 3      -1.42       -2.04       -0.787
## 4      -0.876      -0.628      -0.272
## 5      -0.812       0.785       0.254
## 6      -0.780       1.10        0.535
## 7      -0.716      -1.57       -0.596
## 8      -0.716      -0.157       0.0280
## 9      -0.685       0.628       0.144
## 10     -0.653      -0.157       0.0280
## # ... with 21 more rows
```

The new column `prediction` contains the nearest neighbor predictions. Let us overlay these atop the scatter plot.

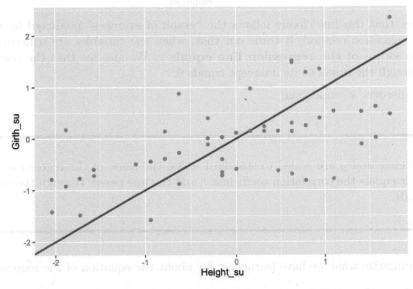

The graph of these predictions is called a **"graph of averages."** If the relationship between the response and predictor variables is roughly linear, then points on the "graph of averages" tend to fall on a line.

What is the equation of that line? It appears that the $y = x$ overestimates observations where Height > 0 and underestimates observations where Height < 0.

How about we try a line where the slope is the correlation coefficient, r, we found earlier? That follows the equation:

$$\text{Girth} = r * \text{Height} = 0.519 * \text{Height}$$

Let us overlay this on the scatter plot using a purple line.

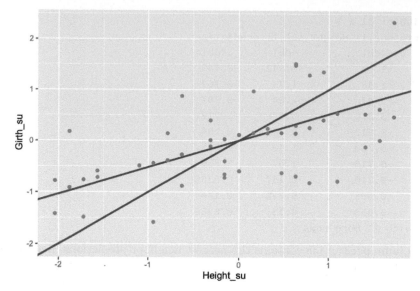

We can see that this line closely follows the "graph of averages" predicted by the nearest neighbor regression method. It turns out that, when the variables are scaled to standard units, **the slope of the regression line equals** r. We also see that the regression line passes through the origin so its intercept equals 0.

Thus, we discover a connection:

When x and y are scaled to standard units, the slope of the regression of y on x equals the correlation coefficient r and the line passes through the origin $(0,0)$.

Let us summarize what we have learned so far about the equation of the regression line.

9.3.3 Equation of the regression line

When working in standard units, the average of x and y are both 0 and the standard deviation of x (SD_x) is 1 and the standard deviation of y is 1 (SD_y). Using what we know, we can recover the equation of the regression line **in original units**. If the slope is r in standard units, then moving 1 unit along the x-axis in standard units is equivalent to moving SD_x units along the x-axis in original units. Similarly, moving r units along the y-axis in standard units is equivalent to moving $r*SD_y$ units along the y-axis in original units. Thus, we have the slope of the regression line in original units:

$$\text{slope} = r * \frac{SD_y}{SD_x}$$

Furthermore, if the line passes through the origin in standard units, then the line in original units passes through the point $(\bar{x}, \bar{y})$, where $\bar{x}$ and $\bar{y}$ is the mean of x and y, respectively. So if the equation of the regression line follows:

$$\text{estimate of } y = \text{slope} * x + \text{intercept}$$

Then the intercept is:

$$\text{intercept} = \bar{y} - \text{slope} * \bar{x}$$

9.3.4 The line of least squares

We have shown that the "graph of averages" can be modeled using the equation of the regression line. However, how do we know that the regression line is the *best* line that passes through the collection of points? We need to be able to *quantify* the amount of error at each point.

For this, we introduce the notion of a *residual*, which we define as the vertical distance from the point to the line (which can be positive or negative depending on the location of the point relative to the line). More formally, we say that a residual e_i has the form:

$$e_i = y_i - (c_0 + c_1 x_i)$$

where plugging into $c_0 + c_1 x_i$, the equation of the line, returns the *predicted* value for some observation i. We usually call this the *fitted* value.

We can annotate the plot with the regression line using the residual amounts.

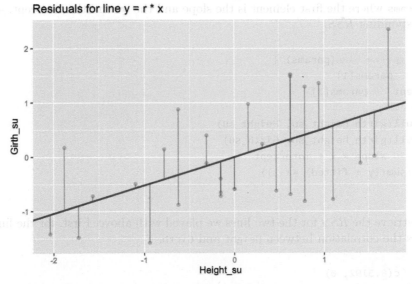

The vertical blue lines show the residual for each point; note how some are small while others are quite large. How do the residuals for this line compare with those for, say, the line $y = x$ or a horizontal line at $y = 0$?

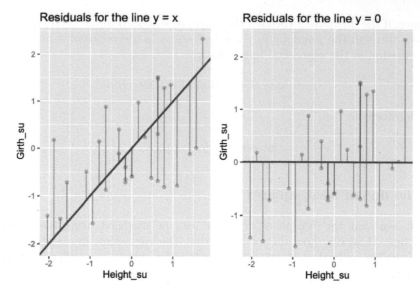

If we compare these plots side-by-side, it becomes clear that the equation $y = r * x$ has smaller residuals overall than the $y = x$ line and the horizontal line that passes through $y = 0$. This observation is key: to get an overall sense of the error in the model, we can *sum* the residuals for each point. However, there is a problem with such an approach. Some residuals can be positive while others can be negative, so a straightforward sum of the residuals will total to 0! Thus, we take the *square* of each residual and then issue the sum.

Our goal can now be phrased as follows: we would like to find the line that *minimizes* the residual sum of squares. We normally refer to this quantity as RSS. Let us write a function that returns the RSS for a given linear model. This function receives a vector called `params` where the first element is the slope and the second the intercept, and returns the corresponding RSS.

```
line_rss <- function(params) {
  slope <- params[1]
  intercept <- params[2]

  x <- pull(girth_height_su, Height_su)
  y <- pull(girth_height_su, Girth_su)
  fitted <- slope * x + intercept
  return(sum((y - fitted) ** 2))
}
```

We can retrieve the RSS for the two lines we played with above. First, for the line $y = r * x$, where r is the correlation between `Height` and `Girth`.

```
params <- c(0.5192, 0)
line_rss(params)
```

```
## [1] 21.91045
```

Next, for the line $y = x$.

```
params <- c(1, 0)
line_rss(params)
```

```
## [1] 28.8432
```

Finally, for the horizontal line passing through $y = 0$. Note how the larger RSS for this model confirms what we saw graphically when plotting the vertical distances.

```
params <- c(0, 0)
line_rss(params)
```

```
## [1] 30
```

We could continue experimenting with values until we found a configuration that seems "good enough". While that would hardly be acceptable to our peers, solving this rigorously involves methods of calculus which are beyond the scope of this text. Fortunately, we can use something called numerical optimization which allows R to do the trial-and-error work for us, "nudging" the above line until it minimizes RSS.

9.3.5 Numerical optimization

The optim function can be used to accomplish the task of finding the linear model that yields the smallest RSS. optim takes as arguments an initial vector of values and a function to be minimized using an initial guess as a starting point.

Let us see an example of it by finding an input that minimizes the output given by a quadratic curve. Consider the following quadratic equation:

```
my_quadratic_equation <- function(x) {
  (x - 2) ** 2 + 3
}
```

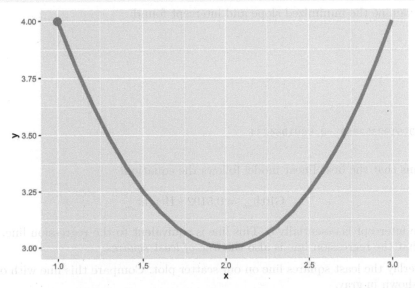

Geometrically, we know that the minimum occurs at the bottom of the "bowl", so the answer is 2.

The orange dot at $x = 1$ shows our first "guess". You can imagine the call optim(initial_guess, my_quadratic_equation) nudging this point around until a minimum is reached. Let us try it.

```
initial_guess <- c(1)   # start with a guess at x = 1
best <- optim(initial_guess, my_quadratic_equation)
```

We can inspect the value found using pluck from purrr.

```
best |>
  pluck("par")
```

```
## [1] 1.9
```

This value closely approximates the expected value of 2.

Let us now use the optim function to find the linear model that yields the smallest RSS. Note that our initial guess now consists of two values, one for the slope and the second for the intercept.

```
initial_guess <- c(1,0)   # start with a guess, the y = x line
best <- optim(initial_guess, line_rss)
```

We can examine the minimized slope and intercept found.

```
best |>
  pluck("par")
```

```
## [1]  0.5192362885 -0.0001084234
```

This means that the best linear model follows the equation:

$$\text{Girth}_{su} = 0.5192 * \text{Height}_{su}$$

where the intercept is essentially 0. This line is equivalent to the regression line. Therefore, we find that the regression line is also the *line of least squares*.

Let us overlay the least squares line on our scatter plot. Compare this line with our original guesses, shown in gray.

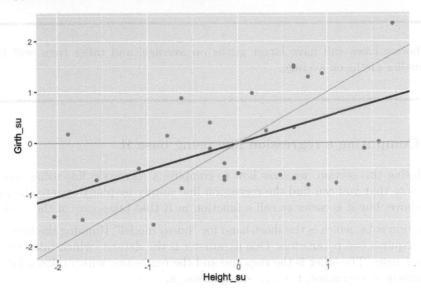

Compare the line we have just found (in purple) with the line $y = x$. For trees with heights that are *less* than 0 *SD*s away from the mean, the regression line predicts a *larger* girth than the $y = x$ line. And, for trees with heights that are *more* than 0 *SD* away from the mean, the regression line predicts a *smaller* girth than the $y = x$ line. The effect should become clearer in the following plot.

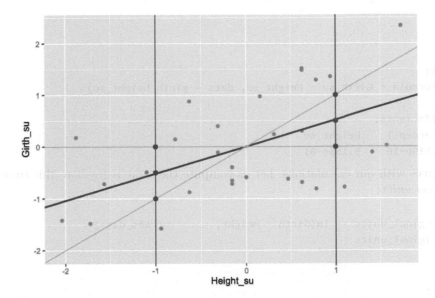

Observe the position of the regression line relative to $y = x$ at each vertical blue line. Note how the regression line approaches the horizontal line that passes through $y = 0$. The phenomena we are observing is called **"regression towards the mean"**, as the regression line prefers to predict extreme points closer to the mean. We can summarize the effect on the data as follows:

Shorter trees will have larger girths on average, and taller trees will have smaller girths on average.

9.3.6 Computing a regression line using base R

Before closing this section, we show how to compute a regression line using functions from base R. Note that you can find the regression line simply by plugging into the formulas we derived above, but it is easier to call a function in R that takes care of the work.

That function is `lm`, which is the short-hand for "linear model". Running the linear regression function requires two parameters. One parameter is the dataset, which takes the form `data = DATA_SET_NAME`. The other is the response and the regressors, which takes a formula of the form `response ~ regressor_1 + ... + regressor_k`.

For a regression of `Girth` on `Height` in standard units, we have the following call.

```
lmod <- lm(Girth_su ~ Height_su, data = girth_height_su)
```

The resulting linear model is stored in a variable `lmod`. We can inspect it by simply typing its name.

```
lmod
```

```
##
## Call:
## lm(formula = Girth_su ~ Height_su, data = girth_height_su)
##
## Coefficients:
## (Intercept)      Height_su
##   -6.519e-16     5.193e-01
```

This agrees with our calculations. Let us compute the same regression, this time in *terms of original units*.

```
lmod_original_units <- lm(Girth ~ Height, data = trees_data)
lmod_original_units
```

```
##
## Call:
## lm(formula = Girth ~ Height, data = trees_data)
##
## Coefficients:
## (Intercept)      Height
##     -6.1884      0.2557
```

So the equation of the line follows:

$$\text{Girth} = 0.2557 * \text{Height} - 6.1884$$

Let us reflect briefly on the meaning of the intercept and the slope. First, the intercept is measured **in inches** as it takes on the units of the response variable. However, its interpretation is nonsensical: a tree with no height is predicted to have a negative diameter.

The slope of the line is a ratio quantity. It measures the increase in the estimated diameter per unit increase in height. Therefore, the slope of this line is **0.2557 inches per foot**. Note that this slope does not say that the diameter of an individual black cherry tree gets wider as it gets taller. The slope tells us the difference in the *average* diameters of two groups of trees that are 1 foot apart in height.

We can also make predictions using the `predict` function.

```
my_pred <- predict(lmod_original_units,
                   newdata = tibble(Height = c(75)))
my_pred
```

```
##        1
## 12.99264
```

So a tree that is 75 feet tall is expected, *on average*, to have a diameter of about 12.9 inches.

An important function for doing regression analysis in R is the function `summary`, which we will not cover. The function reports a wealth of useful information about the linear model, such as significance of the intercept and slope coefficients. However, to truly appreciate and understand what is presented requires a deeper understanding of statistics than what we have developed so far – if this sounds at all interesting, we encourage you to take a course in statistical analysis!

Nevertheless, for the curious reader, we include the incantation to use.

```
summary(lmod)
```

9.4 Using Linear Regression

The last section developed a theoretical foundation for understanding linear regression. We also saw how to fit a regression model using built-in R features such as `lm`. This section will involve linear regression more practically, and we will see how to run a linear regression analysis using tools from a collection of packages called `tidymodels`.

9.4.1 Prerequisites

This section introduces a new meta-package called `tidymodels`. This package can be installed from CRAN using `install.packages`. We will also make use of the tidyverse so let us load this in as usual.

```
library(tidyverse)
library(tidymodels)
library(palmerpenguins)
```

9.4.2   Palmer Station penguins

For the running example in this section, we appeal to the dataset `penguins` from the `palmer-penguins` package, which contains information on different types of penguins in the Palmer Archipelago, Antarctica. The dataset contains 344 observations and `?penguins` can be used to pull up further information. Let us perform basic preprocessing on this dataset by removing any missing values that may be present.

```
my_penguins <- penguins |>
  drop_na()
my_penguins
```

```
## # A tibble: 333 x 7
##    species island     bill_~1 bill_~2 flipp~3 body_~4 sex
##    <fct>   <fct>        <dbl>   <dbl>   <int>   <int> <fct>
##  1 Adelie  Torgersen     39.1    18.7     181    3750 male
##  2 Adelie  Torgersen     39.5    17.4     186    3800 fema~
##  3 Adelie  Torgersen     40.3    18       195    3250 fema~
##  4 Adelie  Torgersen     36.7    19.3     193    3450 fema~
##  5 Adelie  Torgersen     39.3    20.6     190    3650 male
##  6 Adelie  Torgersen     38.9    17.8     181    3625 fema~
##  7 Adelie  Torgersen     39.2    19.6     195    4675 male
##  8 Adelie  Torgersen     41.1    17.6     182    3200 fema~
##  9 Adelie  Torgersen     38.6    21.2     191    3800 male
## 10 Adelie  Torgersen     34.6    21.1     198    4400 male
## # ... with 323 more rows, and abbreviated variable names
## #   1: bill_length_mm, 2: bill_depth_mm,
## #   3: flipper_length_mm, 4: body_mass_g
```

We refer to the preprocessed data as `my_penguins`. Let us visualize the relationship between `bill_length_mm` and `bill_depth_mm`.

```
ggplot(my_penguins) +
  geom_point(aes(x = bill_length_mm, y = bill_depth_mm),
             color = "darkcyan")
```

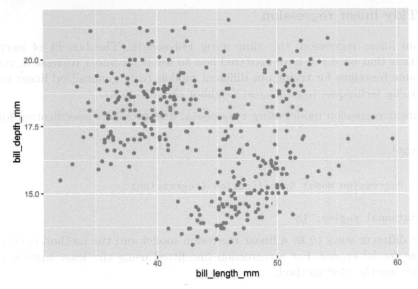

How much correlation is present between the two variables?

```
my_penguins |>
  summarize(r = cor(bill_length_mm,
                    bill_depth_mm))
```

```
## # A tibble: 1 x 1
##        r
##    <dbl>
## 1 -0.229
```

The negative correlation is suspicious considering that we should expect to see a positive correlation after viewing the above the scatter plot. Plotting a histogram for each of these variables reveals a distribution where there appears to be two modes in the data, shown geometrically as two "humps".

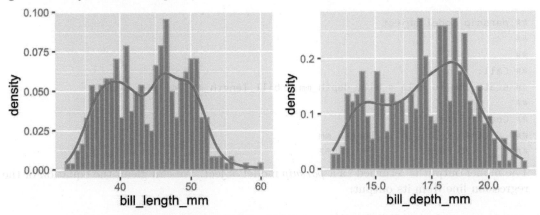

We refer to this as a *bimodal* distribution and here it suggests that there is something about this dataset that we have not yet considered (i.e., the penguin species). For now, we will proceed with the regression analysis.

9.4.3 Tidy linear regression

Let us run linear regression, this time using `tidymodels`. The benefit of learning about `tidymodels` is that once you have mastered how to use it for linear regression, you can then use the same functions for trying out different models (e.g., generalized linear models) and other learning techniques (e.g., nearest neighbor classification).

To fit a linear regression model using `tidymodels`, first provide a specification for it.

```
linear_reg()
```

```
## Linear Regression Model Specification (regression)
##
## Computational engine: lm
```

There are different ways to fit a linear regression model and the method is determined by setting the model *engine*. For a regression line fitted using the least squares method we learned, we use the `"lm"` method.

```
linear_reg() |>
  set_engine("lm") # ordinary least squares
```

```
## Linear Regression Model Specification (regression)
##
## Computational engine: lm
```

We then fit the model by specifying `bill_depth_mm` as the response variable and `bill_length_mm` as the predictor variable.

```
lmod_parsnip <- linear_reg() |>
  set_engine("lm") |>
  fit(bill_depth_mm ~ bill_length_mm, data = penguins)
lmod_parsnip
```

```
## parsnip model object
##
##
## Call:
## stats::lm(formula = bill_depth_mm ~ bill_length_mm, data = data)
##
## Coefficients:
##    (Intercept)   bill_length_mm
##       20.88547         -0.08502
```

The model output is returned as a *parsnip* model object. We can glean the equation of the regression line from its output:

$$\text{Bill Depth} = -0.08502 * \text{Bill Length} + 20.885$$

We can *tidy* the model output using the `tidy` function from the `broom` package (a member of `tidymodels`). This returns the estimates as a *tibble*, a form we can manipulate well in the usual ways.

```
linear_reg() %>%
  set_engine("lm") %>%
  fit(bill_depth_mm ~ bill_length_mm, data = penguins) %>%
  tidy() # broom
```

```
## # A tibble: 2 x 5
##   term            estimate std.error statistic  p.value
##   <chr>              <dbl>     <dbl>     <dbl>    <dbl>
## 1 (Intercept)        20.9     0.844      24.7  4.72e-78
## 2 bill_length_mm  -0.0850    0.0191     -4.46  1.12e- 5
```

Note the estimates for the intercept and slope given in the `estimate` column. The `p.value` column indicates the significance of each term; we will return to this shortly.

We may wish to annotate each observation in the dataset with the predicted (or *fitted*) values and residual amounts. This can be accomplished by *augmenting* the model output. To do this, we extract the `"fit"` element of the model object and then call the function augment from the parsnip package.

```
lmod_augmented <- lmod_parsnip |>
  pluck("fit") |>
  augment()
lmod_augmented
```

```
## # A tibble: 342 x 9
##    .rownames bill_~1 bill_~2 .fitted .resid    .hat .sigma
##    <chr>       <dbl>   <dbl>   <dbl>  <dbl>   <dbl>  <dbl>
## 1  1           18.7    39.1    17.6  1.14   0.00521   1.92
## 2  2           17.4    39.5    17.5 -0.127  0.00485   1.93
## 3  3           18      40.3    17.5  0.541  0.00421   1.92
## 4  5           19.3    36.7    17.8  1.53   0.00806   1.92
## 5  6           20.6    39.3    17.5  3.06   0.00503   1.92
## 6  7           17.8    38.9    17.6  0.222  0.00541   1.93
## 7  8           19.6    39.2    17.6  2.05   0.00512   1.92
## 8  9           18.1    34.1    18.0  0.114  0.0124    1.93
## 9  10          20.2    42      17.3  2.89   0.00329   1.92
## 10 11          17.1    37.8    17.7 -0.572  0.00661   1.92
## # ... with 332 more rows, 2 more variables:
## #   .cooksd <dbl>, .std.resid <dbl>, and abbreviated
## #   variable names 1: bill_depth_mm, 2: bill_length_mm
```

Augmenting model output is useful if you wish to overlay the scatter plot with the regression line.

```
ggplot(lmod_augmented) +
  geom_point(aes(x = bill_length_mm,
                 y = bill_depth_mm),
             color = "darkcyan") +
  geom_line(aes(x = bill_length_mm,
                y = .fitted),
```

```
            color = "purple",
            size = 2)
```

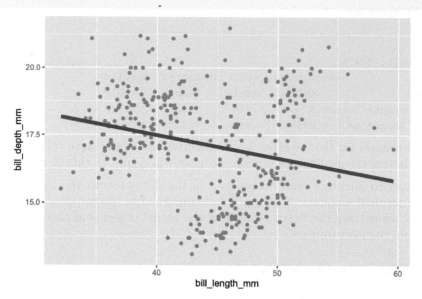

Note that we can obtain an equivalent plot using a smooth geom with the lm method.

```
ggplot(my_penguins,
       aes(x = bill_length_mm, y = bill_depth_mm)) +
  geom_point(color = "darkcyan") +
  geom_smooth(method = "lm", color = "purple", se = FALSE)
```

9.4.4 Including multiple predictors

As hinted earlier, there appear to be issues with the regression model found. We saw a bimodal distribution when visualizing the bill depths and bill lengths, and the negative slope in the regression model has a dubious interpretation.

We know that the dataset is composed of three different *species* of penguins, yet so far we have excluded this information from the analysis. We will now bring in species, a *factor* variable, into the model.

Let us visualize the scatter again and map the color aesthetic to the species variable.

```
ggplot(my_penguins) +
  geom_point(aes(x = bill_length_mm,
                 y = bill_depth_mm,
                 color = species))
```

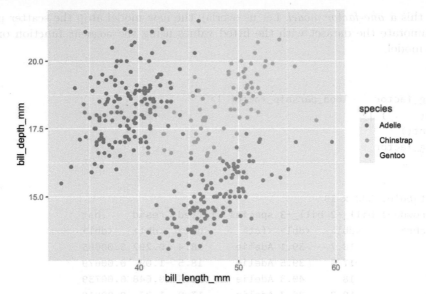

If we were to fit a regression line with respect to each species, we should expect to see a positive slope. Let us try this out by modifying the linear regression specification to include the factor variable species.

```
lmod_parsnip_factor <- linear_reg() |>
  set_engine("lm") %>%
  fit(bill_depth_mm ~ bill_length_mm + species,
      data = penguins)

lmod_parsnip_factor |>
  tidy()
```

```
## # A tibble: 4 x 5
##   term            estimate std.error statistic  p.value
##   <chr>              <dbl>     <dbl>     <dbl>    <dbl>
## 1 (Intercept)        10.6     0.683      15.5  2.43e-41
## 2 bill_length_mm    0.200    0.0175      11.4  8.66e-26
## 3 speciesChinstrap  -1.93     0.224     -8.62  2.55e-16
## 4 speciesGentoo     -5.11     0.191     -26.7  3.65e-85
```

Indeed, we find a positive estimate for the slope, as shown for the bill_length_mm term. However, we now have *three* regression lines, one for each category in the variable species. The slope of each of these lines is the same, however, the intercepts are different. We can write the equation of this line as follows:

$$\text{Depth} = 0.2 * \text{Length} - 1.93 * \text{Chinstrap} - 5.1 * \text{Gentoo} + 10.6$$

The variables Chinstrap and Gentoo in the equation should be treated as Boolean variables. If we want to recover the line for "Gentoo" penguins, set Chinstrap = 0 and Gentoo = 1 to obtain the equation Depth = 0.2 * Length + 5.5. Recovering the line for "Adelie" penguins is curious and requires setting Chinstrap = 0 and Gentoo = 0. This yields the equation Depth = 0.2 * Length + 10.6.

We call this a *one-factor model*. Let us overlay the new model atop the scatter plot. First, let us annotate the dataset with the fitted values using the augment function on the new parsnip model.

```
lmod_aug_factor <- lmod_parsnip_factor |>
  pluck("fit") |>
  augment()
lmod_aug_factor
```

```
## # A tibble: 342 x 10
##    .rowna~1 bill_~2 bill_~3 species .fitted .resid   .hat
##    <chr>      <dbl>   <dbl> <fct>     <dbl>  <dbl>  <dbl>
## 1  1           18.7    39.1 Adelie     18.4  0.292 0.00665
## 2  2           17.4    39.5 Adelie     18.5 -1.09  0.00679
## 3  3           18      40.3 Adelie     18.6 -0.648 0.00739
## 4  5           19.3    36.7 Adelie     17.9  1.37  0.00810
## 5  6           20.6    39.3 Adelie     18.4  2.15  0.00671
## 6  7           17.8    38.9 Adelie     18.4 -0.568 0.00663
## 7  8           19.6    39.2 Adelie     18.4  1.17  0.00668
## 8  9           18.1    34.1 Adelie     17.4  0.691 0.0140
## 9  10          20.2    42   Adelie     19.0  1.21  0.0101
## 10 11          17.1    37.8 Adelie     18.1 -1.05  0.00695
## # ... with 332 more rows, 3 more variables: .sigma <dbl>,
## #   .cooksd <dbl>, .std.resid <dbl>, and abbreviated
## #   variable names 1: .rownames, 2: bill_depth_mm,
## #   3: bill_length_mm
```

We can use the .fitted column in this dataset to plot the new regression line in a line geom layer.

```
ggplot(lmod_aug_factor,
       aes(x = bill_length_mm)) +
  geom_point(aes(y = bill_depth_mm,
                 color = species)) +
  geom_line(aes(y = .fitted,
                group = species),
            color = "purple",
            size = 1)
```

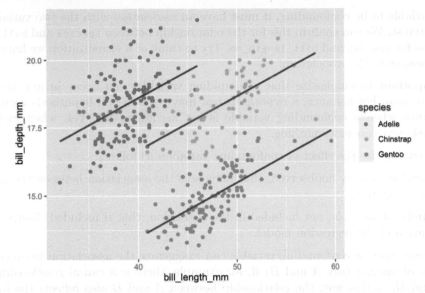

Note how the slope of these lines are the same, but the intercepts in each are different.

9.4.5 Curse of confounding variables

Let us reflect for a moment the importance of species and the impact this variable has when it is incorporated into the regression model. Let us first show the two models again side-by-side.

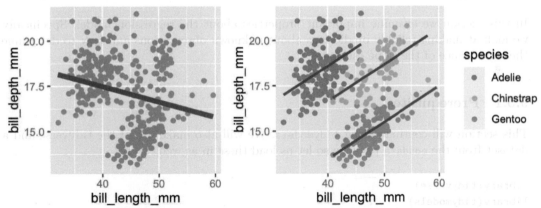

When species is *excluded* from the model, we find a negative estimate for the slope. When species is included, as in the one-factor model, the sign of the slope is flipped. This effect is known as **Simpson's paradox**.

While the inclusion of species seemed obvious for this dataset, we are often unaware of critical variables like these that are "missing" in real-world datasets. These kinds of variables are either neglected from the analysis or were simply not measured. Yet, when included, they can have dramatic effects such as flipping the sign of the slope in a regression model. We call these **confounding variables** because said variables implicate the relationship between two observed variables. For instance, species is *confounding* in the relationship between bill_depth_mm and bill_length_mm.

For a variable to be confounding, it must have an association with the two variables of primary interest. We can confirm this for the relationship between species and bill_depth_mm, as well as for species and bill_length_mm. Try to think of a visualization we have used that can demonstrate the association.

It is important to emphasize that confounding variables can occur in any analysis, not just regression. For instance, a hypothesis test may reject a null hypothesis with very high significance when a confounding variable is not accounted for. Yet, when included, the hypothesis test concludes nothing.

We can summarize the effect of confounding variables as follows:

- A variable, usually unobserved, that influences the association between the variables of primary interest.

- A predictor variable, not included in the regression, that if included changes the interpretation of the regression model.

- In some cases, a confounding variable can exaggerate the association between two variables of interest (say A and B) if, for example, there is a causal relationship between A and B. In this way, the relationship between A and B also reflects the influence of the confounding variable and the association is strengthened because of the effect the confounding variable has on both variables A and B.

9.5 Regression and Inference

In this section we examine important properties about the regression model. Specifically, we look at making reliable predictions about unknown observations and how to determine the significance of the slope estimate.

9.5.1 Prerequisites

This section will continue using tidymodels. We will also make use of the tidyverse and a dataset from the edsdata package so let us load these in as well.

```
library(tidyverse)
library(tidymodels)
library(palmerpenguins)
library(edsdata)
```

For the running example in this section, we examine the athletes dataset from the edsdata package. This is a historical dataset that contains bio and medal count information for Olympic athletes from Athens 1896 to Rio 2016. The dataset is sourced from "120 years of Olympic history: athletes and results"[1] on Kaggle.

Let us focus specifically on athletes that competed in recent Summer Olympics after 2008. We will apply some preprocessing to ensure that each row corresponds to a unique athlete

---

[1] https://www.kaggle.com/datasets/heesoo37/120-years-of-olympic-history-athletes-and-results

as the same athlete can compete in multiple events and in more than one Olympic Games. We accomplish this using the dplyr verb distinct.

```
my_athletes <- athletes |>
  filter(Year > 2008, Season == "Summer") |>
  distinct(ID, .keep_all = TRUE)
my_athletes
```

```
## # A tibble: 3,180 x 15
##       ID Name  Sex     Age Height Weight Team  NOC   Games
##    <dbl> <chr> <chr> <dbl>  <dbl>  <dbl> <chr> <chr> <chr>
## 1     62 "Gio~ M        21    198     90 Italy ITA   2016~
## 2     65 "Pat~ F        21    165     49 Azer~ AZE   2016~
## 3     73 "Luc~ M        27    182     86 Fran~ FRA   2012~
## 4    250 "Sae~ M        26    170     80 Iran  IRI   2016~
## 5    395 "Jen~ F        20    160     62 Cana~ CAN   2012~
## 6    455 "Den~ M        19    161     62 Russ~ RUS   2012~
## 7    465 "Mat~ M        30    197     92 Aust~ AUS   2016~
## 8    495 "Ala~ M        21    188     87 Egypt EGY   2012~
## 9    576 "Ale~ M        23    198     93 Spain ESP   2016~
## 10   608 "Ahm~ M        20    178     68 Jord~ JOR   2016~
## # ... with 3,170 more rows, and 6 more variables:
## #   Year <dbl>, Season <chr>, City <chr>, Sport <chr>,
## #   Event <chr>, Medal <chr>
```

The preprocessed dataset contains 3,180 Olympic athletes.

Let us visualize the relationship between their heights and weights.

```
ggplot(my_athletes) +
  geom_point(aes(x = Height, y = Weight),
             color = "darkcyan")
```

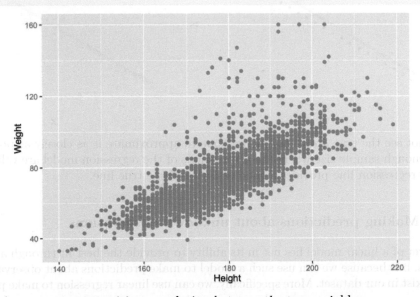

We also observe a strong positive correlation between the two variables.

```
my_athletes |>
  summarize(r = cor(Height, Weight))
```

```
## # A tibble: 1 x 1
##        r
##    <dbl>
## 1 0.790
```

9.5.2   Assumptions of the regression model

Before using inference for regression, we note briefly some assumptions of the regression model. In a simple linear regression model, the regression model assumes that the underlying relation between the response y and the predictor x is perfectly linear and follows a **true line**. We cannot see this line.

The scatter plot that is shown to us is generated by taking points on this line and "pushing" them off the line vertically by some random amount. For each observation, the process is as follows:

- Find the corresponding point on the **true line**
- Make a random draw with replacement from a population of errors that follows a normal distribution with mean 0 (i.e., `rnorm(1, mean = 0)`)
- Draw a point on the scatter with coordinates $(x, y + \text{error})$

Let us demonstrate the effect for two scatter plots with two different sample sizes. The broken dashed line shows the "true" line that generated the scatter. The purple line is the regression line fitted using the least squares method we learned.

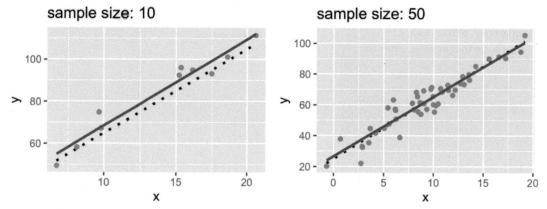

We cannot see the true line, however, we hope to approximate it as closely as possible. For a large enough sample size where the assumptions of the regression model are valid, we find that the regression line provides a good estimate of the true line.

9.5.3   Making predictions about unknown observations

The power of a linear model lies not in its ability to provide the best fit through a collection of points, but because we can use such a model to make predictions about observations that do not exist in our dataset. More specifically, we can use linear regression to make predictions about the expected weight of an Olympic athlete using just one piece of information: their height.

Let us first fit a regression for weight on height using the athlete data.

```
lmod_parsnip <- linear_reg() |>
  set_engine("lm") |>
  fit(Weight ~ Height, data = my_athletes)

lmod_parsnip |>
  tidy()
```

```
## # A tibble: 2 x 5
##   term          estimate std.error statistic p.value
##   <chr>            <dbl>     <dbl>     <dbl>   <dbl>
## 1 (Intercept)    -120.      2.69      -44.8       0
## 2 Height           1.09     0.0150     72.6       0
```

Making a prediction is easy: just plug into the equation of the linear model! For example, what is the expected weight of an Olympic athlete who is 190 centimeters tall?

$$\text{Weight} = 1.0905 * (190) - 120.402 = 86.79$$

So an athlete that is 190 centimeters tall has an expected weight of about 86.8 kilograms.

We can accomplish the work with tidymodels using the function predict. Note that this function receives a dataset in the form of a data frame or tibble, and returns the prediction(s) as a tibble.

```
test_data_tib <- tibble(
  Height = c(190))

lmod_parsnip |>
  predict(test_data_tib)
```

```
## # A tibble: 1 x 1
##   .pred
##   <dbl>
## 1  86.8
```

The prudent reader may have some suspicions about this result – *that's it*? Is this something you can take to the bank?

Indeed, there is a key element missing from the analysis so far: *confidence*. How confident are we in our predictions? We learned that data scientists rarely report singular (or pointwise) estimates because so often they are working with samples of data. The errors are random under the regression model, so the regression line could have turned out differently depending on the scatter plot we get to see. Consequently, the predicted value can change as well.

To combat this, we can provide a *confidence interval* that quantifies the amount of uncertainty in a prediction. We hope to capture the prediction that would be given by the true line with this interval. We learned one way of obtaining such intervals in the previous chapters: resampling.

9.5.4   Resampling a confidence interval

We will use the resampling method we learned earlier to estimate a confidence interval for the prediction we just made. To accomplish this, we will:

- Create a resampled scatter plot by resampling the samples present in my_athletes by means of sampling with replacement.
- Fit a linear model to the bootstrapped scatter plot.
- Generate a prediction from this linear model.
- Repeat the process a large number of times, say, 1,000 times.

We could develop the bootstrap using `dplyr` code as we did before. However, this time we will let `tidymodels` take care of the bootstrap work for us.

First, let us write a function that fits a linear model from a scatter plot and then predicts the expected weight for an athlete who is 190 cm tall.

```
predict190_from_scatter <- function(tib) {
  lmod_parsnip <- linear_reg() |>
    set_engine("lm") |>
    fit(Weight ~ Height, data = tib)

  lmod_parsnip |>
    predict(tibble(Height = c(190))) |>
    pull()
}
```

We use the function `specify` from the `infer` package to specify which columns in the dataset are the relevant response and predictor variables.

```
my_athletes |>
  specify(Weight ~ Height) # infer package
```

```
## Response: Weight (numeric)
## Explanatory: Height (numeric)
## # A tibble: 3,180 x 2
##     Weight Height
##      <dbl>  <dbl>
## 1       90    198
## 2       49    165
## 3       86    182
## 4       80    170
## 5       62    160
## 6       62    161
## 7       92    197
## 8       87    188
## 9       93    198
## 10      68    178
## # ... with 3,170 more rows
```

We form 1,000 resampled scatter plots using the function `generate` with the "bootstrap" setting.

```
resampled_scatter_plots <- my_athletes |>
  specify(Weight ~ Height) |>
  generate(reps = 1000, type = "bootstrap")
resampled_scatter_plots
```

```
## Response: Weight (numeric)
## Explanatory: Height (numeric)
## # A tibble: 3,180,000 x 3
## # Groups:   replicate [1,000]
##    replicate Weight Height
##        <int>  <dbl>  <dbl>
## 1          1     81    181
## 2          1     80    188
## 3          1     78    184
## 4          1     54    162
## 5          1     64    170
## 6          1     67    178
## 7          1     83    187
## 8          1     58    168
## 9          1     67    181
## 10         1     83    190
## # ... with 3,179,990 more rows
```

Note that this function returns a tidy tibble with one resampled observation per row, where the variable `replicate` designates which bootstrap the resampled observation belongs to. This effectively increases the size of the original dataset by a factor of 1,000. Hence, the returned table contains 3.1M entries.

Let us make this more compact by applying a transformation so that we have one *resampled dataset* per row. This can be accomplished using the function `nest` which creates a *nested* dataset.

```
resampled_scatter_plots |>
  nest()
```

```
## # A tibble: 1,000 x 2
## # Groups:   replicate [1,000]
##    replicate data
##        <int> <list>
## 1          1 <tibble [3,180 x 2]>
## 2          2 <tibble [3,180 x 2]>
## 3          3 <tibble [3,180 x 2]>
## 4          4 <tibble [3,180 x 2]>
## 5          5 <tibble [3,180 x 2]>
## 6          6 <tibble [3,180 x 2]>
## 7          7 <tibble [3,180 x 2]>
## 8          8 <tibble [3,180 x 2]>
## 9          9 <tibble [3,180 x 2]>
## 10        10 <tibble [3,180 x 2]>
## # ... with 990 more rows
```

Observe how the number of rows in the nested table (1,000) equals the number of desired bootstraps. We can create a new column that contains a resampled prediction for each scatter plot using the function `predict190_from_scatter` we wrote earlier. We call this function using a `purrr` map.

```
bstrap_predictions <- resampled_scatter_plots |>
  nest() |>
  mutate(prediction = map_dbl(data,
                      predict190_from_scatter))
bstrap_predictions
```

```
## # A tibble: 1,000 x 3
## # Groups:   replicate [1,000]
##    replicate data                     prediction
##        <int> <list>                        <dbl>
## 1          1 <tibble [3,180 x 2]>           86.8
## 2          2 <tibble [3,180 x 2]>           86.9
## 3          3 <tibble [3,180 x 2]>           86.8
## 4          4 <tibble [3,180 x 2]>           87.1
## 5          5 <tibble [3,180 x 2]>           87.1
## 6          6 <tibble [3,180 x 2]>           86.8
## 7          7 <tibble [3,180 x 2]>           86.2
## 8          8 <tibble [3,180 x 2]>           86.9
## 9          9 <tibble [3,180 x 2]>           87.8
## 10        10 <tibble [3,180 x 2]>           87.0
## # ... with 990 more rows
```

We can identify a 95% confidence interval from the 1,000 resampled predictions. We can use the function `get_confidence_interval` from the `infer` package as a shortcut this time. This uses the same percentile method we learned earlier.

```
middle <- bstrap_predictions |>
  get_confidence_interval(level = 0.95)
middle
```

```
## # A tibble: 1 x 2
##   lower_ci upper_ci
##      <dbl>    <dbl>
## 1     86.3     87.3
```

Let us plot a sampling histogram of the predictions and annotate the confidence interval on this histogram.

```
ggplot(bstrap_predictions,
      aes(x = prediction, y = after_stat(density))) +
  geom_histogram(col="grey", fill = "darkcyan", bins = 13) +
  geom_segment(aes(x = middle[1][[1]], y = 0,
                  xend = middle[2][[1]], yend = 0),
                  size = 2, color = "salmon")
```

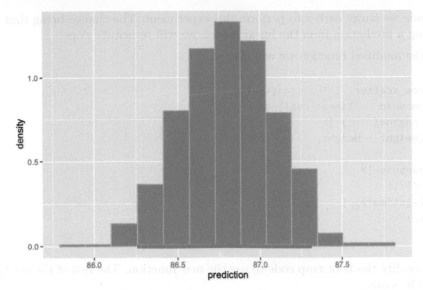

We find that the distribution is approximately normal. Our prediction interval spans between 86.28 and 87.35 kilograms.

We can use R to give us the prediction interval by amending the call to `predict`.

```
lmod_parsnip |>
  predict(test_data_tib,
          type = "conf_int",
          level = 0.95)
```

```
## # A tibble: 1 x 2
##    .pred_lower .pred_upper
##          <dbl>       <dbl>
## 1         86.3        87.3
```

`predict` obtains the interval by means of statistical theory, and we can see that the result is very close to what we found using the the bootstrap. Formally, these intervals go by a special name: *confidence intervals for the mean response*. That's a mouthful!

9.5.5   How significant is the slope?

We showed that the predictions generated by a linear model can vary depending on the sample we have at hand. The slope of the linear model can also turn out differently for the same reasons. This is especially important if our regression model estimates a non-zero slope when the slope of the true line turns out to be 0.

Let us conduct a hypothesis test to evaluate this claim for the athlete data:

- **Null Hypothesis**: Slope of true line is equal to 0.
- **Alternative Hypothesis**: Slope of true line is *not* equal to 0.

We can use a confidence interval to test the hypothesis. In the same way we used bootstrapping to estimate a confidence interval for the mean response, we can apply bootstrapping to estimate the slope of the true line. Fortunately, we need only to make small modifications to the code we wrote earlier to perform this experiment. The change being that instead of generating a prediction from the linear model, we will return its slope.

Here is the modified function we will use.

```
slope_from_scatter <- function(tib) {
  lmod_parsnip <- linear_reg() |>
    set_engine("lm") |>
    fit(Weight ~ Height, data = tib)

  lmod_parsnip |>
    tidy() |>
    pull(estimate) |>
    last() # retrieve slope estimate as a vector
}
```

We also modify the bootstrap code to use the new function. The rest of the bootstrap code remains the same.

```
bstrap_slopes <- my_athletes |>
  specify(Weight ~ Height) |>
  generate(reps = 1000, type = "bootstrap") |>
  nest() |>
  mutate(prediction = map_dbl(data, slope_from_scatter))
bstrap_slopes
```

```
## # A tibble: 1,000 x 3
## # Groups:   replicate [1,000]
##    replicate data                prediction
##        <int> <list>                   <dbl>
## 1          1 <tibble [3,180 x 2]>      1.10
## 2          2 <tibble [3,180 x 2]>      1.07
## 3          3 <tibble [3,180 x 2]>      1.09
## 4          4 <tibble [3,180 x 2]>      1.10
## 5          5 <tibble [3,180 x 2]>      1.07
## 6          6 <tibble [3,180 x 2]>      1.10
## 7          7 <tibble [3,180 x 2]>      1.08
## 8          8 <tibble [3,180 x 2]>      1.06
## 9          9 <tibble [3,180 x 2]>      1.11
## 10        10 <tibble [3,180 x 2]>      1.09
## # ... with 990 more rows
```

Here is the 95% confidence interval.

```
middle <- bstrap_slopes |>
  get_confidence_interval(level = 0.95)
middle
```

```
## # A tibble: 1 x 2
##    lower_ci upper_ci
##       <dbl>    <dbl>
## 1      1.06     1.12
```

Following is the sampling histogram of the bootstrapped slopes with the annotated confidence interval.

```
ggplot(bstrap_slopes,
       aes(x = prediction, y = after_stat(density))) +
   geom_histogram(col="grey", fill = "darkcyan", bins = 13) +
   geom_segment(aes(x = middle[1][[1]], y = 0,
                    xend = middle[2][[1]], yend = 0),
                size = 2, color = "salmon")
```

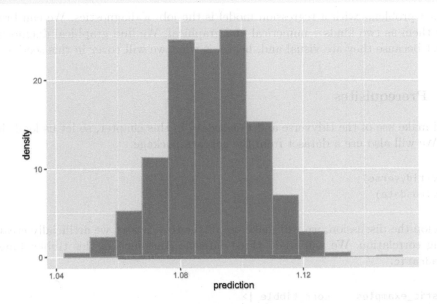

As before, we observe that the distribution is roughly normal. Our 95% confidence interval ranges from about 1.06 to 1.12. According to our null hypothesis, the value 0 does *not* sit on this interval. Therefore, we reject the null hypothesis at the 95% significance level and conclude that the slope estimate obtained is non-zero. The benefit of using the confidence interval is that we also have a range of estimates for what the true slope is.

The normality of this distribution is important since it is the basis for the statistical theory that regression builds on. This means that our interval should be extremely close to what R reports, which is calculated using normal theory. The function confint gives us the answer.

```
lmod_parsnip |>
  pluck("fit") |>
  confint()
```

```
##                    2.5 %      97.5 %
## (Intercept) -125.670130 -115.134110
## Height         1.061099    1.120017
```

Indeed, they are quite close! As a bonus, we also get a confidence interval for the intercept.

9.6   Graphical Diagnostics

An important part of using regression analysis well is understanding how to apply it. Many situations will present itself to you as what seems a golden opportunity for applying linear regression only to find out that the data does not meet any of the assumptions. Do not despair – there are many transformation techniques and diagnostics available that can render linear regression suitable. However, using any of them first requires a realization that there is a problem at hand with the current linear model.

Detecting problems with a regression model is the job of diagnostics. We can broadly categorize them as two kinds – numerical and graphical. We find graphical diagnostics easier to digest because they are visual and, hence, is what we will cover in this section.

9.6.1   Prerequisites

We will make use of the tidyverse and `tidymodels` in this chapter, so let us load these in as usual. We will also use a dataset from the `edsdata` package.

```
library(tidyverse)
library(edsdata)
```

To develop the discussion, we will make use of the toy dataset we artificially created when studying correlation. We will study three datasets contained in this tibble: `linear`, `cone`, and `quadratic`.

```
diagnostic_examples <- corr_tibble |>
  filter(dataset %in% c("linear", "cone", "quadratic"))
diagnostic_examples
```

```
## # A tibble: 1,083 x 3
##          x       y dataset
##      <dbl>   <dbl> <chr>
##  1  0.695   2.26   cone
##  2  0.0824  0.0426 cone
##  3  0.0254 -0.136  cone
##  4  0.898   3.44   cone
##  5  0.413   1.15   cone
##  6 -0.730  -1.57   cone
##  7  0.508   1.26   cone
##  8 -0.0462 -0.0184 cone
##  9 -0.666  -0.143  cone
## 10 -0.723   0.159  cone
## # ... with 1,073 more rows
```

9.6.2   A reminder on assumptions

To begin, let us recap on the assumptions we have made about the linear model.

- The model is of the form $y = c_0 + c_1 x$, that is, the data can be modeled linearly.
- The variance of the residuals is constant, i.e., it fulfills the property of *homoskedasticity*.

Recall that the regression model assumes that the scatter is derived from points that start on a straight line and are then "nudged off" by adding random normal noise.

9.6.3 Some instructive examples

Let us now examine the relationship between y and x across three different datasets. As always, we start with visualization.

```
ggplot(diagnostic_examples) +
  geom_point(aes(x = x, y = y, color = dataset)) +
  facet_wrap( ~ dataset, ncol = 3)
```

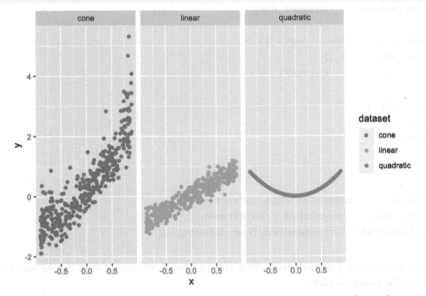

Assuming we have been tasked with performing regression on these three datasets, does it seem like a simple linear model of y on x will get the job done? One way to tell is by using something we have already learned: look at the correlation between the variables for each dataset.

We use dplyr to help us accomplish the task.

```
diagnostic_examples |>
  group_by(dataset) |>
  summarize(r = cor(x, y))
```

```
## # A tibble: 3 x 2
##   dataset          r
##   <chr>        <dbl>
## 1 cone         0.888
## 2 linear       0.939
## 3 quadratic  -0.0256
```

Even without computing the correlation, it should be clear that something is very wrong with the quadratic dataset. We are trying to fit a line to something that follows a curve – a clear violation of our assumptions! The correlation coefficient also confirms that x and y do not have a linear relationship in that dataset.

The situation with the other datasets is more complicated. The correlation coefficients are roughly the same and signal a strong positive linear relationship in both. However, we can see a clear difference in how the points are spread in each of the datasets. Something looks "off" about the cone dataset.

Notwithstanding our suspicions, let us proceed with fitting a linear model for each. Let us use tidymodels to write a function that fits a linear model from a dataset and returns the augmented output.

```
augmented_from_dataset <- function(dataset_name) {
  dataset <- diagnostic_examples |>
    filter(dataset == dataset_name)

  lmod_parsnip <- linear_reg() |>
    set_engine("lm") |>
    fit(y ~ x, data = dataset)

  lmod_parsnip |>
    pluck("fit") |>
    augment()
}
```

```
augmented_cone <- augmented_from_dataset("cone")
augmented_linear <- augmented_from_dataset("linear")
augmented_quadratic <- augmented_from_dataset("quadratic")
```

We need a diagnostic for understanding the appropriateness of our linear model. For that, we turn to the residual plot.

9.6.4   The residual plot

One of the main diagnostic tools we use for studying the fit of a linear model is the residual plot. A *residual plot* can be drawn by plotting the residuals against the fitted values.

This can be accomplish in a straightforward manner using the augmented output. Let us look at the residual plot for augmented_linear.

```
ggplot(augmented_linear) +
  geom_point(aes(x = .fitted, y = .resid), color = "red") +
  geom_hline(yintercept = 0, color = "gray",
             lty = "dashed", size = 1)
```

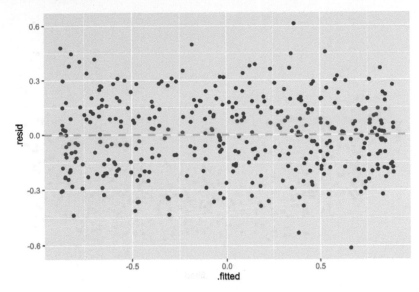

This residual plot tells us that the linear model has a good fit. The residuals are distributed roughly the same around the horizontal line at 0, and the width of the plot is not wider in some parts while narrower in others. The resulting shape is "blobbish nonsense" with no tilt.

Thus, our first observation:

A residual plot corresponding to a good fit shows no pattern. The residuals are distributed roughly the same around the horizontal line passing through 0.

9.6.5 Detecting lack of homoskedasticity

Let us now turn to the residual plot for `augmented_cone`, which we suspected had a close resemblance to `augmented_linear`.

```
ggplot(augmented_cone) +
  geom_point(aes(x = .fitted, y = .resid), color = "red") +
  geom_hline(yintercept = 0, color = "gray",
             lty = "dashed", size = 1)
```

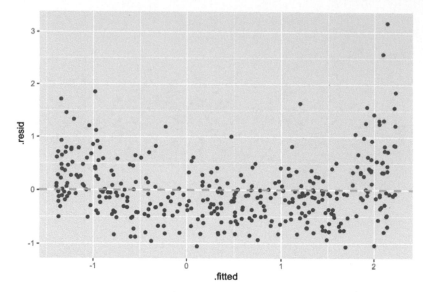

Observe how the residuals fan out at both ends of the residual plot. Meaning, the variance in the size of the residuals is higher in some places while lower in others – a clear violation of the linear model assumptions. Note how this problem would be harder to detect with the scatter plot or correlation coefficient alone.

Thus, our second observation:

A residual plot corresponding to a fit with nonconstant variance shows uneven variation around the horizontal line passing through 0. The resulting shape of the residual plot usually resembles a "cone" or a "funnel".

9.6.6   Detecting nonlinearity

Let us move onto the final model, given by augmented_quadratic.

```
ggplot(augmented_quadratic) +
  geom_point(aes(x = .fitted, y = .resid), color = "red") +
  geom_hline(yintercept = 0, color = "gray",
             lty = "dashed", size = 1)
```

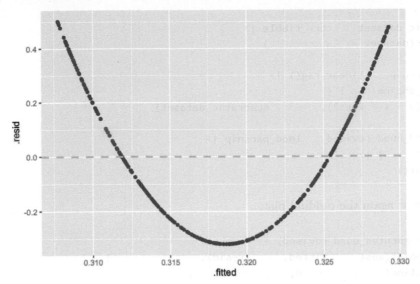

This one shows a striking pattern, which bears the shape of a quadratic curve. That is a clear violation of the model assumptions, and indicates that the variables likely do not have a linear relationship. It would have been better to use a curve instead of a straight line to estimate y using x.

Thus, our final observation:

When a residual plot presents a pattern, there may be nonlinearity between the variables.

9.6.7   What to do from here?

The residual plot is helpful in that it is a tell-tale sign of whether a linear model is appropriate for the data. The next question is, of course, what comes next? If a residual plot indicates a poor fit, data scientists will likely first look to techniques like transformation to see if a quick fix is possible. While such methods are certainly beyond the scope of this text, we can demonstrate what the residual plot will look like after the problem is addressed using these methods.

Let us return to the model given by `augmented_quadratic`. The residual plot signaled a problem of nonlinearity, and the shape of the plot (as well as the original scatter plot) gives a hint that we should try a quadratic curve.

We will modify our model to include a new regressor, which contains the term x^2. Our model will then have the form:

$$y = c_0 + c_1 x + c_2 x^2$$

which is the standard form of a quadratic curve. Let us amend the parsnip model call and re-fit the model.

```
quadratic_dataset <- corr_tibble |>
  filter(dataset == "quadratic")

lmod_parsnip <- linear_reg() |>
    set_engine("lm") |>
    fit(y ~ x + I(x^2), data = quadratic_dataset)

augmented_quad_revised <- lmod_parsnip |>
  pluck("fit") |>
  augment()
```

Let us draw again the residual plot.

```
ggplot(augmented_quad_revised) +
  geom_point(aes(x = .fitted, y = .resid), color = "red") +
  geom_hline(yintercept = 0, color = "gray",
             lty = "dashed", size = 1)
```

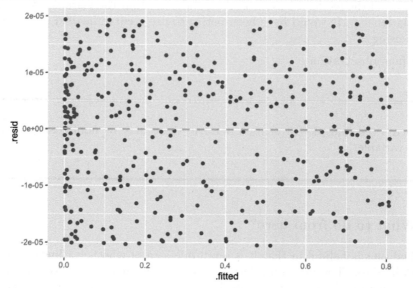

The residual plot shows no pattern whatsoever and, therefore, we can be more confident in this model.

9.7 Exercises

Be sure to install and load the following packages into your R environment before beginning this exercise set.

```
library(tidyverse)
library(edsdata)
library(gapminder)
```

Answering questions for this chapter can be done using the linear modeling functions already available in base R (e.g., lm) or by using the tidymodels package as shown in the textbook. Let us load this package. If you do not have it, then be sure to install it first.

```
library(tidymodels)
```

Question 1 You are deciding what Halloween candy to give out this year. To help make good decisions, you defer to FiveThirtyEight's Ultimate Halloween Candy Power Ranking[2] survey that collected a sample of people's preferences for 269,000 randomly generated matchups.

To measure popularity of candy, you try to describe the popularity of a candy in terms of a single attribute, the amount of sugar a candy has. That way, when you are shopping at the supermarket for treats, you can predict the popularity of a candy just by looking at the amount of sugar it has!

We have collected the results into a tibble candy in the edsdata package. We will use this dataset to see if we can make such predictions accurately using linear regression.

Let's have a look at the data:

```
library(edsdata)
candy
```

```
## # A tibble: 85 x 13
##    competi~1 choco~2 fruity caramel peanu~3 nougat crisp~4
##    <chr>       <dbl>  <dbl>   <dbl>   <dbl>  <dbl>   <dbl>
## 1  100 Grand       1      0       1       0      0       1
## 2  3 Musket~       1      0       0       0      1       0
## 3  One dime        0      0       0       0      0       0
## 4  One quar~       0      0       0       0      0       0
## 5  Air Heads       0      1       0       0      0       0
## 6  Almond J~       1      0       0       1      0       0
## 7  Baby Ruth       1      0       1       1      1       0
## 8  Boston B~       0      0       0       1      0       0
## 9  Candy Co~       0      0       0       0      0       0
## 10 Caramel ~       0      1       1       0      0       0
## # ... with 75 more rows, 6 more variables: hard <dbl>,
## #   bar <dbl>, pluribus <dbl>, sugarpercent <dbl>,
## #   pricepercent <dbl>, winpercent <dbl>, and abbreviated
## #   variable names 1: competitorname, 2: chocolate,
## #   3: peanutyalmondy, 4: crispedricewafer
```

- **Question 1.1** Filter this dataset to contain only the variables winpercent and sugarpercent. Assign the resulting tibble to the name candy_relevant.

- **Question 1.2** Have a look at the data dictionary[3] given for this dataset. What does sugarpercent and winpercent mean? What type of data are these (doubles, factors, etc.)?

Before we can use linear regression to make predictions, we must first determine if the data are roughly linearly associated. Otherwise, our model will not work well.

---

[2]https://fivethirtyeight.com/features/the-ultimate-halloween-candy-power-ranking/
[3]https://github.com/fivethirtyeight/data/tree/master/candy-power-ranking

- **Question 1.3** Make a scatter plot of `winpercent` versus `sugarpercent`. By convention, the variable we will try to predict is on the vertical axis and the other variable – the *predictor* – is on the horizontal axis.

- **Question 1.4** Is the percentile of sugar and the overall win percentage roughly linearly associated? Do you observe a correlation that is positive, negative, or neither? Moreover, would you guess that correlation to be closer to 1, -1, or 0?

- **Question 1.5** Create a tibble called `candy_standard` containing the percentile of sugar and the overall win percentage *in standard units*. There should be two variables in this tibble: `winpercent_su` and `sugarpercent_su`.

- **Question 1.6** Repeat **Question 1.3**, but this time in standard units. Assign your ggplot object to the name `g_candy_plot`.

- **Question 1.7** Compute the correlation r using `candy_standard`. Do *NOT* try to shortcut this step by using `cor()`! You should use the same approach as shown in Section 9.1[4]. Assign your answer to the name r.

- **Question 1.8** Recall that the correlation is the *slope* of the regression line when the data are put in standard units. Here is that regression line overlaid atop your visualization in `g_candy_plot`:

```
g_candy_plot +
  geom_smooth(aes(x = sugarpercent_su, y = winpercent_su),
              method = "lm", se = FALSE)
```

What is the slope of the above regression line in *original units*? Use `dplyr` code and the tibble `candy` to answer this. Assign your answer (a double value) to the name `candy_slope`.

- **Question 1.9** After rearranging that equation, what is the intercept *in original units*? Assign your answer (a double value) to the name `candy_intercept`.

  Hint: Recall that the regression line passes through the point (`sugarpcnt_mean`, `winpcnt_mean`) and, therefore, the equation for the line follows the form (where `winpcnt` and `sugarpcnt` are win percent and sugar percent, respectively):

$$\text{winpcnt} - \text{winpcnt mean} = \texttt{slope} \times (\text{sugarpcnt} - \text{sugarpcnt mean})$$

- **Question 1.10** Compute the predicted win percentage for a candy whose amount of sugar is at the 30th percentile and then for a candy whose sugar amount is at the 70th percentile. Assign the resulting predictions to the names `candy_pred10` and `candy_pred70`, respectively.

The next code chunk plots the regression line and your two predictions (in purple).

```
ggplot(candy,
       aes(x = sugarpercent, y = winpercent)) +
  geom_point(color = "darkcyan") +
  geom_abline(aes(slope = candy_slope,
                  intercept = candy_intercept),
              size = 1, color = "salmon") +
  geom_point(aes(x = 0.3,
```

---

[4]https://ds4world.cs.miami.edu/regression.html#correlation

```
              y = candy_pred30),
        size = 3, color = "purple") +
  geom_point(aes(x = 0.7,
              y = candy_pred70),
        size = 3, color = "purple")
```

- **Question 1.11** Make predictions for the win percentage for each candy in the `candy_relevant` tibble. Put these predictions into a new variable called `prediction` and assign the resulting tibble to a name `candy_predictions`. This tibble should contain three variables: `winpercent`, `sugarpercent`, and `prediction` (which contains the prediction for that candy).

- **Question 1.12** Compute the *residual* for each candy in the dataset. Add the residuals to `candy_predictions` as a new variable called `residual`, naming the resulting tibble `candy_residuals`.

- **Question 1.13** Following is a residual plot. Each point shows one candy and the over- or under-estimation of the predicted win percentage.

```
ggplot(candy_residuals,
       aes(x = sugarpercent, y = residual)) +
  geom_point(color = "darkred")
```

Do you observe any pattern in the residuals? Describe what you see.

- **Question 1.14** In `candy_relevant`, there is no candy whose sugar amount is at the 1st, median, and 100th percentile. Under the regression line you found, what is the predicted win percentage for a candy whose sugar amount is at these three percentiles? Assign your answers to the names `percentile_0_win`, `percentile_median_win`, and `percentile_100_win`, respectively.

- **Question 1.15** Are these values, if any, reliable predictions? Explain why or why not.

Question 2 This question is a continuation of the linear regression analysis of the `candy` tibble in **Question 1**.

Let us see if we can obtain an overall better model by including another attribute in the analysis. After looking at the results from **Question 1**, you have a hunch that whether or not a candy includes chocolate is an important aspect in determining the most popular candy.

This time we will use functions from R and `tidymodels` to perform the statistical analysis, rather than rolling our own regression model as we did in **Question 1**.

- **Question 2.1** Convert the variable `chocolate` in `candy` to a *factor* variable. Then select three variables from the dataset: `winpercent`, `sugarpercent`, and the factor `chocolate`. Assign the resulting tibble to the name `with_chocolate`.

- **Question 2.2** Use a parsnip model to fit a simple linear regression model of `winpercent` on `sugarpercent` on the `with_chocolate` data. Assign the model to the name `lmod_simple`.

  NOTE: The slope and intercept you get should be the same as the slope and intercept you found during lab – just this time we are letting R do the work.

- **Question 2.3** Using a parsnip model again, fit another regression model of `winpercent` on `sugarpercent`, this time adding a new regressor which is the factor `chocolate`. Use the `with_chocolate` data. Assign this model to the name `lmod_with_factor`.

- **Question 2.4** Using the function augment, *augment* the model output from lmod_with_factor so that each candy in the dataset is given information about its predicted (or *fitted*) value, residual, etc. Assign the resulting tibble to the name lmod_augmented.

The following code chunk uses your lmod_augmented tibble to plot the data and overlays the predicted values from lmod_with_factor in the .fitted variable – two different regression lines! The slope of each line is actually the same. The only difference is the intercept.

```
lmod_augmented |>
  ggplot(aes(x = sugarpercent, y = winpercent, color = chocolate)) +
  geom_point() +
  geom_line(aes(y = .fitted))
```

Is the lmod_with_factor model any better than lmod_simple? One way to check is by computing the *RSS* for each model and seeing which model has a lower *RSS*.

- **Question 2.5** The augmented tibble has a variable .resid that contains the residual for each candy in the dataset; this can be used to compute the *RSS*. Compute the *RSS* for each model and assign the result to the appropriate name below.

```
print(paste("simple linear regression RSS :",
            lmod_simple_rss))
print(paste("linear regression with factor RSS :",
            lmod_with_factor_rss))
```

- **Question 2.6** Based on what you found, do you think the predictions produced by lmod_with_factor would be more or less accurate than the predictions from lmod_simple? Explain your answer.

Question 3 This question is a continuation of the linear regression analysis of the candy tibble in **Question 1**.

Before we can be confident in using our linear model for the candy dataset, we would like to know whether or not there truly exists a relationship between the popularity of a candy and the amount of sugar the candy contains. If there is no relationship between the two, we expect the correlation between them to be 0. Therefore, the slope of the regression line would also be 0.

Let us use a hypothesis test to confirm the true slope of regression line. Here is the **null hypothesis** statement:

The true slope of the regression line that predicts candy popularity from the amount of sugar it contains, computed using a dataset that contains the entire population of all candies that have ever been matched up, is 0. Any difference we observed in the slope of our regression line is because of chance variation.

- **Question 3.1** What would be a good **alternative hypothesis** for this problem?

  The following function `slope_from_scatter` is adapted from the textbook. It receives a dataset as input, fits a regression of `winpercent` on `sugarpercent`, and returns the slope of the fitted line:

```
slope_from_scatter <- function(tib) {
  lmod_parsnip <- linear_reg() |>
    set_engine("lm") |>
    fit(winpercent ~ sugarpercent, data = tib)

  lmod_parsnip |>
    tidy() |>
    pull(estimate) |>
    last() # retrieve slope estimate as a vector
}
```

- **Question 3.2** Using the `infer` package, create 1,000 resampled slopes from `candy` using a regression of `winpercent` on `sugarpercent`. You should make use of the function `slope_from_scatter` in your code. Assign your answer to the name `resampled_slopes`.

- **Question 3.3** Derive the approximate 95% confidence interval for the true slope using `resampled_slopes`. Assign this interval to the name `middle`.

- **Question 3.4** Based on this confidence interval, would you accept or reject the null hypothesis that the true slope is 0? Why?

- **Question 3.5** Just in time for Halloween, Reese's Pieces released a new candy this year called Reese's Chunks. We are told its sugar content places the candy at the 64th percentile of sugar within the dataset. We would like to use our linear model `lmod_simple` from **Question 1** to make a prediction about its popularity. However, we know that we can't give an exact estimate because our prediction depends on a sample of 85 different candies being matched up!

  Instead, we can provide an approximate 95% confidence interval for the prediction using the resampling approach in Section 9.3. Recall that such an interval goes by a special name: *confidence interval for the mean response*.

  Suppose we find this interval to be [48.7, 55.8]. Does this interval cover around 95 percent of the candies in `candy` whose sugar amount is at the 64th percentile? Why or why not?

Question 4 Linear regression may not be the best method for describing the relationship between two variables. We would like to have techniques that can help us decide whether or not to use a linear model to predict one variable from another.

If a regression fits a scatter plot well, then the residuals from our regression model should show no pattern when plotted against the predictor variable. This is called the *residual plot*. Section 9.4[5] shows how we can use `ggplot` to generate a residual plot from a `parsnip` linear model.

```
library(lterdatasampler)
```

---

[5]https://ds4world.cs.miami.edu/regression.html#the-residual-plot

The tibble `and_vertebrates` from the package `lterdatasampler` contains length and weight observations for Coastal Cutthroat Trout and two salamander species (Coastal Giant Salamander, and Cascade Torrent Salamander) in HJ Andrews Experimental Forest, Willamette National Forest, Oregon. See the dataset description[6] for more information.

`and_vertebrates`

```
## # A tibble: 32,209 x 16
##      year sitecode section reach  pass unitnum unittype
##     <dbl> <chr>    <chr>   <chr> <dbl>   <dbl> <chr>
##  1   1987 MACKCC-L CC      L         1       1 R
##  2   1987 MACKCC-L CC      L         1       1 R
##  3   1987 MACKCC-L CC      L         1       1 R
##  4   1987 MACKCC-L CC      L         1       1 R
##  5   1987 MACKCC-L CC      L         1       1 R
##  6   1987 MACKCC-L CC      L         1       1 R
##  7   1987 MACKCC-L CC      L         1       1 R
##  8   1987 MACKCC-L CC      L         1       1 R
##  9   1987 MACKCC-L CC      L         1       1 R
## 10   1987 MACKCC-L CC      L         1       1 R
## # ... with 32,199 more rows, and 9 more variables:
## #   vert_index <dbl>, pitnumber <dbl>, species <chr>,
## #   length_1_mm <dbl>, length_2_mm <dbl>, weight_g <dbl>,
## #   clip <chr>, sampledate <date>, notes <chr>
```

Let us try to predict the weight (in grams) of Coastal Giant Salamanders based on their snout-vent length (in millimeters).

- **Question 4.1** Filter the tibble `and_vertebrates` to contain only those observations that pertain to Coastal Giant Salamanders. Assign the resulting tibble to the name `and_salamanders`.

- **Question 4.2** Generate a scatter plot of `weight_g` versus `length_1_mm`.

- **Question 4.3** Generate the residual plot for a linear regression of `weight_g` on `length_1_mm`.

- **Question 4.4** Following are some statements that can be made about the above residual plot. For each of these statements, state whether or not the statement is correct and explain your reasoning.

    - The residuals are distributed roughly the same around the horizontal line passing through 0. Because there is no visible pattern in the residual plot, the linear model is a good fit.
    - The residual plot shows uneven variation around the horizontal line passing through 0.
    - The residual plot shows a pattern, which points to nonlinearity between the variables.

- **Question 4.5** For the problem(s) you found in **Question 4.3**, try applying transformations as shown in Section 9.4[7] to see if they can be corrected. If you did not find any problems, suggest some ways to improve the linear model.

---

[6]https://lter.github.io/lterdatasampler/articles/and_vertebrates_vignette.html
[7]https://ds4world.cs.miami.edu/regression.html#what-to-do-from-here

Question 5 The tibble `mango_exams` from the `edsdata` package contains exam scores from four different offerings of a course taught at the University of Mango. The course contained two midterm assessments and a final exam administered at the end of the semester. Here is a preview of the data:

```
library(edsdata)
mango_exams
```

```
## # A tibble: 440 x 4
##     Year Midterm1 Midterm2 Final
##    <dbl>    <dbl>    <dbl> <dbl>
##  1  2001     80.1     90.8  96.7
##  2  2001     87.5     78.3  93.3
##  3  2001     29.4     30    46.7
##  4  2001     97.5     96.7  98.3
##  5  2001     79.4     75.5  66.7
##  6  2001     82.5     57.5  80
##  7  2001     97.2     83.8  75
##  8  2001     80       85    80
##  9  2001     38.8     60    51.7
## 10  2001     67.5     62.5  66.7
## # ... with 430 more rows
```

- **Question 5.1** Following are some questions you would like to address about the data:

  - Is there a difference in Final exam performance between the offerings in `2001` and `2002`?
  - How high will student scores on the Final be on average, given a Midterm 1 score of 75?
  - What is the estimated range of mean Midterm 2 scores in the population of University of Mango students who have taken the course?

  Here are four techniques we have learned:

  - Hypothesis test
  - Bootstrapping
  - Linear Regression
  - Central Limit Theorem

  Choose the *best* one to address each of the above questions. **For each question, select only one technique.**

- **Question 5.2** Using an appropriate geom from `ggplot2`, visualize the distribution of scores in `Midterm1`. Then generate another visualization showing the distribution of scores in `Final`. Do the distributions appear to be symmetrical and roughly normally distributed? Explain your reasoning.

- **Question 5.3** Visualize the relationship between `Midterm1` and `Final` using an appropriate geom with `ggplot2`. Then answer in English: do these variables appear to be associated? Are they *linearly* related?

- **Question 5.4** Fit a regression line to this scatter plot. Write down the equation of this line. Does the intercept have a sensible interpretation? Then, augment your visualization from **Question 5.3** with this line.

Note: it is tempting to use `geom_smooth` but `geom_smooth` computes its own linear regression; you must find another way so that the visualization shows *your* regression line.

- **Question 5.5** Generate a residual plot for this regression. Does the association between final and midterm scores appear linear? Or does the plot suggest problems with the regression? Why or why not?

- **Question 5.6** You want to know how students who scored a 80 on the first midterm will perform on the final exam, on average. Assuming the regression model holds for this data, what could plausibly be the predicted final exam score?

- **Question 5.7** Following are some statements about the prediction you just generated. Which of these are valid statements that can be made about this prediction? In English, explain why or why not each is a valid claim.

    - This is the score an individual student can expect to receive on the final exam after scoring a 80 on the first midterm.
    - A possible interpretation of this prediction is that a student who scored a 80 on the first midterm cannot possibly score a 90 or higher on the final exam.
    - This is an estimate of the height of the true line at $x = 80$. Therefore, we can confidently report this result.
    - A prediction cannot be determined using a linear regression model fitted on this data.

- **Question 5.8** An alternative to transformation is to set aside data that appears inconsistent with the rest of the dataset and apply linear regression only to the subset. For instance, teaching style and content may change with each course offering and it can be fruitful to focus first on those offerings that share similar characteristics (e.g., the slope and intercept of the fitted regression line on each individual offering is similar). We may also consider cut-offs and set aside exam scores that are too low or high.

    Using `dplyr` and your findings so far, craft a subset of `mango_exams` scores that seems appropriate for a linear regression of `Final` on `Midterm1`. Explain why your subset makes sense. Then perform the linear regression using this subset and show `ggplot2` visualizations that demonstrate whether the model fitted on this subset is any better than the model fitted on the full data.

Question 6 This question is a continuation of **Question 7** from Chapter 3. There is evidence suggesting[8] that mean annual temperature is a primary factor for the change observed in lake ice formation. Let us explore the role of air temperature as a *confounding variable* in the relationship between year and ice cover duration. We will bring in another data source: daily average temperature data from 1869 to 2019 in Madison, Wisconsin. These data are available in the tibble `ntl_airtemp`, also sourced from the `lterdatasampler` package.

- **Question 6.1** Form a tibble named `by_year_avg` that reports the mean annual temperature. According to the documentation, data prior to (and including) 1884 should be filtered as we are told data for these dates contain biases.

- **Question 6.2** Create a tibble named `icecover_with_temp` that contains **both** the ice cover duration (from `ntl_icecover`) and the air temperature data (`by_year_avg`). The resulting tibble should not contain any data before 1885.

- **Question 6.3** Generate a scatter plot showing ice duration versus mean annual temperature.

---

[8]http://hpkx.cnjournals.com/uploadfile/news_images/hpkx/2019-03-14/s41558-018-0393-56789.pdf

- **Question 6.4** Generate a line plot showing mean annual air temperature versus year.

- **Question 6.5** Fit a regression model for `ice_duration` on year. Then fit another regression model for `ice_duration` on year and `avg_air_temp_year`. Note if the sign of the estimate given for year changes when including the additional `avg_air_temp_year` regressor.

- **Question 6.6** We have now visualized the relationship between mean annual air temperature and year and between mean annual air temperature and ice duration. We also generated two regression models where the only difference is inclusion of the variable `avg_air_temp_year`. Based on these, which of the following conclusions can be correctly drawn about the data? Explain which visualization and/or model gives evidence for each statement.

  - There is a positive correlation between year and mean annual temperature.
  - There is a negative correlation between ice duration and mean annual temperature.
  - Mean annual air temperature is a confounding variable in the relationship between year and ice cover duration.
  - The relationship between year and ice cover duration is spurious.

10

Text Analysis

A lot of the data that data scientists deal with daily tends to be *quantitative*, that is, data that is numerical. We have seen many examples of this throughout the text: the miles per gallon of car makes, the diameter of oak trees, the height and yardage of top football players, and the number of minutes flight were delayed for departure. We also applied methods from statistical analysis like *resampling* to estimate how much a value can vary and *regression* to make predictions about unknown quantities. All of this has been in the context of data that is quantitative.

Data scientists prefer quantitative data because computing and reasoning with them is straightforward. If we wish to develop an understanding of a numerical sample at hand, we can simply compute a statistic such as its mean or median or visualize its distribution using a histogram. We could also go further and look at confidence intervals to quantify the uncertainty in any statistic we may be interested in.

However, these methods are no longer useful when the data handed to us is *qualitative*. Qualitative data comes in many different kinds – like speech recordings and images – but perhaps the most notorious among them: *textual* data. With text data, it becomes impossible to go straight to the statistic. For instance, can you say what the *mean* or *median* is of this sentence: "I found queequeg's arm thrown over me"? Probably not!

If we cannot compute anything from the text directly, how can we possibly extract any kind of insight? The trick: transform it! This idea should not seem unfamiliar. The "Getting Started" online chapter gave us a hint when we plotted word relationships in Herman Melville's *Moby Dick*. There, we transformed the text into *word counts* and recorded the number of times a word occurred in each chapter which allowed us to visualize word relationships. Word counts are quantitative, which means we know how to work with it very well.

This chapter turns to such transformations and builds up our understanding of how to deal with and build models for text data so that we may gain some insight from it. We begin with the idea of *tidy text*, an extension of tidy data, which sets us up for a study in frequency analysis. We then turn to an advanced technique called *topic modeling* which is a powerful modeling tool that gives us a sense of the "topics" that are present in a collection of text documents.

These methods have important implications in an area known as the Digital Humanities[1], where a dominant line of its research is dedicated to the study of text by means of computation.

---

[1] https://en.wikipedia.org/wiki/Digital_humanities

10.1   Tidy Text

We begin this chapter by studying what tidy text looks like, which is an extension of the tidy data principles we saw earlier when working with `dplyr`. Recall that tidy data has four principles:

1. Each *variable* forms a column.
2. Each *observation* forms a row.
3. Each value must have its own cell.
4. Each type of observational unit forms a table.

Much like what we saw when keeping tibbles tidy, tidy text provides a way to deal with text data in a straightforward and consistent manner.

10.1.1   Prerequisites

We will continue making use of the tidyverse in this chapter , so let us load that in as usual. We will also require two new packages this time:

- The `tidytext` package, which contains the functions we need for working with tidy text and to make text tidy.
- The `gutenbergr` package, which allows us to search and download public domain works from the Project Gutenberg[2] collection.

```
library(tidyverse)
library(tidytext)
library(gutenbergr)
```

10.1.2   Downloading texts using `gutenbergr`

To learn about tidy text, we need a subject for study. We will return to Herman Melville's epic novel *Moby Dick*[3], which we worked with briefly in the online "Getting Started" chapter. Since this work is in the public domain, we can use the `gutenbergr` package to download the text from the Project Gutenberg database and load it into R.

We can use the function `gutenberg_works` to confirm that *Moby Dick*[4] is indeed in the database and fetch its corresponding Gutenberg ID.

```
gutenberg_works(title == "Moby Dick")
```

```
## # A tibble: 0 x 8
## # ... with 8 variables: gutenberg_id <int>, title <chr>,
## #    author <chr>, gutenberg_author_id <int>,
## #    language <chr>, gutenberg_bookshelf <chr>,
## #    rights <chr>, has_text <lgl>
```

---

[2]http://www.gutenberg.org/
[3]http://www.gutenberg.org/ebooks/15
[4]http://www.gutenberg.org/ebooks/15

Using its ID 15, we can proceed with retrieving the text into a variable called `moby_dick`.
Observe how the data returned is in the form of a tibble – a format we are well familiar
with – with one row per each line of the text.

```r
moby_dick <- gutenberg_download(gutenberg_id = 15,
                mirror = "http://mirrors.xmission.com/gutenberg/")
moby_dick
```

```
## # A tibble: 22,243 x 2
##    gutenberg_id text
##           <int> <chr>
##  1           15 "Moby-Dick"
##  2           15 ""
##  3           15 "or,"
##  4           15 ""
##  5           15 "THE WHALE."
##  6           15 ""
##  7           15 "by Herman Melville"
##  8           15 ""
##  9           15 ""
## 10           15 "Contents"
## # ... with 22,233 more rows
```

The function `gutenberg_download` will do its best to strip any irrelevant header or footer
information from the text. However, we still see a lot of preface material like the table of
contents which is also unnecessary for analysis. To remove these, we will first split the text
into its corresponding chapters using a call to the mutate `dplyr` verb. While we are at it,
let us also drop the column `gutenberg_id`.

```r
by_chapter <- moby_dick |>
  select(-gutenberg_id) |>
  mutate(document = 'Moby Dick',
         linenumber = row_number(),
         chapter = cumsum(str_detect(text, regex('^CHAPTER '))))
by_chapter
```

```
## # A tibble: 22,243 x 4
##    text                 document   linenumber chapter
##    <chr>                <chr>           <int>   <int>
##  1 "Moby-Dick"          Moby Dick           1       0
##  2 ""                   Moby Dick           2       0
##  3 "or,"                Moby Dick           3       0
##  4 ""                   Moby Dick           4       0
##  5 "THE WHALE."         Moby Dick           5       0
##  6 ""                   Moby Dick           6       0
##  7 "by Herman Melville" Moby Dick           7       0
##  8 ""                   Moby Dick           8       0
##  9 ""                   Moby Dick           9       0
## 10 "Contents"           Moby Dick          10       0
## # ... with 22,233 more rows
```

There is a lot going on in this `mutate` call. Let us unpack the important parts:

- The `str_detect` checks if the pattern `CHAPTER` occurs at the beginning of the line, which returns `TRUE` if found and `FALSE` otherwise. This is done using the regular expression `^CHAPTER` (note the white space at the end). To understand this expression, compare the following examples and try to explain why only the first example (a proper line signaling the start of a new chapter) yields a match.

```
regex <- "^CHAPTER "
```

```
str_view_all("CHAPTER CXVI. THE DYING WHALE", regex) # match
```

```
## [1] | <CHAPTER >CXVI. THE DYING WHALE
```

```
str_view_all("Beloved shipmates, clinch the last verse of the
first chapter of Jonah—", regex)   # no match
```

```
## [1] | Beloved shipmates, clinch the last verse of the
##      | first chapter of Jonah—
```

- The function `cumsum` is a cumulative sum over the boolean values returned by `str_detect` indicating a match. The overall effect is that we can assign a row which chapter it belongs to.

Eliminating the preface material becomes straightforward as we need only to filter any rows pertaining to chapter 0.

```
by_chapter <- by_chapter |>
  filter(chapter > 0)
by_chapter
```

```
## # A tibble: 21,580 x 4
##    text                       docum~1 linen~2 chapter
##    <chr>                      <chr>     <int>   <int>
##  1 "CHAPTER I. LOOMINGS"      Moby D~     664       1
##  2 ""                         Moby D~     665       1
##  3 ""                         Moby D~     666       1
##  4 "Call me Ishmael. Some years a~ Moby D~  667    1
##  5 "little or no money in my purs~ Moby D~  668    1
##  6 "on shore, I thought I would s~ Moby D~  669    1
##  7 "of the world. It is a way I h~ Moby D~  670    1
##  8 "regulating the circulation. W~ Moby D~ 671     1
##  9 "the mouth; whenever it is a d~ Moby D~  672    1
## 10 "I find myself involuntarily p~ Moby D~ 673     1
## # ... with 21,570 more rows, and abbreviated variable
## #   names 1: document, 2: linenumber
```

Looks great! We are now ready to define what tidy text is.

10.1.3 Tokens and the principle of tidy text

The basic meaningful unit in text analysis is the *token*. It is usually a word, but it can be more or less granular depending on the context, e.g., sentence units or vowel units. For us, the token will always represent the word unit.

Tokenization is the process of splitting text into tokens. Here is an example using the first few lines from *Moby Dick*.

```
some_moby_df <- by_chapter |>
  slice(4:6)
some_moby_df
```

```
## # A tibble: 3 x 4
##    text                         docum~1 linen~2 chapter
##    <chr>                        <chr>     <int>   <int>
## 1 Call me Ishmael. Some years ago~ Moby D~    667       1
## 2 little or no money in my purse,~ Moby D~    668       1
## 3 on shore, I thought I would sai~ Moby D~    669       1
## # ... with abbreviated variable names 1: document,
## #    2: linenumber
```

```
tokenized <- some_moby_df |>
  pull(text) |>
  str_split(" ")
tokenized
```

```
## [[1]]
##  [1] "Call"              "me"
##  [3] "Ishmael."          "Some"
##  [5] "years"             "ago—never"
##  [7] "mind"              "how"
##  [9] "long"              "precisely—having"
##
## [[2]]
##  [1] "little"      "or"          "no"      "money"
##  [5] "in"          "my"          "purse,"  "and"
##  [9] "nothing"     "particular"  "to"      "interest"
## [13] "me"
##
## [[3]]
##  [1] "on"       "shore,"   "I"    "thought" "I"
##  [6] "would"    "sail"     "about" "a"      "little"
## [11] "and"      "see"      "the"  "watery" "part"
```

Note how in each line the text has been split into tokens, and we can access any of them using list and vector notation.

```
tokenized[[1]][3]
```

```
## [1] "Ishmael."
```

Text that is tidy is a table with one token per row. Assuming that the text is given to us in tabular form (like `text_df`), we can break text into tokens and transform it to a tidy structure in one go using the function `unnest_tokens`.

```
tidy_df <- some_moby_df |>
  unnest_tokens(word, text)
tidy_df
```

```
## # A tibble: 40 x 4
##    document  linenumber chapter word
##    <chr>          <int>   <int> <chr>
##  1 Moby Dick        667       1 call
##  2 Moby Dick        667       1 me
##  3 Moby Dick        667       1 ishmael
##  4 Moby Dick        667       1 some
##  5 Moby Dick        667       1 years
##  6 Moby Dick        667       1 ago
##  7 Moby Dick        667       1 never
##  8 Moby Dick        667       1 mind
##  9 Moby Dick        667       1 how
## 10 Moby Dick        667       1 long
## # ... with 30 more rows
```

Note how each row of this table contains just one token, unlike `text_df` which had a row per line. When text is in this form, we say it follows a **one-token-per-row** structure and, therefore, is tidy.

10.1.4 Stopwords

Let us return to the full `by_chapter` tibble and make it tidy using `unnest_tokens`.

```
tidy_moby <- by_chapter |>
  unnest_tokens(word, text)
tidy_moby
```

```
## # A tibble: 212,610 x 4
##    document  linenumber chapter word
##    <chr>          <int>   <int> <chr>
##  1 Moby Dick        664       1 chapter
##  2 Moby Dick        664       1 i
##  3 Moby Dick        664       1 loomings
##  4 Moby Dick        667       1 call
##  5 Moby Dick        667       1 me
##  6 Moby Dick        667       1 ishmael
##  7 Moby Dick        667       1 some
##  8 Moby Dick        667       1 years
##  9 Moby Dick        667       1 ago
## 10 Moby Dick        667       1 never
## # ... with 212,600 more rows
```

With our text in this form, we can start answering some basic questions about the text. For example: what are the most popular words in *Moby Dick*? We can answer this by piping `tidy_moby` into the function `count`, which lets us count the number of times each word occurs.

```
tidy_moby |>
  count(word, sort = TRUE)
```

```
## # A tibble: 17,450 x 2
##    word       n
##    <chr>  <int>
##  1 the    14170
##  2 of      6463
##  3 and     6327
##  4 a       4620
##  5 to      4541
##  6 in      4077
##  7 that    2934
##  8 his     2492
##  9 it      2393
## 10 i       1980
## # ... with 17,440 more rows
```

The result is disappointing: top-ranking words that appear are so obivous and do not clue us as to the language used in *Moby Dick*. It obstructs from any kind of analysis being made.

These common words (e.g., "this", "his", "that", "in") that appear in almost every written English sentence are known as *stopwords*. It is a typical preprocessing step in text analysis studies to remove such stopwords before proceeding with the analysis.

The tibble `stop_words` is a table provided by `tidytext` that contains a list of English stopwords.

```
stop_words
```

```
## # A tibble: 1,149 x 2
##    word        lexicon
##    <chr>       <chr>
##  1 a           SMART
##  2 a's         SMART
##  3 able        SMART
##  4 about       SMART
##  5 above       SMART
##  6 according   SMART
##  7 accordingly SMART
##  8 across      SMART
##  9 actually    SMART
## 10 after       SMART
## # ... with 1,139 more rows
```

Using the function `anti_join` (think: the opposite of a join), we can filter any rows in `tidy_moby` that match with a stopword in the tibble `stop_words`.

```
tidy_moby_filtered <- tidy_moby |>
  anti_join(stop_words)
tidy_moby_filtered
```

```
## # A tibble: 84,049 x 4
##    document  linenumber chapter word
##    <chr>          <int>   <int> <chr>
##  1 Moby Dick        664       1 chapter
##  2 Moby Dick        664       1 loomings
##  3 Moby Dick        667       1 call
##  4 Moby Dick        667       1 ishmael
##  5 Moby Dick        667       1 ago
##  6 Moby Dick        667       1 mind
##  7 Moby Dick        667       1 precisely
##  8 Moby Dick        668       1 money
##  9 Moby Dick        668       1 purse
## 10 Moby Dick        669       1 shore
## # ... with 84,039 more rows
```

Observe how the total number of rows has decreased dramatically. Let us redo the most popular word list again.

```
tidy_moby_filtered |>
  count(word, sort = TRUE)
```

```
## # A tibble: 16,864 x 2
##    word        n
##    <chr>   <int>
##  1 whale    1029
##  2 sea       435
##  3 ahab      431
##  4 ship      430
##  5 ye        424
##  6 head      336
##  7 time      332
##  8 captain   306
##  9 boat      289
## 10 white     279
## # ... with 16,854 more rows
```

Much better! We can even visualize the spread using a bar geom in ggplot2.

```
tidy_moby_filtered |>
  count(word, sort = TRUE) |>
  filter(n > 200) |>
  mutate(word = reorder(word, n)) |>
  ggplot() +
  geom_bar(aes(x=n, y=word), stat="identity")
```

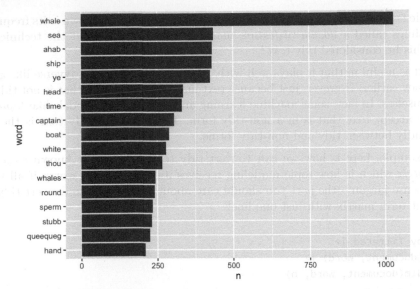

From this visualization, we can clearly see that the story has a lot to do with whales, the sea, ships, and a character named "Ahab". Some readers may point out that "ye" should also be considered a stopword, which raises an important point about standard stopword lists: they do not do well with lexical variants. For this, a custom list should be specified.

10.1.5 Tidy text and non-tidy forms

Before we end this section, let us further our understanding of tidy text by comparing it to other non-tidy forms.

This first example should seem familiar. Does it follow the one-token-per-word structure? If not, how is this table structured?

```
## # A tibble: 3 x 4
##    text                            docum~1 linen~2 chapter
##    <chr>                           <chr>     <int>   <int>
## 1 Call me Ishmael. Some years ago~ Moby D~     667       1
## 2 little or no money in my purse,~ Moby D~     668       1
## 3 on shore, I thought I would sai~ Moby D~     669       1
## # ... with abbreviated variable names 1: document,
## #   2: linenumber
```

Here is another example:

```
some_moby_df |>
  unnest_tokens(word, text) |>
  count(document, word) |>
  cast_dfm(document, word, n)
```

```
Document-feature matrix of: 1 document, 36 features (0.00% sparse) and 0 docvars.
            features
docs          a about ago and call having how i in interest
  Moby Dick 1    1   1   2   1      1   1 2 1         1
[ reached max_nfeat ... 26 more features ]
```

This table is structured as one token per *column*, where the value shown is its frequency. This is sometimes called a *document-feature matrix* (do not worry about the technical jargon). Would this be considered tidy text?

Tidy text is useful in that it plays well with other members of the tidyverse like `ggplot2`, as we just saw earlier. However, just because text may come in a form that is not tidy does not make it useless. In fact, some machine learning and text analysis models like *topic modeling* will only accept text that is in a non-tidy form. The beauty of tidy text is the ability to move fluidly between tidy and non-tidy forms.

As an example, here is how we can convert tidy text to a format known as a *document term matrix* which is how topic modeling receives its input. Don't worry if all that seems like nonsense jargon – the part you should care about is that we can convert tidy text to a document term matrix with just one line!

```
tidy_moby_filtered |>
  count(document, word) |>
  cast_dtm(document, word, n)
```

```
## <<DocumentTermMatrix (documents: 1, terms: 16864)>>
## Non-/sparse entries: 16864/0
## Sparsity           : 0%
## Maximal term length: 20
## Weighting          : term frequency (tf)
```

Disclaimer: the document term matrix expects word frequencies so we actually need to pipe into `count` before doing the `cast_dtm`.

10.2   Frequency Analysis

In this section we use tidy text principles to carry out a first study in text analysis: frequency (or word count) analysis. While looking at word counts may seem like a simple idea, they can be helpful in exploring text data and informing next steps in research.

10.2.1   Prerequisites

We will continue making use of the tidyverse in this chapter, so let us load that in as usual. Let us also load in the `tidytext` and `gutenbergr` packages.

```
library(tidyverse)
library(tidytext)
library(gutenbergr)
```

10.2.2 An oeuvre of Melville's prose

We will study Herman Melville's works again in this section. However, unlike before, we will include many more texts from his oeuvre of prose. We will collect: *Moby Dick, Bartleby, the Scrivener: A Story of Wall-Street, White Jacket*, and *Typee: A Romance of the South Seas*.

```
melville <- gutenberg_download(c(11231, 15, 10712, 1900),
                mirror = "http://mirrors.xmission.com/gutenberg/")
```

We will tidy it up as before and filter the text for stopwords. We will also add one more step to the preprocessing where we extract words that are strictly alphabetical. This is accomplished with a `mutate` call using `str_extract` and the regular expression `[a-z]+`.

```
tidy_melville <- melville |>
  unnest_tokens(word, text) |>
  anti_join(stop_words) |>
  mutate(word = str_extract(word, "[a-z]+"))
```

We will not concern ourselves this time with dividing the text into chapters and removing preface material. But it would be helpful to add a column to `tidy_melville` with the title of the work a line comes from, rather than a ID which is hard to understand.

```
tidy_melville <- tidy_melville |>
  mutate(title = recode(gutenberg_id,
                '15' = 'Moby Dick',
                '11231' = 'Bartleby, the Scrivener',
                '10712' = 'White Jacket',
                '1900' = 'Typee: A Romance of the South Seas'))
```

As a quick check, let us count the number of words that appear in each of the texts.

```
tidy_melville |>
  group_by(title) |>
  count(word, sort = TRUE) |>
  summarize(num_words = sum(n)) |>
  arrange(desc(num_words))
```

```
## # A tibble: 4 x 2
##   title                              num_words
##   <chr>                                  <int>
## 1 Moby Dick                              86233
## 2 White Jacket                           56449
## 3 Typee: A Romance of the South Seas     43060
## 4 Bartleby, the Scrivener                 5025
```

Moby Dick is a mammoth of a book (178 pages!) so it makes sense that it would rank highest in the list in terms of word count.

10.2.3   Visualizing popular words

Let us find the most popular words in each of the titles.

```
tidy_melville <- tidy_melville |>
  group_by(title) |>
  count(word, sort = TRUE) |>
  ungroup()
tidy_melville
```

```
## # A tibble: 42,166 x 3
##    title        word         n
##    <chr>        <chr>    <int>
##  1 Moby Dick    whale     1243
##  2 Moby Dick    ahab       520
##  3 Moby Dick    ship       520
##  4 White Jacket war        481
##  5 Moby Dick    sea        454
##  6 Moby Dick    ye         439
##  7 White Jacket captain    413
##  8 Moby Dick    head       348
##  9 White Jacket ship       345
## 10 Moby Dick    boat       336
## # ... with 42,156 more rows
```

This lends itself well to a bar geom in `ggplot`. We will select out around the 10 most popular, which correspond to words that occur over 300 times.

```
tidy_melville |>
  filter(n > 250) |>
  mutate(word = reorder(word, n)) |>
  ggplot() +
  geom_bar(aes(x=n, y=word, fill=title), stat="identity")
```

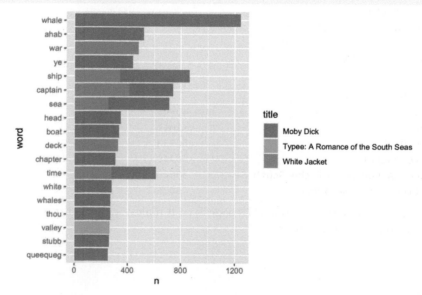

Something is odd about this plot. These top words are mostly coming from *Moby Dick*! As we just saw, *Moby Dick* is the most massive title in our collection so any of its popular words would dominate the overall popular list of words in terms of word count.

Instead of looking at word counts, a better approach is to look at word *proportions*. Even though the word "whale" may have over 1200 occurrences, the proportion in which it appears may be much less when compared to other titles.

Let us add a new column containing these proportions in which a word occurs with respect to the total number of words in the corresponding text.

```
tidy_melville_prop <- tidy_melville |>
  group_by(title) |>
  mutate(proportion = n / sum(n)) |>
  ungroup()
tidy_melville_prop
```

```
## # A tibble: 42,166 x 4
##    title         word       n proportion
##    <chr>         <chr>   <int>      <dbl>
##  1 Moby Dick     whale    1243    0.0144
##  2 Moby Dick     ahab      520    0.00603
##  3 Moby Dick     ship      520    0.00603
##  4 White Jacket  war       481    0.00852
##  5 Moby Dick     sea       454    0.00526
##  6 Moby Dick     ye        439    0.00509
##  7 White Jacket  captain   413    0.00732
##  8 Moby Dick     head      348    0.00404
##  9 White Jacket  ship      345    0.00611
## 10 Moby Dick     boat      336    0.00390
## # ... with 42,156 more rows
```

Let us redo the plot.

```
tidy_melville_prop |>
  filter(proportion > 0.005) |>
  mutate(word = reorder(word, proportion)) |>
  ggplot() +
  geom_bar(aes(x=proportion, y=word, fill=title),
           stat="identity")
```

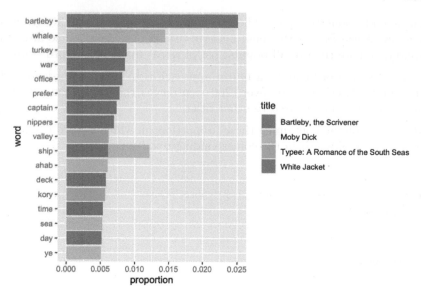

Interesting! In terms of proportions, we see that Melville uses the word "bartleby" much more in *Bartleby, the Scrivener* than he does "whale" in *Moby Dick*. Moreover, *Moby Dick* no longer dominates the popular words list and, in fact, it turns out that *Bartleby, the Scrivener* contributes the most highest-ranking words from the collection.

10.2.4 Just how popular was *Moby Dick*'s vocabulary?

A possible follow-up question is whether the most popular words in *Moby Dick* also saw significant usage across other texts in the collection. That is, for the most popular words that appear in *Moby Dick*, how often do they occur in the other titles in terms of word proportions? This would suggest elements in those texts that are laced with some of the major thematic components in *Moby Dick*.

We first extract the top 10 word proportions from *Moby Dick* to form a "popular words" list.

```
top_moby <- tidy_melville |>
  filter(title == "Moby Dick") |>
  mutate(proportion = n / sum(n)) |>
  arrange(desc(proportion)) |>
  slice(1:10) |>
  select(word)
top_moby
```

```
## # A tibble: 10 x 1
##    word
##    <chr>
##  1 whale
##  2 ahab
##  3 ship
##  4 sea
```

```
##  5 ye
##  6 head
##  7 boat
##  8 time
##  9 captain
## 10 chapter
```

We compute the word proportions with respect to each of the titles and then join the
top_moby words list with `tidy_melville` to extract only the top *Moby Dick* words from the
other three texts.

```
top_moby_words_other_texts <- tidy_melville |>
  group_by(title) |>
  mutate(proportion = n / sum(n)) |>
  inner_join(top_moby, by="word") |>
  ungroup()
top_moby_words_other_texts
```

```
## # A tibble: 32 x 4
##    title         word      n proportion
##    <chr>         <chr>  <int>      <dbl>
##  1 Moby Dick     whale   1243    0.0144
##  2 Moby Dick     ahab     520    0.00603
##  3 Moby Dick     ship     520    0.00603
##  4 Moby Dick     sea      454    0.00526
##  5 Moby Dick     ye       439    0.00509
##  6 White Jacket  captain  413    0.00732
##  7 Moby Dick     head     348    0.00404
##  8 White Jacket  ship     345    0.00611
##  9 Moby Dick     boat     336    0.00390
## 10 Moby Dick     time     334    0.00387
## # ... with 22 more rows
```

Now, the plot. Note that the `factor` in the y aesthetic mapping allows us to preserve the
order of popular words in top_moby so that we can observe an upward trend in the *Moby
Dick* bar heights.

```
ggplot(top_moby_words_other_texts) +
  geom_bar(aes(x=proportion,
               y=factor(word, level=pull(top_moby, word)),
               fill=title),
           position="dodge",stat="identity") +
  labs(y="word")
```

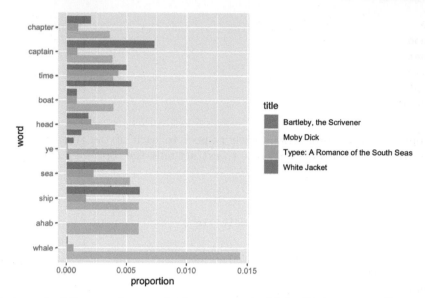

We see that most of the popular words that appear in *Moby Dick* are actually quite unique. The words "whale", "ahab", and even dialects like "ye" appear almost exclusively in *Moby Dick*. There are, however, some notable exceptions, e.g., the words "captain" and "time" appear much more in other titles than they do in *Moby Dick*.

10.3 Topic Modeling

We end this chapter with a preview of an advanced technique in text analysis known as *topic modeling*. In technical terms, topic modeling is an unsupervised method of classification that allows users to find clusters (or "topics") in a collection of documents even when it is not clear to us how the documents should be divided. It is for this reason that we say topic modeling is "unsupervised." We do not say how the collection should be organized into groups; the algorithm simply learns how to without any guidance.

10.3.1 Prerequisites

Let us load in the main packages we have been using throughout this chapter. We will also need one more package this time, `topicmodels`, which has the functions we need for topic modeling.

```
library(tidyverse)
library(tidytext)
library(gutenbergr)
library(topicmodels)
```

We will continue with our running example of Herman Melville's *Moby Dick*. Let us load it in.

```
melville <- gutenberg_download(15,
            mirror = "http://mirrors.xmission.com/gutenberg/")
```

We will tidy the text as usual and, this time, partition the text into chapters.

```
melville <- melville |>
  select(-gutenberg_id) |>
  mutate(title = 'Moby Dick',
         linenumber = row_number(),
         chapter = cumsum(str_detect(text, regex('^CHAPTER ')))) |>
  filter(chapter > 0)
```

10.3.2 Melville in perspective: the American Renaissance

Scope has been a key element of this chapter's trajectory. We began exploring tidy text with specific attention to Herman Melville's *Moby Dick* and then "zoomed out" to examine his body of work at large by comparing four of his principal works with the purpose of discovering elements that may be in common among the texts. We will go even more macroscale in this section by putting Melville in perspective with some of his contemporaries. We will study Nathaniel Hawthorne's *The Scarlet Letter* and Walt Whitman's *Leaves of Grass*.

Since topic modeling is the subject of this section, let us see if we can apply this technique to cluster documents according to the author who wrote them. The documents will be the individual chapters in each of the books and the desired clusters the three works: *Moby Dick*, *The Scarlet Letter*, and *Leaves of Grass*.

We begin as always: loading the texts, tidying them, splitting according to chapter, and filtering out preface material. Let us start with Whitman's *Leaves of Grass*.

```
whitman <- gutenberg_works(title == "Leaves of Grass") |>
  gutenberg_download(meta_fields = "title",
    mirror = "http://mirrors.xmission.com/gutenberg/")
```

```
whitman <- whitman |>
  select(-gutenberg_id) |>
  mutate(linenumber = row_number(),
         chapter = cumsum(str_detect(text, regex('^BOOK ')))) |>
  filter(chapter > 0)
```

Note that we have adjusted the regular expression here so that it is appropriate for the text. On to Hawthorne's *The Scarlet Letter*.

```
hawthorne <- gutenberg_works(title == "The Scarlet Letter") |>
  gutenberg_download(meta_fields = "title",
    mirror = "http://mirrors.xmission.com/gutenberg/")
```

```
hawthorne <- hawthorne |>
  select(-gutenberg_id) |>
  mutate(linenumber = row_number(),
         chapter = cumsum(str_detect(text, regex('^[XIV]+\\.')))) |>
  filter(chapter > 0)
```

Now that the three texts are available in tidy form, we can merge the three tibbles into one by stacking the rows. We call the merged tibble books.

```
books <- bind_rows(whitman, melville, hawthorne)
books
```

```
## # A tibble: 46,251 x 4
##    text                          title linen~1 chapter
##    <chr>                         <chr>   <int>   <int>
##  1 "BOOK I.   INSCRIPTIONS"      Leav~      23       1
##  2 ""                            Leav~      24       1
##  3 ""                            Leav~      25       1
##  4 ""                            Leav~      26       1
##  5 ""                            Leav~      27       1
##  6 "One's-Self I Sing"           Leav~      28       1
##  7 ""                            Leav~      29       1
##  8 "  One's-self I sing, a simple s~ Leav~  30       1
##  9 "  Yet utter the word Democratic~ Leav~  31       1
## 10 ""                            Leav~      32       1
## # ... with 46,241 more rows, and abbreviated variable
## #   name 1: linenumber
```

As a quick check, we can have a look at the number of chapters in each of the texts.

```
books |>
  group_by(title) |>
  summarize(num_chapters = max(chapter))
```

```
## # A tibble: 3 x 2
##   title             num_chapters
##   <chr>                    <int>
## 1 Leaves of Grass             34
## 2 Moby Dick                  135
## 3 The Scarlet Letter          24
```

10.3.3   Preparation for topic modeling

Before we can create a topic model, we need to do some more preprocessing. Namely, we need to create the documents that will be used as input to the model.

As mentioned earlier, these documents will be every chapter in each of the books, which we will give a name like Moby Dick_12 or Leaves of Grass_1. This information is already available in books in the columns title and chapter, but we need to unite the two columns together into a single column. The dplyr function unite() will do the job for us.

```
chapter_documents <- books |>
  unite(document, title, chapter)
```

We can then tokenize words in each of the documents as follows.

```
documents_tokenized <- chapter_documents |>
  unnest_tokens(word, text)
documents_tokenized
```

```
## # A tibble: 406,056 x 3
##    document          linenumber word
##    <chr>                  <int> <chr>
##  1 Leaves of Grass_1         23 book
##  2 Leaves of Grass_1         23 i
##  3 Leaves of Grass_1         23 inscriptions
##  4 Leaves of Grass_1         28 one's
##  5 Leaves of Grass_1         28 self
##  6 Leaves of Grass_1         28 i
##  7 Leaves of Grass_1         28 sing
##  8 Leaves of Grass_1         30 one's
##  9 Leaves of Grass_1         30 self
## 10 Leaves of Grass_1         30 i
## # ... with 406,046 more rows
```

We will filter stopwords as usual, and proceed with adding a column containing word counts.

```
document_counts <- documents_tokenized |>
  anti_join(stop_words) |>
  count(document, word, sort = TRUE) |>
  ungroup()
document_counts
```

```
## # A tibble: 112,834 x 3
##    document              word         n
##    <chr>                 <chr>    <int>
##  1 Moby Dick_32          whale      102
##  2 Leaves of Grass_17    pioneers    54
##  3 Leaves of Grass_31    thee        54
##  4 Leaves of Grass_31    thy         51
##  5 Moby Dick_16          captain     49
##  6 Leaves of Grass_32    thy         48
##  7 Moby Dick_36          ye          48
##  8 Moby Dick_9           jonah       48
##  9 Moby Dick_42          white       46
## 10 The Scarlet Letter_17 thou        46
## # ... with 112,824 more rows
```

This just about completes all the tidying we need. The final step is to convert this tidy tibble into a document-term-matrix (DTM) object.

```
chapters_dtm <- document_counts |>
  cast_dtm(document, word, n)
chapters_dtm
```

```
## <<DocumentTermMatrix (documents: 193, terms: 24392)>>
## Non-/sparse entries: 112834/4594822
## Sparsity          : 98%
## Maximal term length: 20
## Weighting         : term frequency (tf)
```

10.3.4   Creating a three-topic model

We are now ready to create the topic model. Ours is a three-topic model since we expect a topic for each of the books. We will use the function LDA() to create the topic model, where LDA is an acronym that stands for Latent Dirichlet Allocation. LDA is one such algorithm for creating a topic model.

The best part about this step: we can create the model in just one line!

```
lda_model <- LDA(chapters_dtm, k = 3, control = list(seed = 50))
```

Note that the k argument given is the desired number of clusters.

10.3.5   A bit of LDA vocabulary

Before we get to seeing what our model did, we need to cover some basics on LDA. While the mathematics of this algorithm is beyond the scope of this text, learning some of its core ideas are important for understanding its results.

LDA follows three key ideas:

- **Every document is a mixture of topics**, e.g., it may be 70% topic "Moby Dick", 20% topic "The Scarlet Letter", and 10% topic "Leaves of Grass".
- **Every topic is a mixture of words**, e.g., we would expect a topic "Moby Dick" to have words like "captain", "sea", and "whale".
- **LDA is a "fuzzy clustering" algorithm**, that is, it is possible for a word to be generated by multiple topics, e.g., "whale" may be generated by both the topics "Moby Dick" and "The Scarlet Letter".

Finally, keep in mind the following definitions:

β (or "beta") is the probability that a *word* is generated by some topic N. These are also sometimes called *per-topic-per-word* probabilities.

γ (or "gamma") is the probability that a *document* is generated by some topic N. These are also sometimes called *per-topic-per-document* probabilities.

Generally speaking, the more words in a document that are generated by a topic gives "weight" to the document's γ so that the document is also generated by that same topic. We can say then that β and γ are associated.

10.3.6 Visualizing top per-word probabilities

Let us begin by unpacking the β probabilities. At the moment we have an LDA object, which needs to be transformed back into a tidy tibble so that we can begin examining it. We can do this using the function `tidy` and referencing the β matrix.

```
chapter_beta <- tidy(lda_model, matrix = "beta")
chapter_beta
```

```
## # A tibble: 73,176 x 3
##    topic term           beta
##    <int> <chr>         <dbl>
##  1     1 whale     1.38e-  8
##  2     2 whale     1.36e-  2
##  3     3 whale     8.07e-  5
##  4     1 pioneers  1.31e-108
##  5     2 pioneers  1.06e-119
##  6     3 pioneers  9.91e-  4
##  7     1 thee      2.99e-  3
##  8     2 thee      1.62e-  3
##  9     3 thee      4.86e-  3
## 10     1 thy       2.68e-  3
## # ... with 73,166 more rows
```

The format of this tidy tibble is one-topic-per-term-per-row. While the probabilities shown are tiny, we can see that the term "whale" has the greatest probability of being generated by topic 2 and the lowest by topic 3. "pioneers" has even tinier probabilities, but when compared relatively we see that it is most likely to be generated by topic 1.

What are the top terms within each topic? Let us use `dplyr` to retrieve the top 5 terms with the highest β values in each topic.

```
top_terms <- chapter_beta |>
  group_by(topic) |>
  arrange(topic, -beta) |>
  slice(1:5) |>
  ungroup()
top_terms
```

```
## # A tibble: 15 x 3
##    topic term      beta
##    <int> <chr>    <dbl>
## 1      1 hester   0.0109
## 2      1 thou     0.00757
## 3      1 pearl    0.00672
## 4      1 child    0.00593
## 5      1 minister 0.00469
## 6      2 whale    0.0136
## 7      2 ahab     0.00570
## 8      2 ship     0.00566
## 9      2 ye       0.00557
## 10     2 sea      0.00548
## 11     3 love     0.00507
## 12     3 life     0.00491
## 13     3 thee     0.00486
## 14     3 day      0.00478
## 15     3 soul     0.00463
```

It will be easier to visualize this using `ggplot`. Note that the function `reorder_within()` is not one we have used before. It is a handy function that allows us to order the bars *within* a group according to some other value, e.g., its β value. For an explanation of how it works, we defer to this helpful blog post[5] written by Julia Silge.

```
top_terms |>
  mutate(term = reorder_within(term, beta, topic)) |>
  ggplot() +
  geom_bar(aes(beta, term, fill = factor(topic)),
           stat="identity", show.legend = FALSE) +
  facet_wrap(~topic, scales = "free") +
  scale_y_reordered()
```

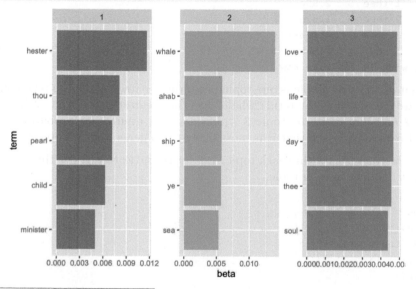

We conclude that our model is a success! The words "love", "life", and "soul" correspond closest to Whitman's *Leaves of Grass*; the words "whale", "ahab", and "ship" with Melville's *Moby Dick*; and the words "hester", "pearl", and "minister" with Hawthorne's *The Scarlet Letter*.

As we noted earlier, this is quite a success for the algorithm since it is possible for some words, say "whale", to be in common with more than one topic.

10.3.7 Where the model goes wrong: per-document misclassifications

Each document in our model corresponds to a single chapter in a book, and each document is associated with some topic. It would be interesting then to see how many chapters are associated with its corresponding book, and which chapters are associated with something else and, hence, are "misclassified." The per-document probabilities, or γ, can help address this question.

We will transform the LDA object into a tidy tibble again, this time referencing the γ matrix.

```
chapters_gamma <- tidy(lda_model, matrix = "gamma")
chapters_gamma
```

```
## # A tibble: 579 x 3
##     document            topic      gamma
##     <chr>               <int>      <dbl>
##  1 Moby Dick_32             1 0.0000171
##  2 Leaves of Grass_17       1 0.0000253
##  3 Leaves of Grass_31       1 0.0000421
##  4 Moby Dick_16             1 0.0000173
##  5 Leaves of Grass_32       1 0.0000138
##  6 Moby Dick_36             1 0.0000284
##  7 Moby Dick_9              1 0.577
##  8 Moby Dick_42             1 0.368
##  9 The Scarlet Letter_17    1 1.00
## 10 Leaves of Grass_34       1 0.00000791
## # ... with 569 more rows
```

Let us separate the document "name" back into its corresponding title and chapter columns. We will do so using the separate() dplyr verb.

```
chapters_gamma <- chapters_gamma |>
  separate(document, c("title", "chapter"), sep = "_",
           convert = TRUE)
chapters_gamma
```

```
## # A tibble: 579 x 4
##     title            chapter topic      gamma
##     <chr>              <int> <int>      <dbl>
##  1 Moby Dick             32     1 0.0000171
##  2 Leaves of Grass       17     1 0.0000253
##  3 Leaves of Grass       31     1 0.0000421
```

```
##   4 Moby Dick                  16        1 0.0000173
##   5 Leaves of Grass            32        1 0.0000138
##   6 Moby Dick                  36        1 0.0000284
##   7 Moby Dick                   9        1 0.577
##   8 Moby Dick                  42        1 0.368
##   9 The Scarlet Letter         17        1 1.00
## 10 Leaves of Grass            34        1 0.00000791
## # ... with 569 more rows
```

We can find the topic that is most associated with each chapter by taking the topic with the highest γ value. For instance:

```
chapters_gamma |>
  filter(title == "The Scarlet Letter", chapter == 17)
```

```
## # A tibble: 3 x 4
##    title                chapter topic    gamma
##    <chr>                  <int> <int>    <dbl>
## 1 The Scarlet Letter        17     1 1.00
## 2 The Scarlet Letter        17     2 0.0000287
## 3 The Scarlet Letter        17     3 0.0000287
```

That γ for this chapter corresponds to topic 3, so we will take this to be its "classification" or label".

Let us do this process for all of the chapters.

```
chapter_label <- chapters_gamma |>
  group_by(title, chapter) |>
  slice_max(gamma) |>
  ungroup()
chapter_label
```

```
## # A tibble: 193 x 4
##    title                chapter topic gamma
##    <chr>                  <int> <int> <dbl>
##  1 Leaves of Grass           1     3 1.00
##  2 Leaves of Grass           2     3 1.00
##  3 Leaves of Grass           3     3 1.00
##  4 Leaves of Grass           4     3 1.00
##  5 Leaves of Grass           5     3 1.00
##  6 Leaves of Grass           6     3 1.00
##  7 Leaves of Grass           7     3 1.00
##  8 Leaves of Grass           8     3 1.00
##  9 Leaves of Grass           9     3 1.00
## 10 Leaves of Grass          10     3 1.00
## # ... with 183 more rows
```

How were the documents labeled? The dplyr verb `summarize` can tell us.

```
labels_summarized <- chapter_label |>
  group_by(title, topic) |>
```

```
  summarize(num_chapters = n()) |>
  ungroup()
labels_summarized
```

```
## # A tibble: 5 x 3
##   title              topic num_chapters
##   <chr>              <int>        <int>
## 1 Leaves of Grass        3           34
## 2 Moby Dick              1           12
## 3 Moby Dick              2          122
## 4 Moby Dick              3            1
## 5 The Scarlet Letter     1           24
```

This lends itself well to a nice bar geom with `ggplot`.

```
ggplot(labels_summarized) +
  geom_bar(aes(x = num_chapters,
               y = title, fill = factor(topic)),
           stat = "identity") +
  labs(fill = "topic")
```

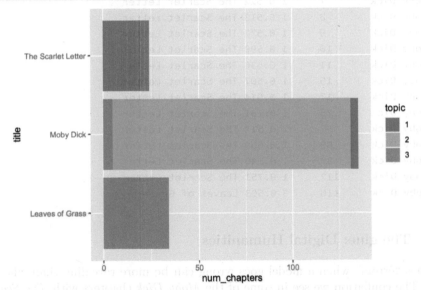

We can see that all of the chapters in *The Scarlet Letter* and *Leaves of Grass* were sorted out completely into its associated topic. While the same is overwhelmingly true for *Moby Dick*, this title did have some chapters that were misclassified into one of the other topics. Let us see if we can pick out which chapters these were.

To find out, we will label each book with the topic that had a majority of its chapters classified as that topic.

```
book_labels <- labels_summarized |>
  group_by(title) |>
  slice_max(num_chapters) |>
  ungroup() |>
```

```
    transmute(label = title, topic)
book_labels
```

```
## # A tibble: 3 x 2
##    label                topic
##    <chr>                <int>
## 1 Leaves of Grass           3
## 2 Moby Dick                 2
## 3 The Scarlet Letter        1
```

We can find the misclassifications by joining `book_labels` with the `chapter_label` table and then filtering chapters where the label does not match the title.

```
chapter_label |>
  inner_join(book_labels, by = "topic") |>
  filter(title != label)
```

```
## # A tibble: 13 x 5
##       title      chapter topic gamma label
##       <chr>        <int> <int> <dbl> <chr>
## 1  Moby Dick          7     1 0.522 The Scarlet Letter
## 2  Moby Dick          8     1 0.513 The Scarlet Letter
## 3  Moby Dick          9     1 0.577 The Scarlet Letter
## 4  Moby Dick         10     1 0.590 The Scarlet Letter
## 5  Moby Dick         11     1 0.536 The Scarlet Letter
## 6  Moby Dick         15     1 0.507 The Scarlet Letter
## 7  Moby Dick         17     1 0.514 The Scarlet Letter
## 8  Moby Dick         25     1 0.781 The Scarlet Letter
## 9  Moby Dick         82     1 0.517 The Scarlet Letter
## 10 Moby Dick         89     1 0.638 The Scarlet Letter
## 11 Moby Dick         90     1 0.546 The Scarlet Letter
## 12 Moby Dick        112     1 0.751 The Scarlet Letter
## 13 Moby Dick        116     3 0.503 Leaves of Grass
```

10.3.8   The glue: Digital Humanities

Sometimes "errors" when a model goes wrong can be more revealing than where it found success. The conflation we see in some of the *Moby Dick* chapters with *The Scarlet Letter* and *Leaves of Grass* points to a connection among the three authors. In fact, Melville and Hawthorne shared such a close writing relationship[6] that it was Hawthorne's influence on Melville that led him to evolve *Moby Dick* from adventure tale into a creative, philosophically rich story. So it is interesting to see so many of the *Moby Dick* chapters mislabeled as *The Scarlet Letter*.

This connection between data and interpretation is the impetus for a field of study known as Digital Humanities[7] (DH). While the text analysis techniques we have covered in this chapter can produce exciting results, they are worth little without context. That context, the authors would argue, is made possible by the work of DH scholars.

---

[6]https://scholarworks.boisestate.edu/cgi/viewcontent.cgi?article=1164&context=mcnair_journal
[7]https://en.wikipedia.org/wiki/Digital_humanities

10.3.9 Further reading

This chapter has served as a preview of the analyses that are made possible by text analysis tools. If you are interested in taking a deeper dive into any of the methods discussed here, we suggest the following resources:

- **"Text Mining with R: A Tidy Approach"**[8] by Julia Silge and David Robinson. The tutorials in this book inspired most of the code examples seen in this chapter. The text also goes into much greater depth than what we have presented here. Check it out if you want to take your R text analysis skills to the next level!

- **Distant Reading**[9] by Franco Moretti. This is a seminal work by Moretti, a literary historian, which expounds on the purpose and tools for literary history. He attempts to redefine the lessons of literature with confident reliance on computational tools. It has been consequential in shaping the methods that form research in Digital Humanities today.

10.4 Exercises

Answering questions for the exercises in Chapter 10 requires some additional packages beyond the tidyverse. Let us load them. Make sure you have installed them before loading them.

```
library(tidytext)
library(gutenbergr)
library(topicmodels)
```

Question 1. Jane Austen Letters Jane Austen (December 17, 1775 to July 18, 1817) was an English novelist. Her well-known novels include "Pride and Prejudice" and "Sense and Sensibility". The Project Gutenberg has all her novels as well as a collection of her letters. The collection, "The Letters of Jane Austen," is from the compilation by her great nephew "Edward Lord Bradbourne".

```
gutenberg_works(author == "Austen, Jane")
```

The project ID for the letter collection is 42078. Using the ID, load the project text as `austen_letters`.

```
austen_letters <- gutenberg_download(gutenberg_id = 42078,
  "http://mirrors.xmission.com/gutenberg/")
austen_letters
```

The tibble has only two variables, `gutenberg_id`, which is 42078 for all rows, and `text`, which is the line-by-line text.

---

[8]https://www.tidytextmining.com/index.html
[9]https://www.google.com/books/edition/Distant_Reading/YKMCy9I3PG4C?hl=en

Let us examine some rows of the tibble. First, unlike "Moby Dick," each letter in the collection appears with Greek numerals as the header. The sixth line of the segment below shows the first letter with the header "I."

```
(austen_letters |> pull(text))[241:250]
```

Next, some letters contain a footnote, which is not part of the original letter. The fifth and the seventh lines of the segment below show the header for a footnote and a footnote with a sequential number "[39]"

```
(austen_letters |> pull(text))[6445:6454]
```

Footnotes can appear in groups. The segment below shows two footnotes. Thus, the header is "FOOTNOTES:" instead of "FOOTNOTE:".

```
(austen_letters |> pull(text))[3216:3225]
```

The letter sequence concludes with the header "THE END." as shown below.

```
(austen_letters |> pull(text))[8531:8540]
```

- **Question 1.1** Suppose we have the the following vector `test_vector`:

```
test_vector <- c( "I.", "    I.", "VII.",
                  "THE END.", "FOOTNOTE:",
                  "FOOTNOTES:",
                  "ds", "    world")
```

Let us filter out the unwanted footnote headers from this vector. Create a regular expression `regex_foot` that detects any line starting with "FOOTNOTE". Test the regular expression on `test_vector` using `str_detect` function and ensure that your regular expression functions correctly.

- **Question 1.2** Using the regular expression `regex_foot`, remove all lines matching the regular expression in the tibble `austen_letters`. Also, remove the variable `gutenberg_id`. Store the result in `letters`. Check out the rows 351-360 of the original and then in the revised version.

- **Question 1.3** Next, find out the location of the start line and the end line. The start line is the one that begins with "I.", and the end line is the one that begins with "THE END.". Create a regular expression `regex_start` for the start and a regular expression `regex_end` for the end. Test the expression on `test_vector`.

- **Question 1.4** Apply the regular expressions to the variable `text` of `letters` using the `stringr` function `str_which`. The result of the first is the start line. The result of the second minus 1 is the very end. Store these indices in `start_no` and `end_no`, respectively.

The following code chunk filters the text tibble to select only those lines between `start_no` and `end_no`.

```
letters_clean <- letters |>
  slice(start_no:end_no)
letters_clean
```

- **Question 1.5** Now, as with the textbook example using "Moby Dick", accumulate the lines that correspond to each individual letter. The following regular expression can be used to detect a letter index.

```
regex_index <- regex('^[IVXLCDM]++\\.')
str_which(letters$text, regex_index)
```

Apply this regular expression, as we did for "Moby Dick", to letters_clean. Add the letter index as letter and the row_number() as linenumber. Store the result in letters_with_num.

Question 2. This is a continuation from the previous question. The previous question has prepared us to investigate the letters in letters_with_num.

- **Question 2.1** First, extract tokens from letters_with_num using unnest_tokens, and store it in letter_tokens. Then, from letter_tokens remove stop_words using anti_join and store it in a name letter_tokens_nostop.

- **Question 2.2** Let us obtain the counts of the words in these two tibbles using count. Store the result in letter_tokens_ranked and letter_tokens_nostop_ranked, respectively.

- **Question 2.3** As in the textbook, use a lower bound of 60 to collect the words and show a bar plot.

- **Question 2.4** Now generate word counts with respect to each letter. The source is letter_tokens_nostop. The counting is by executing count(letter, word, sort = TRUE). Store the result in word_counts.

- **Question 2.5** Generate, from the letter-wise word counts word_counts, a document-term matrix letters_dtm.

- **Question 2.6** From the document-term matrix, generate an LDA model with 2 classes. Store it in lda_model.

- **Question 2.7** Transform the model output into a tibble. Use the function tidy() and set the variable "beta" using the matrix entries.

- **Question 2.8** Select the top 15 terms from each topic as we did for "Moby Dick". Store it in top_terms.

- **Question 2.9** Now plot the top terms as we did in the textbook.

Question 3. The previous attempt to create topic models may not have worked well, possibly because of the existence of frequent non-stop-words that may dominate the term-frequency matrix. Here we attempt to revise the analysis after removing such non-stop-words.

- **Question 3.1** We generated a ranked tibble of words, letter_tokens_nostop_ranked.

```
letter_tokens_nostop_ranked
```

The first few words on the list appear generic, so let us remove these words from consideration. Let us form a character vector named `add_stop` that captures the first 10 words that appear in the tibble `letter_tokens_nostop_ranked`. You can use the `slice()` and `pull()` dplyr verbs to accomplish this.

- **Question 3.2** Let us remove the rows of `word_counts` where the `word` is one of the words in `add_stop`. Store it in `word_counts_revised`.

- **Question 3.3** Generate, from the letter-wise word counts `word_counts_revised`, a document-term matrix `letters_dtm_revised`.

- **Question 3.4** From this document-term matrix, generate an LDA model with 2 classes. Store it in `lda_model_revised`.

- **Question 3.5** Transform the model output back into a tibble. Use the function `tidy()` and set the variable "beta" using the matrix entries.

- **Question 3.6** Select the top 15 terms and store it in `top_terms_revised`.

- **Question 3.7** Now plot the top terms of the two classes. Also, show the plot without excluding the common words. In that matter, we should be able to compare the two topic models side by side.

- **Question 3.8** What differences do you observe between the two topics in the bar plot for `top_terms_revised`? Are these differences more or less apparent (or about the same) when comparing the differences in the original bar plot for `top_terms`?

Question 4. Chesterton Essays G. K. Chesterton (May 29, 1874 – June 14, 1936) was a British writer, who is the best known for his "Father Brown" works. The Gutenbrerg Project ID 8092 is a collection of his essays "Tremendous Trifles".

- **Question 4.1** Load the work in `trifles`. Then remove the unwanted variable and store the result in `trifles0`.

- **Question 4.2** Each essay in the collection has a Greek number header. As we did before, using the Greek number header to capture the start of an essay and add that index as the variable `letter`. Store the mutated table in `tifles1`.

- **Question 4.3** First, extract tokens from `trifles1` using `unnest_tokens`, and store it in `trifles_tokens`. Then, from `trifles_tokens` remove `stop_words` using `anti_join` and store it in `trifles_tokens_nostop`.

- **Question 4.4** Let us obtain the counts of the words in the two token lists using `count`. Store the result in `trifles_tokens_ranked` and `trifles_tokens_nostop_ranked`, respectively.

- **Question 4.5** Use a lower bound of 35 to collect the words and show a bar plot.

- **Question 4.6** Now generate word counts letter-wise. The source is `trifles_tokens_nostop`. The counting is by executing `count(letter, word, sort = TRUE)`. Store the result in `trifles_word_counts`.

- **Question 4.7** Generate, from the letter-wise word counts `trifles_word_counts`, a document-term matrix `trifles_dtm`.

- **Question 4.8** From the document-term matrix, generate an LDA model with 4 classes. Store it in `trifles_lda_model`.

- **Question 4.9** Take the model and turn it into a data frame. Use the function `tidy()` and set the variable "beta" using the matrix entries.

- **Question 4.10** If you run the top-term map, you notice that the words are much similar among the four classes. So, let us skip the first 5 and select the words that are ranked 6th to the 15th. Select the top 10 terms from each topic and store it in `trifles_top_terms`.

- **Question 4.11** Now plot the top terms.

- **Question 4.12** What differences do you observe, if any, among the above four topics?

Question 5 In this question, we explore the relationship between the Chesterton essays and the Jane Austen letter collection. This question assumes you have already formed the tibbles `word_counts_revised` and `trifles_word_counts`.

- **Question 5.1** Using `bind_rows`, merge the datasets `word_counts_revised` and `trifles_word_counts`. However, before merging, discard the `letter` variable present in each tibble and create a new column `author` that gives the author name together with each word count. Assign the resulting tibble to the name `merged_frequencies`.

- **Question 5.2** The current word counts given in `merged_frequencies` are with respect to each letter/essay, but we would like to obtain these counts with respect to each *author*. Using `group_by` and `summarize` from `dplyr`, obtain updated counts for each word by summing the counts over its respective texts. The resulting tibble should contain three variables: `author`, `word`, and `n` (the updated word count).

- **Question 5.3** Instead of reporting word counts, we would like to report word *proportions* so that we can make comparisons between the two authors. Create a new variable `prop` that, with respect to each *author*, reports the proportion of times a word appears over the total count of words for that author. The resulting tibble should contain three variables: `author`, `word`, and `prop`. Assign the resulting tibble to the name `freq_by_author_prop`.

- **Question 5.4** Apply a pivot transformation so that three variables materialize in the `freq_by_author_prop` tibble: `word`, `G. K. Chesterton` (giving the word proportion for Chesterton essays), and `Jane Austen` (giving the word proportion for Jane Austen letters). Drop any resulting missing values after the transformation. Assign the resulting tibble to the name `freq_by_author_prop_long`.

- **Question 5.5** Using `freq_by_author_prop_long`, fit a linear regression model of the G. K. Chesterton proportions on the Jane Austen proportions.

- **Question 5.6** How significant is the estimated slope of the regression line you found? Use `confint()` with the linear model you developed.

- **Question 5.7** The following scatter plot shows the Chesterton word proportions against the Jane Austen word proportions; the color shown is the absolute difference between the two. Also given is a dashed line that follows $y = x$.

  Amend this `ggplot` visualization by adding another geom layer that visualizes the equation of the linear model you found.

```
ggplot(freq_by_author_prop_long,
       aes(x = `Jane Austen`, y = `G. K. Chesterton`,
           color = abs(`G. K. Chesterton` - `Jane Austen`))) +
  geom_abline(color = "gray", lty = 2) +
  geom_text(aes(label = word), check_overlap = TRUE) +
  theme(legend.position="none")
```

- **Question 5.8** What does it mean for words to be close to the $y = x$ line? Also, briefly comment on the relationship between the regression line you found and the $y = x$ line – what does it mean that the slope of your line is relatively smaller?

Index